Multiscale Modelling in Biomedical Engineering

**IEEE Press**
445 Hoes Lane
Piscataway, NJ 08854

# Multiscale Modelling in Biomedical Engineering

*Dimitrios I. Fotiadis*
*University of Ioannina, Ioannina, Greece*

*Antonis I. Sakellarios*
*University of Ioannina, Ioannina, Greece*

*Vassiliki T. Potsika*
*University of Ioannina, Ioannina, Greece*

IEEE Press Series in Biomedical Engineering

IEEE PRESS

WILEY

Published by John Wiley & Sons, Inc., Hoboken, New Jersey.
Published simultaneously in Canada.



For general information on our other products and services or for technical support, please contact our Customer Care Department within the United States at (800) 762-2974, outside the United States at (317) 572-3993 or fax (317) 572-4002.

Wiley also publishes its books in a variety of electronic formats. Some content that appears in print may not be available in electronic formats. For more information about Wiley products, visit our web site at www.wiley.com.

*Library of Congress Cataloging-in-Publication Data Applied for:*
[Hardback ISBN: 9781119517344]

Cover Design: Wiley
Cover Image: © santoelia/Shutterstock

Set in 9.5/12.5pt STIXTwoText by Straive, Pondicherry, India

# Contents

# Author Biographies

Dimitrios I. Fotiadis is a professor of biomedical engineering at the Department of Materials Science and Engineering at the University of Ioannina, the director of the Unit of Medical Technology and Intelligent Information Systems, and an affiliated researcher at the Biomedical Research Institute – FORTH and the Michailideion Cardiology Center. He is the editor in chief of the *IEEE Journal of Biomedical and Health Informatics.* His research interests include multiscale modeling, biomedical informatics, medical and biological data engineering, wearable and implantable devices, and machine/deep learning.

Antonis I. Sakellarios is currently an associate researcher and adjunct lecturer in the Department of Materials Science and Engineering of University of Ioannina and in the Foundation for Research and Technology – Greece, Department of Biomedical Research. He has published over 45 journal articles, 100 conference papers, and 4 book chapters.

Vassiliki T. Potsika is a senior researcher at the Unit of Medical Technology and Intelligent Information Systems, Department of Materials Science and Engineering, University of Ioannina, Greece, and project manager for relevant R&D-funded projects. She is the managing editor of *IEEE Journal of Biomedical and Health Informatics*. Her research interests include biomechanics, bone mechanics, ultrasonic evaluation of fractured and osteoporotic bones, wave scattering in composite materials, computational modeling of cardiovascular diseases, biomedical signal processing and rehabilitation engineering, and robotics.

# Preface

Nowadays, multiscale modeling is considered a fundamental numerical modeling approach in several engineering disciplines such as materials science, biomedical engineering, and fluid mechanics. The application of numerical methods in biomedical research has increased tremendously in the last two decades due to the exponential growth in computer power and the evolution of imaging modalities. The aim of this book is to describe the basic engineering principles, and how multiscale modeling can enhance clinical practice, support clinical trials, improve patient treatment options, and thus ameliorate the quality of life by focusing on biological systems that are the most complex structures in scientific research. It provides fundamental knowledge from the molecular to the cellular functions, as well as at the tissue and organ levels. Considering the significant diversity of the topic, this book approaches the implementation of multiscale modeling to biological systems focusing on the musculoskeletal and cardiovascular systems. For a complete understanding of pathologies such as osteoporosis, fracture healing, and atherosclerosis, the musculoskeletal and cardiovascular systems should be considered as complex systems combining different modeling scales.

More specifically, Chapter 1 deals with the role of systems biology in biomedical engineering and presents the fundamental principles in multiscale modeling, as well as the most popular multiscale methods and approaches. Systems biology refers to biology at a global scale where biological functions are analyzed as a result of complex mechanisms that evolve at different scales, from the nanoscale (molecular) to macroscale (organ). This chapter explains how multiscale modeling and simulation can play a key role in the description, prediction, and better understanding of those mechanisms in a quantitative and integrative manner.

In Chapter 2, biomedical imaging principles and applications related to multiscale modeling are presented by providing in-depth coverage from the engineering point of view. It presents the fundamentals and applications of primary medical imaging techniques, such as magnetic resonance imaging, ultrasound, nuclear medicine, X-ray/computed tomography, and molecular imaging.

Additionally, the physical principles, equipment design, data acquisition, image reconstruction techniques, and clinical applications of each modality are discussed. Recent hybrid developments such as multislice spiral computed tomography or fused and combined imaging techniques are also presented.

An introduction to molecular mechanics is presented in Chapter 3 and its fundamental principles are presented. Advanced mathematical modeling, simulation, and data analysis methods are described and applied to biological problems at the molecular level. The crucial role of molecular bioengineering in areas critical to human health is explained focusing on the effect of computational modeling at molecular level on cell biology, diagnostic medicine, preventive medicine, and therapeutic medicine.

Computational modeling at the cell level is presented in Chapter 4. An introduction to cell mechanics is presented and fundamental principles in cell material properties, composition, and structure are described. Cell mechanics is considered a research domain that plays a significant role in cellular and tissue biology, from tissue and organ development to wound healing and cancer cell metastasis and migration. To study the mechanical behavior of cells, computational models have been presented, which aim to explain experimental observations by providing a framework of underlying cellular mechanisms.

Chapter 5 deals with computational modeling at the tissue level. The mechanical and structural properties of the four main categories of human tissues are presented that are the epithelial, connective, muscle, and nervous tissues. Computational modeling approaches are presented as a promising means aiming to provide insight into the complex nature of the different tissue types and provide quantitative and qualitative information on tissue physics for clinical and scientific purposes.

Starting from the molecular level in Chapter 3, we move from the cellular and tissue levels in Chapters 4 and 5 to Chapter 6 describing the organ level. Each organ performs a particular function in the body and is made up of different tissues. The eleven distinct organ systems of humans are presented that form the basis of anatomy, and the implementation of computational modeling approaches is discussed as a means to provide supplementary information to experimental and clinical research.

Then, in Chapter 7, the mechanotransduction perspective, recent progress, and future challenges are described. Mechanotransduction provides a clear bridge between experimental results performed at separate scales. In particular, it involves quantitative measurements versus more qualitative approaches used by the "general" cell biology making connections that do not follow a simple linear chain of events. The effects of mechanotransduction include growth and remodeling, which are typically considered as unique processes, representing mass–volume changes due to bulk material deposition or resorption versus structural changes including trabecular or fiber realignment, respectively. The development of models of these processes requires both constitutive formulation and computational implementation of the constitutive model, typically within a finite volume of finite difference.

Chapters 8 and 9 focus on multiscale modeling approaches of the musculoskeletal and cardiovascular systems, respectively. Chapter 8 describes the main components of the musculoskeletal system and their hierarchical structure and presents cornerstone research in the area of computational modeling in the macro, micro, and nano scales. The basic principles of the theory of elasticity and mechanics of the soft tissues of the musculoskeletal system are presented as fundamental knowledge to establish a computational model. With the assistance of multidimensional computational models, the unknown mechanisms underlying pathologies of the musculoskeletal system can be tested and clinicians can conduct optimal treatment strategies.

Chapter 9 aims to explain how computational modeling has been used to understand the mechanisms of atherosclerosis and to provide insights into the atherosclerosis process in arteries and the development of de novo plaques. The current approaches to plaque growth modeling, as well as a new multilevel modeling approach, which is based on the major mechanisms of plaque growth, are presented. The model can be used as a decision support tool for the medical doctor or the researcher to predict regions that are prone to plaque development. Additionally, decision support for diagnosis is also provided by calculating noninvasively the fractional flow reserve (FFR) index. Finally, it is discussed how treatment decision support can be achieved by modeling the stent positioning and deployment in the arteries.

Chapter 10 presents risk prediction approaches in chronic diseases as a possible top level of a multiscale computational approach in the domain of predictive medicine. In recent years, we observe an increasing acquisition and accumulation of various types of medical data, which, however, have not been used successfully to fully understand the mechanisms or pathways of diseases. Data are also collected in Real World Settings, such as electronic health records (EHRs), wearable systems, and registries. Machine learning methods have been employed for the development of decision support systems to make prediction, risk assessment, diagnosis, prognosis, and treatment management, especially for chronic diseases. This top level of multiscale modeling could be the development of predictive models for the prediction of disease outcomes such as events or for patients' risk stratification.

Chapter 11 deals with future trends. It presents some future and prospective developments in biomedical engineering considering multiscale modeling. More and more data are collected and acquired generating large databases. However, accurate development of the multiscale models for personalized application requires additional data. Virtual populations are considered as a way to generate and use data for such purposes. Another future perspective that is discussed in this chapter is the use of digital twins in healthcare. Due to the increased use of machine learning and artificial intelligence approaches, the potential integration of multiscale modeling with machine learning is also presented.

This book is intended for pregraduate and postgraduate students, as well as for researchers in the domains of biology, biomechanics, biomedical engineering, biomedical informatics, materials science, and medicine. Our aim is twofold: (i) to establish a good background in the principles of mathematics, physics, and engineering, as well as biology and physiology and (ii) to provide knowledge on advanced computational approaches and the status of the current state of the art in multiscale modeling.

Antonis I. Sakellarios, Vassiliki T. Potsika, Dimitrios I. Fotiadis
Ioannina, Greece

# List of Abbreviations

| Abbreviation | Explanation |
|---|---|
| **2D** | Two-dimensional |
| **3D** | Three-dimensional |
| **ABM** | Agent-based models |
| **ACS** | Activation cycle switch |
| **ACW** | Antral contraction waves |
| **AIx** | Augmentation index |
| **ALE** | Arbitrary Lagrange–Euler |
| **ANN** | Artificial neural network |
| **AP** | Action potential |
| **BEM** | Boundary element method |
| **BMI** | Body mass index |
| **BMImean** | Mean body mass index |
| **BMP** | Bone morphogenetic protein |
| **BMU** | Basic multicellular unit |
| **CAD** | Computer-aided design |
| **CFD** | Computational fluid dynamics |
| **CL** | Culprit lesions |
| **CMBE** | Cellular and molecular bioengineering |
| **CNS** | Central nervous system |
| **Creat** | Creatinine |
| **CSK** | Actin cytoskeleton |
| **CT** | Computed tomography |

(*Continued*)

(Continued)

| Abbreviation | Explanation |
|---|---|
| **CTCA** | Computed tomography coronary angiography |
| **CTDI** | Computed tomography dose index |
| **CV** | Cardiovascular |
| **CVD** | Cardiovascular disease |
| **DBP** | Diastolic blood pressure |
| **DE** | Difference equations |
| **DES** | 1) Drug-eluting stent<br>2) Discrete event simulation |
| **DMD** | Duchenne muscular dystrophy |
| **DNA** | Deoxyribonucleic acid |
| **DS** | Dynamic system |
| **DSM** | Detrusor smooth muscle |
| **DT** | Digital twin |
| **DXA** | Dual-energy X-ray absorptiometry |
| **ECG** | Electrocardiogram |
| **ECM** | Extracellular matrix |
| **EEG** | Electroencephalography |
| **EHR** | Electronic health record |
| **eNOS** | Endothelial nitric oxide synthases |
| **ESS** | Endothelial shear stress |
| **EVD** | Eigenvalue decomposition |
| **FCBF** | Fast correlation-based filter |
| **FDG** | Fluorescent analogue of glucose |
| **FDM** | Finite difference method |
| **FE** | Finite elements |
| **FEM** | Finite element method |
| **FES** | Functional electrical stimulation |
| **FFNN** | Feed-forward neural network |
| **FFR** | Fractional flow reserve |
| **FFRCT** | Fractional flow reserve from computed tomography |
| **FLAME** | Flexible large agent-based modeling environment |
| **fMRI** | Functional magnetic resonance imaging |
| **FNNs** | Feedforward neural networks |

| Abbreviation | Explanation |
|---|---|
| **FRPVE** | Fiber-reinforced poro-viscoelastic |
| **FSI** | Fluid–structure interaction |
| **FVM** | Finite volume method |
| **GAN** | Generative adversarial network |
| **GFR** | Growth factor receptor |
| **GLS** | Generalized least squares |
| **Glyc** | Glycemia |
| **GOF** | Goodness of fit |
| **HDL** | High-density lipoprotein |
| **Hmean** | Mean body height |
| **ICA** | Angiography |
| **IMAG** | Interagency modeling and analysis group |
| **ISR** | In-stent restenosis |
| **IVUS** | Intravascular ultrasound |
| **KL** | Kullback–Leibler |
| **LDL** | Low-density lipoprotein |
| **LINC** | Linker of nucleoskeleton and cytoskeleton |
| **LLLT** | Low-level light therapy |
| **LOC** | Lab-on-a-chip |
| **LV** | Linear viscoelastic |
| **MAPEL** | Mechanistic axes population ensemble linkage |
| **MCP-1** | Monocyte chemoattractant protein |
| **Mech-ABM** | Mechano-agent-based model |
| **MIF** | Migration inhibitory factor |
| **ML** | Machine learning |
| **MMP** | Matrix metalloproteinase |
| **MPR** | Multiplanar reformations |
| **MRA** | Magnetic resonance angiography |
| **MRI** | Magnetic resonance imaging |
| **MSC** | Mesenchymal stem cell |
| **MSCT** | Multislice CT scanners |
| **MVND** | Multivariate normal distributions |
| **NLV** | Nonlinear viscoelastic |
| **NMR** | Nuclear magnetic resonance |

*(Continued)*

(Continued)

| Abbreviation | Explanation |
|---|---|
| **NN** | Neural network |
| **OCT** | Optical coherence tomography |
| **OLS** | Ordinary least squares |
| **PAC** | Pia-arachnoid complex |
| **PCT** | Perfusion CT |
| **PDEs** | Partial differential equations |
| **PDGF** | Platelet-derived growth factor |
| **PDMS** | Polydimethylsiloxane |
| **PET** | Positron emission tomography |
| **PK/PD** | Pharmacokinetic/Pharmacodynamics |
| **PNS** | Peripheral nervous system |
| **POC** | Point-of-care |
| **PP** | Pulse pressure |
| **pQCT** | peripheral quantitative computed tomography |
| **PyEL** | Python edge loading |
| **QCT** | Quantitative computed tomography |
| **QLV** | Quasi-linear viscoelastic |
| **RDE** | Reaction-diffusion equation |
| **RFA** | Radiofrequency ablation |
| **RLS** | Recursive least squares |
| **RP** | Rapid prototyping |
| **sAPs** | spontaneously evoked action potentials |
| **SBP** | Systolic blood pressure |
| **SCI** | Spinal cord injury |
| **SCPC** | Spinal-cord-pia-arachnoid construct |
| **SMC** | Smooth muscle cell |
| **SPECT** | Single photon emission computed tomography |
| **STN** | Signal transduction network |
| **SVM** | Support vector machines |
| **TAZ** | Transcriptional co-activator |
| **TBI** | Traumatic brain injury |
| **TC** | Total cholesterol |
| **TIMP** | Tissue inhibitor of metalloproteinase |

| Abbreviation | Explanation |
|---|---|
| **tNIRS** | transcranial near-infrared stimulation |
| **TRH** | Thyrotropin-releasing hormone |
| **TSH** | Thyroid-stimulating hormone |
| **US** | Ultrasound |
| **VEGFR** | Vascular endothelial growth factor and its receptor |
| **VSEPR** | Valence shell electron pair repulsion |
| **WLS** | Weighted least squares |
| **XAI** | Explainable artificial intelligence |
| **YAP** | Yes-associated protein |

# List of Terms

**Biological systems** It is a complex network that connects several biologically relevant entities. Biological organization spans several scales and is determined based on different structures depending on what the system is.

**Bone remodeling** Bone remodeling (or bone metabolism) is a lifelong process where mature bone tissue is removed from the skeleton (a process called bone resorption) and new bone tissue is formed (a process called ossification or new bone formation). These processes control the reshaping or replacement of bone following injuries like fractures and also microdamage, which occurs during normal activity. Remodeling also responds to functional demands of mechanical loading.

**Cell differentiation** It is the process by which dividing cells change their functional or phenotypical type. All cells presumably derive from stem cells and obtain their functions as they mature.

**Cell migration** It is a central process in the development and maintenance of multicellular organisms. Tissue formation during embryonic development, wound healing, and immune responses all require the orchestrated movement of cells in particular directions to specific locations. Cells often migrate in response to specific external signals, including chemical signals and mechanical signals.

**Cell proliferation** It is the process by which a cell grows and divides to produce two daughter cells. Cell proliferation leads to an exponential increase in cell number and is therefore a rapid mechanism of tissue growth. Cell proliferation requires both cell growth and cell division to occur at the same time, such that the average size of cells remains constant in the population. Cell division can occur without cell growth, producing many progressively smaller cells (as in cleavage of the zygote), while cell growth can occur without cell division to produce a single larger cell (as in growth of neurons).

**Chemical bonds** They hold molecules together and create temporary connections that are essential to life. Types of chemical bonds include covalent, ionic, and hydrogen bonds and London dispersion forces.

**Computed tomography** It combines a series of X-ray images taken from different angles around your body and uses computer processing to create cross-sectional images (slices) of the bones, blood vessels, and soft tissues inside your body.

**Drug delivery** It refers to approaches, formulations, manufacturing techniques, storage systems, and technologies involved in transporting a pharmaceutical compound to its target site to achieve a desired therapeutic effect.

**Enzyme** It is a biological catalyst and is almost always a protein. It speeds up the rate of a specific chemical reaction in the cell.

**Flexible large agent-based modeling environment** FLAME (flexible large-scale agent-based modeling environment) is a very general system for building detailed agent-based models that generate highly efficient simulation software that can run on any computing platform.

**Gene expression** It is the process by which the information encoded in a gene is used either to make RNA molecules that code for proteins or to make noncoding RNA molecules that serve other functions.

**Gene networks** A gene regulatory network is a set of genes, or parts of genes, that interact with each other to control a specific cell function. Gene regulatory networks are important in development, differentiation, and responding to environmental cues.

**General Data Protection Regulation (GDPR)** It is a regulation in EU law on data protection and privacy in the European Union (EU) and the European Economic Area (EEA). The GDPR is an important component of EU privacy law and of human rights law, in particular Article 8(1) of the Charter of Fundamental Rights of the European Union.

**Hardness** It is the resistance of a material to localized plastic deformation.

**Health Insurance Portability and Accountability (HIPPA)** The Health Insurance Portability and Accountability Act of 1996 (HIPAA) is a federal law in the United States that required the creation of national standards to protect sensitive patient health information from being disclosed without the patient's consent or knowledge. The US Department of Health and Human Services (HHS) issued the HIPAA Privacy Rule to implement the requirements of HIPAA. The HIPAA Security Rule protects a subset of information covered by the Privacy Rule.

**Heat conduction** It is the transfer of internal thermal energy by the collisions of microscopic particles and movement of electrons within a body. The microscopic particles in heat conduction can be molecules, atoms, and electrons. Internal energy includes kinematic and potential energy of microscopic particles.

**Hill equation** In biochemistry and pharmacology, the Hill equation refers to two closely related equations that reflect the binding of ligands to macromolecules, as a function of the ligand concentration.

**Hodgkin–Huxley model** The Hodgkin–Huxley model, or conductance-based model, is a mathematical model that describes how action potentials in neurons are initiated and propagated.

**Homogenization model** The numerical homogenization method accurately considers the geometry and spatial distribution of the phases and also precisely estimates the propagation of damage to accurately predict the failure strength.

**Hyperelasticity** They exhibit highly nonlinear elastic response when subjected to very large strains.

***In vitro*** *In vitro* (meaning in glass, or in the glass) studies are performed with microorganisms, cells, or biological molecules outside their normal biological context.

***In vivo*** Studies that are *in vivo* (Latin for "within the living," often not italicized in English) are those in which the effects of various biological entities are tested on whole, living organisms or cells, usually animals, including humans and plants, as opposed to a tissue extract or dead organism.

**Implant** It is a medical device manufactured to replace a missing biological structure, support a damaged biological structure, or enhance an existing biological structure. Medical implants are human-made devices, in contrast to a transplant, which is a transplanted biomedical tissue. The surface of implants that contacts the body might be made of a biomedical material such as titanium, silicone, or apatite depending on what is the most functional. In some cases, implants contain electronics, e.g. artificial pacemaker and cochlear implants. Some implants are bioactive, such as subcutaneous drug delivery devices in the form of implantable pills or drug-eluting stents.

***In silico*** In biology and other experimental sciences, an *in silico* experiment is one performed on a computer or via computer simulation.

**Isotropic material** Isotropic materials are materials whose properties remain the same when tested in different directions. Isotropic materials differ from anisotropic materials, which display varying properties when tested in different directions. Common isotropic materials include glass, plastics, and metals.

**Lab on a chip** It is a device that integrates one or several laboratory functions on a single integrated circuit (commonly called a "chip") of only millimeters to a few square centimeters to achieve automation and high-throughput screening.

**Lewis structures** These are also called Lewis dot diagrams, electron dot structures, or Lewis electron dot structures, which represent atoms and their positions in the molecular structure with their chemical symbols.

**Lipidomics** It is the study of reaction pathways involved in lipid metabolism within biological systems.

**Magnetic resonance imaging** It is a medical imaging technique that uses a magnetic field and computer-generated radio waves to create detailed images of the organs and tissues in your body.

**Mechanosensors** It is a sensory neuron that responds to mechanical stimuli.

**Metabolomics** It is the large-scale study of small molecules, commonly known as metabolites, within cells, biofluids, tissues, or organisms. Collectively, these small molecules and their interactions within a biological system are known as the metabolome.

**Microfluidics** It is both the science that studies the behavior of fluids through microchannels and the technology of manufacturing microminiaturized devices containing chambers and tunnels through which fluids flow or are confined. Microfluidics deal with very small volumes of fluids, down to femtoliters (fL), which is a quadrillionth of a liter.

**Molecular bonds** A molecular or covalent bond is formed when atoms bond by sharing pairs of electrons.

**Molecular structure** It describes the location of the atoms, not the electrons. We differentiate between these two situations by naming the geometry that includes all electron pairs and the electron-pair geometry. The structure that includes only the placement of the atoms in the molecule is called the molecular structure.

**Nanotechnology** It is the manipulation of matter on a near-atomic scale to produce new structures, materials, and devices. The technology promises scientific advancement in many sectors such as medicine, consumer products, energy, materials, and manufacturing.

**Non-Newtonian fluid** A non-Newtonian fluid is a fluid whose flow (viscosity) properties differ from those of Newtonian fluids, described by Sir Isaac Newton. Non-Newtonian fluids have viscosities that change according to the amount of force that is applied to the fluid. The viscosity changes as the force applied changes.

**Orthotropic material** In materials science and solid mechanics, orthotropic materials have material properties at a particular point, which differ along three orthogonal axes, where each axis has twofold rotational symmetry. These directional differences in strength can be quantified with Hankinson's equation.

**Pharmacodynamics** It is the study of how the drug affects the organism.

**Pharmacokinetics** It is the study of how an organism affects a drug.

**Photoelectric effect** It is the emission of electrons when electromagnetic radiation, such as light, hits a material.

**Photon** A photon (from Ancient Greek φῶς, φωτός (phôs, phōtós) "light") is an elementary particle that is a quantum of the electromagnetic field, including electromagnetic radiation such as light and radio waves, and the force carrier for the electromagnetic force.

**Polarity** It is, in chemical bonding, the distribution of electrical charge over the atoms joined by the bond.

**Poroelastic medium** It is a field in materials science and mechanics that studies the interaction between fluid flow and solid's deformation within a linear porous medium, and it is an extension of elasticity and porous medium flow (diffusion equation). The deformation of the medium influences the flow of the fluid and vice versa.

**Porosity** It is a measure of the void (i.e. empty) spaces in a material and is a fraction of the volume of voids over the total volume, between 0 and 1, or as a percentage between 0% and 100%.

**Positron emission tomography** It is an applied method of medical imaging based on two fundamental natural phenomena: (i) the phenomenon of radioactivity $\beta^+$, and (ii) the phenomenon of positron-electron annihilation.

**Protein activity** A critical function of proteins is their activity as enzymes, which are needed to catalyze almost all biological reactions.

**Proteomics** It is the systematic, large-scale analysis of proteins. It is based on the concept of the proteome as a complete set of proteins produced by a given cell or organism under a defined set of conditions.

**Pseudoplastic fluid** It is a fluid that increases viscosity as force is applied. A typical example is a suspension of cornstarch in water with a concentration of one to one. This cornstarch behaves like water when no force is applied; however, it is solidified as force is applied.

**Radiation** It is energy that comes from a source and travels through space at the speed of light. This energy has an electric field and a magnetic field associated with it and has wave-like properties. You could also call radiation "electromagnetic waves."

**Reynolds number** It is the ratio of inertial forces to viscous forces within a fluid that is subjected to relative internal movement due to different fluid velocities. A region where these forces change behavior is known as a boundary layer, such as the bounding surface in the interior of a pipe.

**Single photon emission computed tomography** It is a nuclear medicine tomographic imaging technique using gamma rays.

**Stent** It is a tiny tube that your doctor can insert into a blocked passageway to keep it open. The stent restores the flow of blood or other fluids, depending on where it is placed.

**Stiffness** It is defined as the resistance to a force causing a member to bend.

**Strain** Proportional deformation.

**Stress** Force per unit area.

**Transcriptomics** It is the study of the transcriptome – the complete set of RNA transcripts that are produced by the genome, under specific circumstances or in a specific cell – using high-throughput methods, such as microarray analysis.

**Ultrasonic waves** Ultrasonic waves are sound waves whose frequencies are higher than those of waves normally audible to the human ear. The angular frequencies of the ultrasonic waves produced in laboratories lie from about $10^5\ s^{-1}$ to about $3 \times 10^9\ s^{-1}$, the former value representing the limit of audibility of the human ear.

**Ultrasound imaging** It uses high-frequency sound waves to view inside the body.

**Viscoelastic materials** They combine two different properties. The term "viscous" implies that they deform slowly when exposed to an external force. The term "elastic" implies that once a deforming force has been removed the material will return to its original configuration.

**Viscosity** It is a measure of its resistance to deformation at a given rate.

**Weibel model** It is a model of the respiratory airway tree. In it, each parent airway, starting with the trachea, splits into two daughter airways. Based on measurements from several cadavers, characteristic branching angles, airway diameters, and lengths for the different airway generations are prescribed.

**Windkessel models** The four elements of the Windkessel models are aortic compliance, aortic impedance (resistance and inductance of the aorta), resistance of the peripheral arteries, and compliance of the peripheral arteries.

**Wound healing** It is a complex biological process that consists of hemostasis, inflammation, proliferation, and remodeling. Large numbers of cell types – including neutrophils, macrophages, lymphocytes, keratinocytes, fibroblasts, and endothelial cells – are involved in this process.

**X-ray radiography** It uses a very small dose of ionizing radiation to produce pictures of the body's internal structures. X-rays are the oldest and most frequently used form of medical imaging. They are often used to help diagnose fractured bones, look for injury or infection, and to locate foreign objects in soft tissue.

**Young's modulus** It is a measure of the ability of a material to withstand changes in length when under lengthwise tension or compression. Sometimes referred to as the modulus of elasticity, Young's modulus is equal to the longitudinal stress divided by the strain.

# 1

# Systems Biology and Multiscale Modeling

## 1.1 Introduction

The biological systems are characterized by a significant complexity ideally described using networks and pathways as well as their potential interconnections between their parts or with other external systems. They present temporal and spatial dimensions, which enable the definition of the system's evolution, growth, and development. Furthermore, the biological systems consist of several scales of representation starting from the cell level (e.g. gene pathways, molecular pathways, protein translation, and function) to higher levels such as tissue or organ level. The modeling of such systems is based on the proper definition of the outcomes and goals and according to them on the selection of the necessary subsets of features, which consist of the specific component parts which are necessary for the modeling goals' achievement. Then, the model of the system comprises the set of features and processes requiring some inputs and providing some outputs. First, the inputs are possible external "forces" which affect our set of features or even noise, which does not affect it. Outputs are the responses of the set of features and processes to the input stimuli, as this is observed from outside of the system.

The usual representation of a system model is provided by a mathematical model, which is defined by one or more mathematical equations and the necessary operations among them. The major advantage of using mathematical formulation for the representation of a biological system is that it is based on mathematical theorems and laws, which enable the implementation of simple as well as highly complex models empowering the proper evaluation of our hypothesis. Usually, differential equations are used for the development of biological mathematical models, but in other cases, simpler ones based on algebra or

*Multiscale Modelling in Biomedical Engineering*, First Edition. Dimitrios I. Fotiadis, Antonis I. Sakellarios, and Vassiliki T. Potsika.

Published 2023 by John Wiley & Sons, Inc.

geometry models can be employed. After the definition of a mathematical model, which describes a biological system, it is possible to perform the computational simulation. This means that the computational model simulates the biological system and its functions as those have been defined by the mathematical equations. Besides the computational models, other approaches in biological systems are: the stochastic models, which employ probabilistic mathematical laws and deterministic models whose components, mathematical variables, and parameters are represented by symbols with unique and not random values. The deterministic models are used to present the relationships between large numbers of entities, such as the molecules. Finally, the compartment models are distinguished by discrete boundaries between components called compartments.

## 1.2 Systems Biology

Systems biology is an evolving research field with many definitions, though all of them encompass a biological system, which describes some conditions, pathways, or mechanisms studied at one or more scales. The scales may have two orientations: (i) the first one is the categorization according to the organism (from molecular to organismic scale), and (ii) the second is the temporal scale (e.g. from nanoseconds to years) meaning the time scale under which the process is performed. Additionally, all definitions of systems biology conclude with the same aim: a better description and understanding of the biological system as a whole. Although the trend in computational biology is to employ systems biology for the description of a biological system, the concept is not new. On the other hand, it is being used and implemented for over a century. Schmitt and Schmitt [1] and Rashevsky [2] attempted to describe biological systems using mathematics and physics. Initially, the concepts were applied to neurophysiological systems [3]. The developed models were at organ-system level and they were considered as systems physiology models, while independent studies presenting systems pharmacology models were also reported. Pharmacokinetics and pharmacodynamics were modeled at all levels from cells to tissues and organs in order to describe the interaction of the body with the drug as well as its consequent effect on the body.

After the 1990s, an acceleration of big data collection has been observed meaning that omics types of data are collected. Such data include transcriptomics, lipidomics, proteomics, and metabolomics and they provide the opportunity to develop systems biology models at the microscopic level usually called molecular systems biology. In that case, the interaction between different components and molecules is usually described by creating dynamical networks and pathways. The pathways and networks provide an overall perspective of the mechanism and can describe all the interactions between these molecules as well as the consequences in the case of external factors and forces such as a mutation of a gene and the infection from a pathogenic organism. The first studies, which were based on omics data used available algorithms and methodologies for the development of systems biology models. In most cases, these approaches were adequate, especially in the case of simple systems or in the case of modeling one level or at least for the interaction between two scales.

The collection of big data leads to the development of models in the field of genetics. Computational models not only support the implementation of gene networks but they are also used for the understanding of their functionality [4–7]. In a similar manner, systems biology is used to understand the mechanisms of absorption, distribution, excretion, and toxicity of substances and/or molecules in order to prevent their potential negative effects before their use in clinical practice [8–10]. For that purpose, many algorithms and techniques have been employed such as Pathway Assist™ (http://www.ariadnegenomics.com/), PathArt™ (http://jubilantbiosys.com/), MetaCore™ (http://www.genego.com), and Pathways Analysis™ (http://www.ingenuity.com/). Other methodologies include clustering of gene-expression data and generation of interaction networks [11], superparamagnetic clustering [12], simulated annealing [13], probabilistic graphical models [14], and Monte Carlo optimization [15].

## 1.3 Systems Biology Modeling Goals

One of the main goals of systems biology modeling is to develop models of biological systems described mechanistically and/or mathematically to understand the biological details and interactions. Such models target usually at simulating the biological experiment by predicting its outcome, and in the case of accurate predictions, an important step is the understanding of the biological system's mechanisms [16–19]. As it is expected, the goals of systems biology range from the scale or level of the models. Thus, it is possible to identify very complex aims in the case of microscale modeling, where interactions between molecular components exist and their expressions and concentrations are required as boundary and initial conditions in order to describe adequately the biosystem in the specific scale or in the interaction with other scales. Similarly, simpler aims can be defined for: (i) macroscale models, where the biosystem can be described by simple mathematical equations, differential, or even algebraic and (ii) microscale models, where a simple chemical reaction is enough to describe the dynamics of the system. Ideally, the modeling goals are satisfied when a multidisciplinary approach is adapted which integrates the base theory with the basic experiment and the corresponding mathematical representation. All together support each other in a circular way: theory is necessary to define the experiment, which provides data for

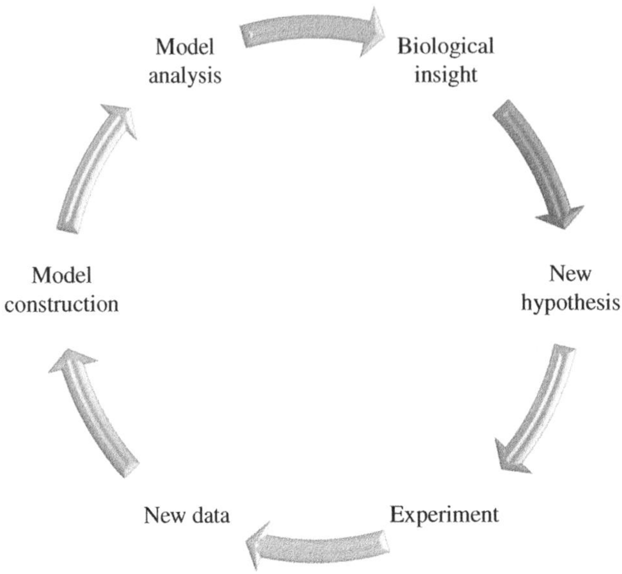


**Figure 1.1** The iterative cycle of wet and dry laboratory research.

the development of the mathematical model but the mathematical model is validated by the experiment, which refines the current theory and determines new. Integrative systems biology involves the iterative cycle of wet and dry laboratory research (Figure 1.1).

Other systems biology modeling goals are a better understanding of the interaction between various systems inside an organization. Such interactions include signaling pathways, biochemical pathways, and gene networks. The homeostatic interactions define the functional states between the systems and may be responsible for pathologic conditions. Thus, the developed models aim to identify the role of each component or feature inside the network or pathway and the degree to which it may affect the organization. These models can describe the cell cycle from the division to its apoptosis or differentiation. Other models focus on the description of the transcription of genes and translation of proteins under specific regulatory pathways, e.g. under the regulation by an enzyme and under specific conditions such as increased body temperature. More complex biosystems include energy generation and intercellular communication which is very common in neurological applications and models.

Another major goal of systems biology modeling is to test a specific biological hypothesis about a biological function. Such models usually attempt to identify the behavior or response of the system to stimuli even internal or external. Moreover, they aim to predict the interaction between other organisms or to identify the effect of pathologic and abnormal conditions in the system. Such models are usual in pharmacodynamics and contribute to the development of new treatment approaches for pathologic conditions and diseases [20, 21]. Traditionally, drug development is based on the identification of the substance, which interacts with the compounds; their experimental and using animal testing; and, finally, their testing in patients in clinical studies. Unfortunately, potential side effects are discovered at a late stage, e.g. after their use in patients. On the other hand, using systems biology models in pharmacology or even in toxicology provides the in silico clinical trials and obviously benefits the proper identification of side effects at earlier stages and, in many cases, even before the animal testing. In this way, the benefit is considered huge not only in terms of socioeconomic factors by reducing the costs of clinical trials and experiments in animals but also socially by improving the healthcare of patients by reducing the potential side effects. Furthermore, a reduction in the population participating in the clinical trials is achieved.

## 1.4 Systems Biology Modeling Approach

Modeling biological processes often requires accounting for action and feedback involving a wide range of spatial and temporal scales. Biological systems are organized at scales of many orders of magnitude in space and time. Space spanning ranges from the molecular scale ($10^{-10}$ m) to the living organism scale (1 m), and time ranges from nanoseconds ($10^{-9}$ s) to years ($10^{8}$ s) (Figure 1.2).

Besides the multidisciplinary character of a systems biology model, the scales of the biosystem also define the type of approach which will be employed. For example, the typical modeling approaches for microscale biosystem modeling are the reaction kinetics using ordinary differential equations (ODEs), the lattice reaction-annihilation processes, and others. These approaches can be used for the modeling of molecular and subcellular processes such as mutations, gene alterations, signaling, metabolic pathways, and parts of the cell cycle. At the mesoscopic scale, the approaches include cell-level ODEs, cellular

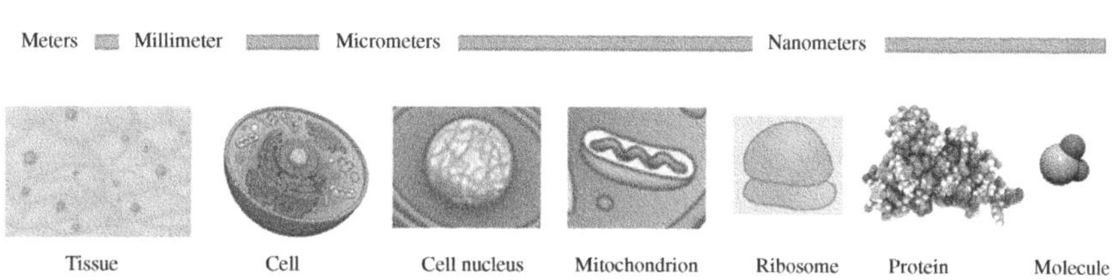


**Figure 1.2** The space scales in biological systems.

automata, and evolution rules in order to model cell–cell interactions and cell–matrix interactions such as the phenomena of angiogenesis, the immune response, the local remodeling of the Extracellular matrix (ECM), etc. Finally, at the macroscopic scale, the typical approaches are usually based on the solution of partial differential equations (PDEs) such as the reaction-diffusion, the continuous mechanics, and the convection equations. These models are used for the simulation not only of processes at the tissue level, e.g. diffusion of nutrients, cell migration, and invasion, but also of the blood flow dynamics, plaque growth modeling, and bone dynamics, which will be described in detail in the next chapters (Chapters 3–9). Table 1.1 presents the modeling approaches for each scale and potential applications in systems biology. It is worth noting that interactions between these scales and approaches can be defined.

From the aforementioned, it is clear that there is a strong relation of the biological scale with the chosen experimental approach and the corresponding modeling approach. Figure 1.3 presents the relation of the modeling and experimental approach in correspondence to the biological scale. However, this scheme lacks one major characteristic of systems biology modeling concerning the interaction between the different scales (spatial and temporal) providing the ability to develop multiscale models able to simulate and describe complex phenomena in the biological systems.

To this end, Figure 1.4 presents a schematic illustration of the biological scales of significant relevance for cancer modeling including atomic, molecular, microscopic (tissue/multicellular), and macroscopic (organ) scales [22–24]. Compared to Figure 1.2, different scales represent both different spatial and temporal ranges. Multiscale cancer modeling requires the establishment of the linking between those scales.

**Table 1.1** Modeling approaches and typical examples per scale in systems biology modeling.

| Scale | Typical modeling approaches | Examples |
|---|---|---|
| Microscopic scale | Reaction kinetics using ODEs, the lattice reaction-annihilation processes | Mutations, gene alterations, signaling, metabolic pathways and parts of the cell cycle, mitosis, apoptosis, etc. |
| Mesoscopic scale | Cell level ODEs, cellular automata, and evolution rules | Angiogenesis, the immune response, the local remodeling of the ECM, etc. |
| Macroscopic scale | Reaction-diffusion, continuous mechanics, convection equations | Diffusion of nutrients, cell migration and invasion, cardiovascular dynamics, bone dynamics, etc. |

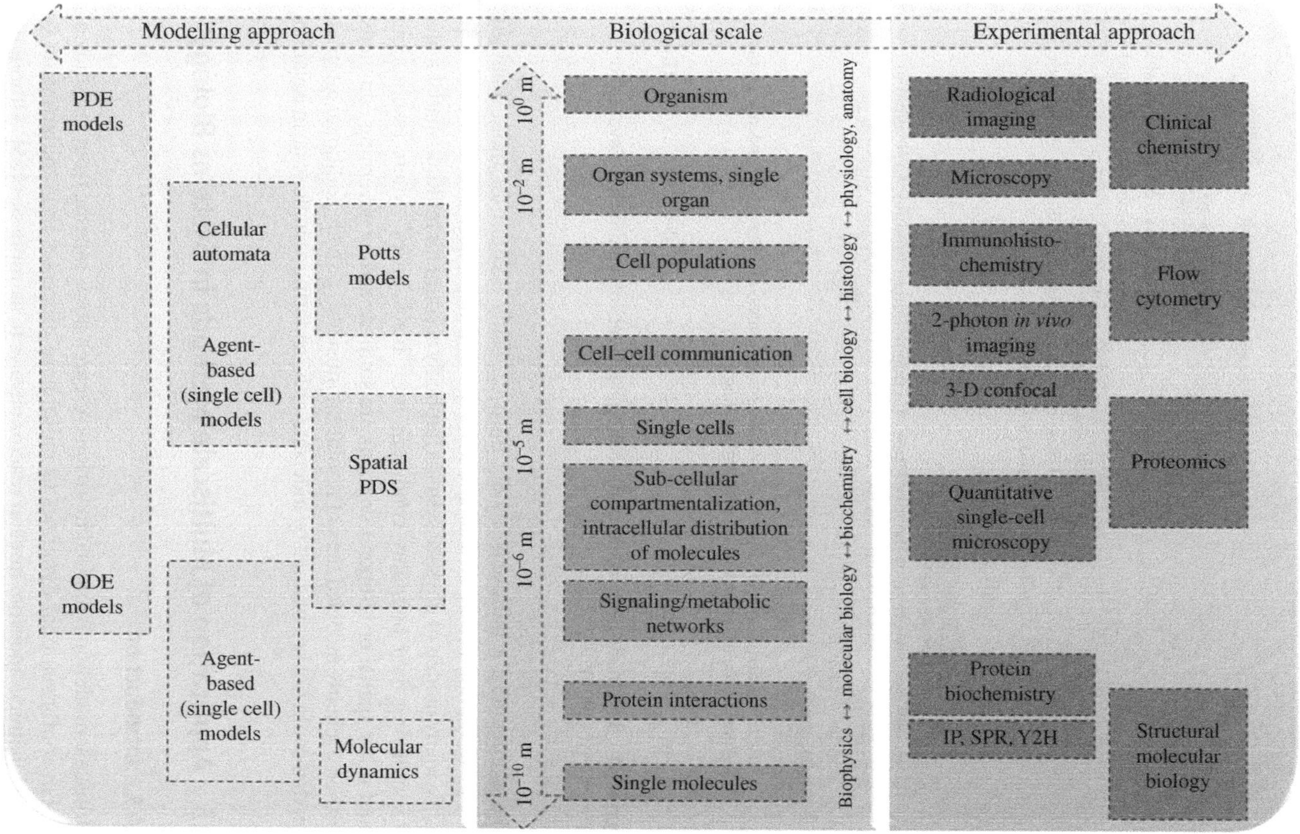


**Figure 1.3** The relation of the modeling and experimental approach according to the biological scale.

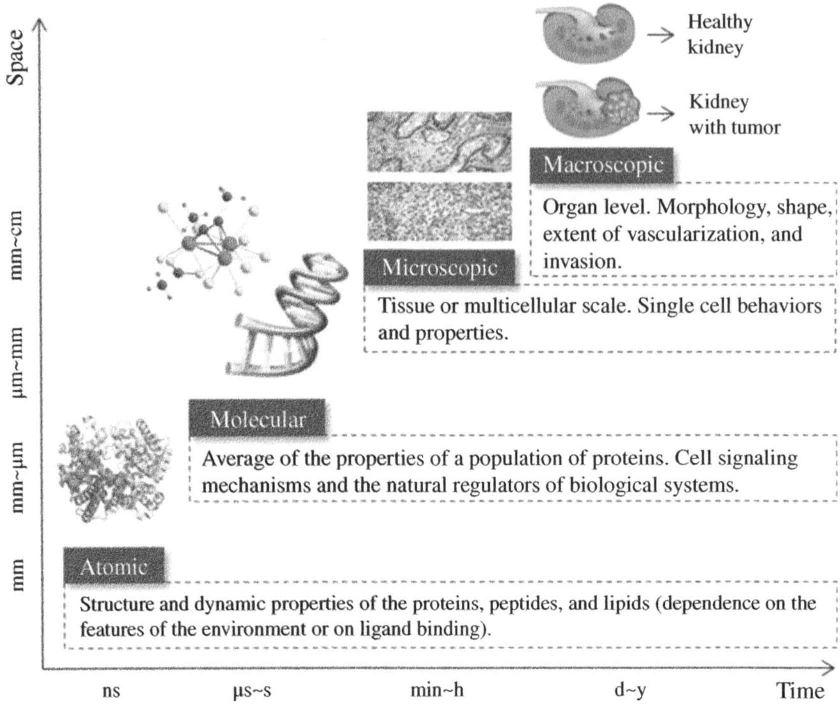


**Figure 1.4** Schematic illustration of the biological levels of significant relevance for kidney cancer modeling.

Fundamentally, a multiscale model must explicitly account for more than one level of resolution across measurable domains of time, space, and/or function. To clarify, many models of physical systems implicitly account for multiple spatial scales by simplifying their boundary conditions into "black boxes" where assumptions about other spatial or temporal domains are summarized by governing equations. Further, explicitly modeled tiers of resolution must also provide additional information that could not be obtained by independently exploring a single scale in isolation.

## 1.5 Application of Multiscale Methods in Systems Biology

### 1.5.1 Introduction

Multiscale modeling is applied in various research areas ranging from the study of protein conformational dynamics to multiphase processes in granular media or hemodynamics, and from nuclear reactor physics to astrophysics [25]. Although a significant diversity is observed in research applications, there are many common principles and challenges such as the need for the development of advanced tools for programming and executing multiscale simulations, the validation of the numerical methods and results using experimental procedures, the selection of the proper numerical method according to the computational cost, and the biological substance or mathematical and physical aspects of the examined problem. In addition, despite the area of application, it is widely accepted that computer simulations are usually more cost-effective, efficient, and time-consuming compared to laboratory experiments and clinical studies.

According to Hoekstra et al. [26], a major question which arises is the determination of the concepts which orchestrate the multiscale modeling approaches which are inherent in multiscale systems. Traditional modeling approaches are focused on one scale. Considering the example of solid media, engineers may be interested in the macro-scale behavior of solids using continuum models and represent the atomistic effects by constitutive relations, while physicists may be more interested in the behavior of solids at the atomic or electronic level, often working under the assumption that the relevant processes are homogeneous at the macroscopic scale [27]. Under this assumption, civil engineers are able to design structures (buildings, bridges, etc.), without the need to deepen the origin of the interactions between the atoms in the materials. On the other hand, physicists can give insight on the evolution of phenomena at a fundamental level, but they may face several difficulties when dealing with an engineering problem at the macroscale level.

The common aim of mathematics, engineering, and systems biology is to achieve a thorough understanding of biological systems at different hierarchical levels. The Interagency Modelling and Analysis Group (IMAG) has suggested the following definition for the term "multiscale modeling" [28–32]:

"Multiscale, biomedical modeling uses mathematics and computations to represent and simulate a physiological system at more than one biological scale. Biological scales include atomic, molecular, molecular complexes, subcellular, cellular, multicell systems, tissue, organ, multiorgan systems, organism, population, and behavior. These multiscale biomedical models may also include dynamical processes which span multiple time and length scales."

A holistic understanding of many biological processes requires multiscale models which capture the relevant properties on all these scales [28–32]. It can be questioned whether the identification of general laws is relevant as a research aim for biology, but universal design principles, without doubt, play a critical role in engineering approaches that inspire a large part of systems biology [33].

For example, cancer is considered a complex, heterogeneous disease, characterized by many interaction processes evolving in multiple scales in time and space that act in parallel to drive cancer formation, progression, invasion, and metastasis [34]. These processes range from molecular reactions to cell–cell interactions, to tumor growth and invasion on the tissue scale, and even to larger scales, such as the physiology, pathophysiology, and population scales. In addition, many cancer properties (e.g. size, cell density, extracellular ligands, cellular receptors, mutation type(s), phenotypic distribution, vasculature status, blood vessel permeability, and treatment prognosis) are dynamic and patient-dependent, changing and evolving with both time and treatments (e.g. cell death rate may vary over time when the patient is subjected to chemotherapy). All these dynamically changing cancer properties make the development of effective cancer therapies extremely difficult. Computational models which include patient-specific parameters could be a supplementary tool to current statistical approaches to enhance personalized medicine and prediction of complex behaviors of cancer.

## 1.6 The Use of Systems Biology and Multiscale Modeling in Biomedical and Medical Science

At the organism level, an "infinite" number of processes is happening throughout its life. Moreover, these processes may be complex and define the interactions between pathways of the same level or multilevel interactions. The basic research in biomedical and medical science aims to identify the role of each process in order to clarify the causative mechanisms and pathways, which promote disease evolution. The overall target of such research is to provide the knowledge for the diagnosis, prognosis, and prediction of events related to the disease.

Computational modeling has an incremental role in understating the mechanisms, which underlie the disease in order to provide predictors potentially used for the diagnosis, prognosis, and prediction. For example, machine learning techniques and systems biology models are being implemented nowadays for risk stratification in cardiovascular disease [35]. In the same field, computational modeling is used for the estimation of hemodynamics and the calculation of variables such as endothelial shear stress or lipid accumulation for the prediction of regions, which are prone to atherosclerotic plaque growth [36–38]. Such models are implemented in many other diseases such as in cancer and oncology research [39], in arthritis [40], brain [41], etc. [42, 43].

## 1.7 Application of Computational Methods in Biomedical Engineering

A biological system is a set of self-organized, differentiated components (elements) that interact pair-wise among themselves through various networks and media, isolated from other sets by boundaries called teguments and whose relation to other systems can be described as a closed loop in a steady state [44].

Advanced computational modeling is essential to understand the complex mechanisms that couple material, structural, and topological hierarchy, merging phenomena of different nature, size, and time scales in hierarchical materials. Numerical modeling also allows extensive parametric studies for the optimization of material properties and arrangement, avoiding time-consuming and complex experimental trials, and providing guidance in the fabrication of novel advanced materials [45].

The terms "top-down" and "bottom-up" are used often in computational modeling, referring to the relative degree of detail in the model. More specifically, the term "top-down" is used to describe numerical approaches at the macroscopic level and "bottom-up" at the microscopic level. Both, however, are usually involved in the modeling process, and deciding on a level of complexity is ultimately dependent on modeling goals and data availability. Top-down modeling approaches capture the overall dynamical features of the system, expressed as a collection of interacting subsystems, and then modeling the subsystems. Bottom-up modeling focuses initially on each subsystem and then the overall system is modeled through their interconnection [46].

This section mainly presents the finite element method (FEM), the finite difference method (FDM), and the boundary element method (BEM), which are the most popular computational methods to carry out numerical simulations in a wide variety of engineering applications [47–52]. Initially, this section presents the fundamental principles, key terms, and nomenclature for the application of computational modeling in biomedical engineering. Then, a description of the main features of FEM, BEM, and FDM is presented.

### 1.7.1 Fundamental Principles

Two of the most basic and frequently used terms in computational studies are "model" and "system." A **model** is a hypothetical description or representation of a complex entity or process [46]. A **system** is a collection of objects, usually interconnected or interacting in a coordinated way. Thus, a **system model** is a collection of objects, or component parts, normally interconnected in some way.

**Models of systems**, or **system models**, also known as structured models or structural models, are usually based on physical (e.g. biophysical or biochemical) principles and hypotheses, descriptive information about how a system is structured and how it functions. For our purposes, this includes models based on physical laws and their consequences (e.g. law of mass action, mass-balance relationships, cell transduction processes, and Newton's second law). Systems have inputs and outputs, as illustrated in Figure 1.5. They represent essential features, satisfy goals, include simplifications, and usually have inputs (stimuli) and outputs (responses). Inputs are stimuli generated externally to the system which enter into the system and influence it. Outputs are system responses to input

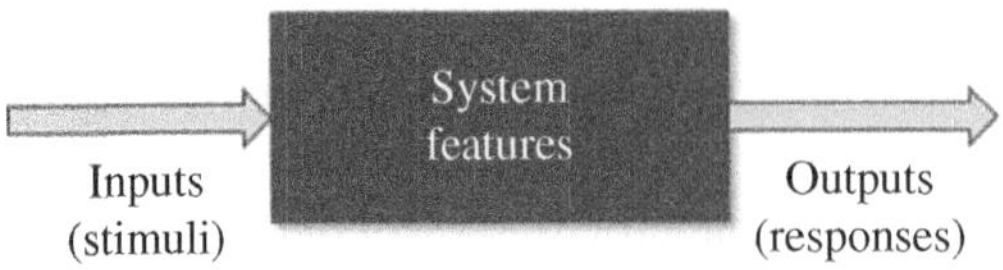


**Figure 1.5** Characteristics of system models.

stimuli. As systems can generally have more than one output, modeled outputs of interest are typically selected according to the aims of the modeling process. Therefore, system models can be defined by the relationships describing object connections and interactions of the inputs and outputs with system features.

**Simulation** means representation of the features of an object or a system using numerical methods. Each modeling process is a type of simulation. Mathematical modeling refers to the implementation of the equations of the system model on a computer aiming to: (i) obtain a solution to the examined problem and (ii) evaluate the model and study its properties or its predictive value. Simulation models are also called *in numero* models or *in silico* models or computational models.

**Computational continuum mechanics** deals with media at the macroscopic scale. Continuum models are established in which the microstructure of a medium is homogenized by phenomenological averaging. **Solid and fluid mechanics** are two of the most traditional areas of application.

**Structural mechanics** is considered a branch of solid mechanics due to the fact that structures are fabricated with solid media. The difference between these two terms can be explained as follows: computational solid mechanics refers to an applied-sciences approach, while computational structural mechanics focuses on the implementation of technological means to the analysis and design of structures.

**Computational fluid mechanics** is a branch of fluid mechanics that uses numerical analysis and data structures to solve and analyze problems that involve fluid flows. These problems deal with the equilibrium and motion of fluids – liquids and gases. Hydrodynamics, aerodynamics, and atmospheric physics are considered traditional subareas of application.

**Explicit and implicit methods** are different approaches applied in computer simulations of physical processes to obtain numerical approximations for the solution of time-dependent ordinary and PDEs. More specifically, explicit methods describe the state of a system at a later time from the state of the system at the current time. On the other hand, implicit methods lead to a solution by solving an equation involving both the current state of the system and the latter one. The engineering problem of heat transfer through an insulated rod is an indicative case study in implicit multiscale modeling. Whether solved using continuous PDEs or a discrete finite element approach, all solutions to this problem rely on carefully defining spatial boundary conditions, the fundamental laws of thermodynamics of a closed system, and material properties such as a thermal conductivity coefficient. Implicit methods require extra computational time and can be much harder to implement. Implicit methods are preferred in many problems which may arise in practice when the use of an explicit method requires impractically small time steps to control the error in the final result. In these cases, to ensure the required accuracy, an implicit method with larger time steps is used as it needs much less computation time. This implies that the selection of an explicit or implicit method depends on the problem to be solved.

Continuum mechanics problems may be subdivided into **statics** and **dynamics** according to whether inertial effects are taken into account or not. In **statics**, inertial forces are ignored or neglected. A **dynamic system (DS) model** is a model of a system, with inputs, outputs, and equations describing system motion in space and time. In other words, in order to know the motion of a DS in the "future," it is a prerequisite to know the status of the system "now." In physics and engineering, this distinction translates into a DS having the property that it can store energy in some form [46]. These problems may be subclassified into time invariant and quasi-static. **Time invariant** problems do not need to be considered explicitly (any time-like response-ordering parameter will do). In **quasi-static** problems (e.g. creep flow, fatigue cycling), a more realistic estimation of time is required but inertial forces are ignored as long as motions remain slow. In dynamics, the time dependence is explicitly considered because the calculation of inertial (and/or damping) forces requires derivatives with respect to the actual time to be taken.

**Static problems** can be classified into **linear and nonlinear statics**. Linear static analysis deals with static problems in which the response is linear in the cause-and-effect sense (e.g. if the applied forces on an object are doubled, the displacements and internal stresses are also doubled). Problems outside this domain are classified as nonlinear.

A final classification of computational solid and structural mechanics for static analysis is based on the **discretization method** by which the continuum mathematical model is discretized in space, i.e. converted to a discrete model of a finite number of degrees of freedom and may include the FEM, BEM, FDM, finite volume method (FVM), spectral method, and mesh-free method. For linear problems, FEMs currently dominate the scene, while BEMs are also widely used. For nonlinear problems, the use of FEM is dominant.

Figure 1.6 presents the solution to problems in biomedical engineering using computational means. **Differential equation models** model systems over continuous-time domains. Differential equations are algebraic or transcendental (sine, cosine, etc.) equalities which include terms with either differentials (e.g. $dx$) or derivatives (e.g. $dx/dt$), defined continuously over one or more independent

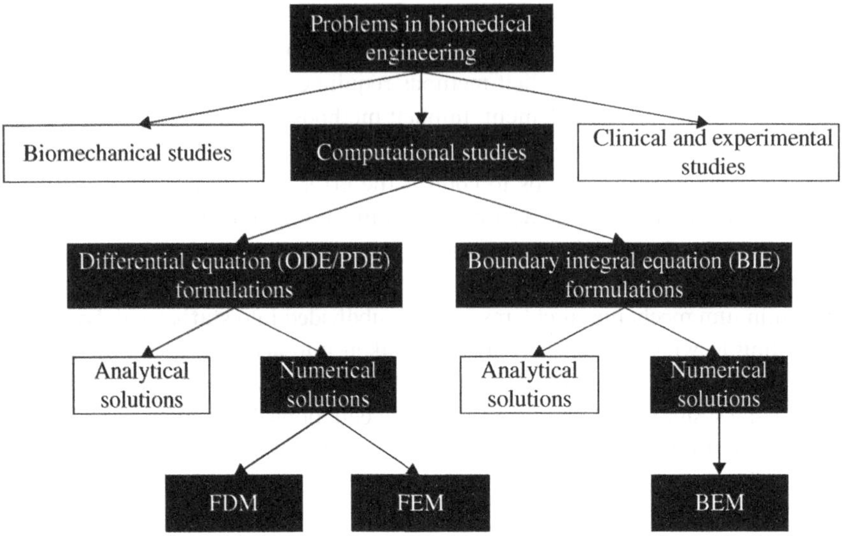


**Figure 1.6** The solution of problems in biomedical engineering using FDM, FEM, and BEM.

variables (e.g. $t$, in which $t$ denotes time). **Difference equations** (DE) are algebraic or transcendental equalities involving discrete sequences of values of the dependent variable(s) of the model, corresponding to discrete sequences of values of one or more of the independent variable(s) of the model. The description of the laws of physics for space- and time-dependent problems is usually expressed in terms of PDEs. Linear systems can often be represented by linear differential (ODE) or DE models, but these are not the only equations representing linear systems. For example, algebraic equations, like Newton's second law, and integral equations can also be linear. There is a close connection between differential and integral equations, and some problems may be formulated either way (e.g. Green's function, Fredholm theory, and Maxwell's equations).

**FEM** is a popular computational method for the solution of various problems in the domains of mathematics, physics, and engineering. It is widely applied to several areas including structural analysis, fluid flow, heat transfer, and mass transport. For the analytical solution of these problems, the solution of boundary value problems for PDEs is usually a prerequisite. The formulation of the problem results in a system of algebraic equations. The method leads to approximate values of the unknowns at discrete number of points over the domain. In order to solve a large problem, the problem is subdivided into smaller, simpler parts that are called **finite elements (FE).** The simple equations which model these FE are then integrated into a larger system of equations which models the problem as a whole. Section 2.3.2 provides a more detailed description on the analysis with FEM, which is often referred to as finite element analysis (FEA).

**FDMs** are numerical methods for solving differential equations by approximating them with DE, in which finite differences approximate the derivatives. FDMs are thus discretization methods. Nowadays, FDMs are the dominant approach to numerical solutions of PDEs. **FDMs**, which focus on the direct discretization of conservation laws, are favored in highly nonlinear problems of fluid mechanics. Spectral methods are based on global transformations, based on eigen decomposition of the governing equations, that map the physical computational domain to transform spaces where the problem can be efficiently solved. A recent newcomer to the scene of biomedical engineering are the mesh-free methods. These are FDMs on arbitrary grids constructed using a subset of FE techniques.

**BEMs** are computational methods of solving linear PDEs which have been formulated as integral equations (i.e. in boundary integral form). BEM applications cover fluid mechanics, acoustics, electromagnetics, fracture mechanics, and contact mechanics. BEM is often more efficient than other methods, including finite elements, in terms of computational resources for problems where there is a small surface/volume ratio. Conceptually, it works by constructing a "mesh" over the modeled surface. BEM is one of the most effective methods for numerical simulation of contact problems, in particular for simulation of adhesive contacts. However, for many problems the BEM is significantly less efficient than volume-discretisation methods (FEM, FDM, FVM).

Figure 1.7 presents the main stages of computer-based simulations showing the interaction between the physical system, the mathematical model, the discrete model, and discrete solution. Figure 1.7 introduces the terms "discrete" and

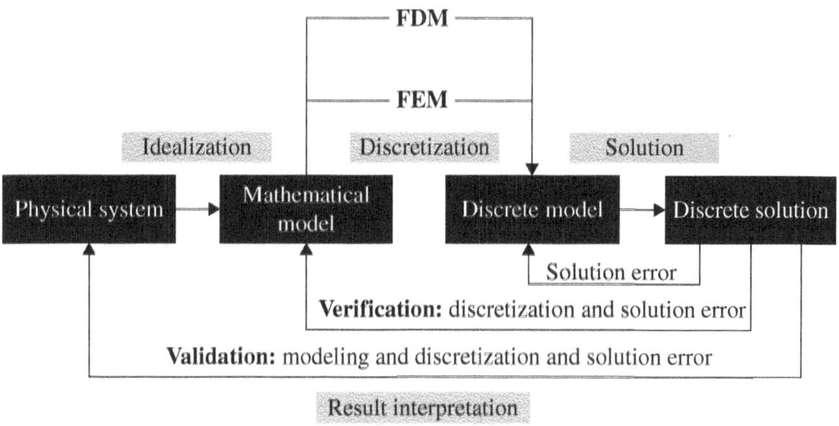


**Figure 1.7** The main stages of computer-based simulations.

"continuous." The term **continuous system** refers to a system in which the state variable(s) change continuously over time (e.g. the amount of water flow over a dam). Continuous systems are also known as analog systems. In a **discrete system**, the state variable(s) change only at a discrete set of points in time (e.g. customers arrive at $3:15, 3:23, 4:01$, etc). Because discrete systems have a countable number of states, they may be described with precise mathematical models. As computers are often used to model discrete systems and continuous systems as well, several methods have been developed to represent real-world continuous systems as discrete systems (e.g. sampling a continuous signal at discrete time intervals). A combination of multiple computational techniques, including both continuous and discrete systems, is optimal for efficiently capturing information across biological scales.

Additional classification **into deterministic and stochastic models** divides systems based on whether they contain a degree of "randomness" that allows for multiple solutions to the same initial conditions. System dynamics usually are represented with **deterministic** (not random) variables, typically as differential or DEs. States and outputs of **stochastic DS models** are stochastic processes (also called random processes) evolving in time and governed by probability distributions, as well as by the basic system dynamics equations depicting biosystem structure. They are especially useful for describing the evolution of biosystem dynamics when the number of objects represented by state variables is very small (e.g. for intracellular processes where a number of molecules interact with a similar number of others).

Each spatial scale can be summarized by the biological functions occupying that tier of resolution, allowing for modeling techniques to be implemented based on how well they represent these functions [53]. Figure 1.8 presents a conceptual map of modeling techniques divided into continuous and discrete categories across spatial scales for which they are most appropriate.

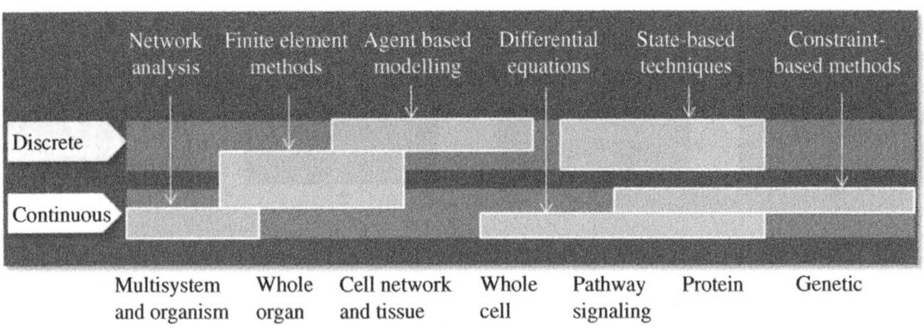


**Figure 1.8** Conceptual map of modeling techniques divided into continuous and discrete categories across spatial scales for which they are most appropriate. *Source:* Adapted from Walpole et al. [53].

**Network analyses** include discrete state-based techniques (e.g. Markov chains and Boolean networks) as well as continuous systems biology approaches. These methods are well suited to modeling the smallest tiers of resolution (genomics, proteomics, and metabolomics) [53].

**Agent-based modeling** has become a very popular and powerful tool for representing discrete stochastic biological processes as either compartmentalized or spatially defined models. These models include geometries in one-, two-, and three-dimensional configurations and may be scaled such that each fundamental agent is as large (groups of organisms) or as small (subcellular membrane components) as it is desired.

**Constraint-based modeling** aims to reduce the number of possible flux profiles and identify one that best predicts the metabolic phenotype of the organism under specified genetic and environmental conditions. Constraints limit the number of possible flux profiles for a given organism to some finite number and additional constraints further reduce the flux space and allow us to make even more accurate *in silico* predictions of metabolic phenotype. Potential constraints are generally assigned to the following four categories: physico-chemical constraints (e.g. mass and energy balance), environmental constraints (e.g. temperature, pH, substrate availability), spatial/topological constraints (e.g. organelle compartmentalization), and self-regulatory constraints [54, 55].

### 1.7.2 Finite Element Method

The FEM is an analytical engineering tool originated in the 1960s by the aerospace and nuclear power industries to find usable, approximate solutions to problems with many complex variables [47]. FEM employs piecewise approximation in which the continuum (domain) of interest is divided into several subregions called finite elements. Initially, the solution of each FE is obtained independently and at the next stage, the overall solution for the continuum is derived through the combination of these individual FE results. The basic procedure in the isoparametric finite element formulation is to express the element coordinates and element displacements in the form of interpolations using the natural coordinate system of the element [56]. Figure 1.9 presents an indicative scheme of the application of FEM in biomedical engineering (e.g. for the modeling of long bones), including indicative input data, system features, and output data as described in Figure 1.4.

The accuracy of FEM depends on the element's geometry, geometry distortion, shape function, laws used in developing the governing equation, and materials to be analyzed. Basically, there are three groups of elements and those are: (i) line elements (for one-dimensional analysis), (ii) planar elements (also known as membrane elements, for two-dimensional analyses), and (iii) solid elements (for

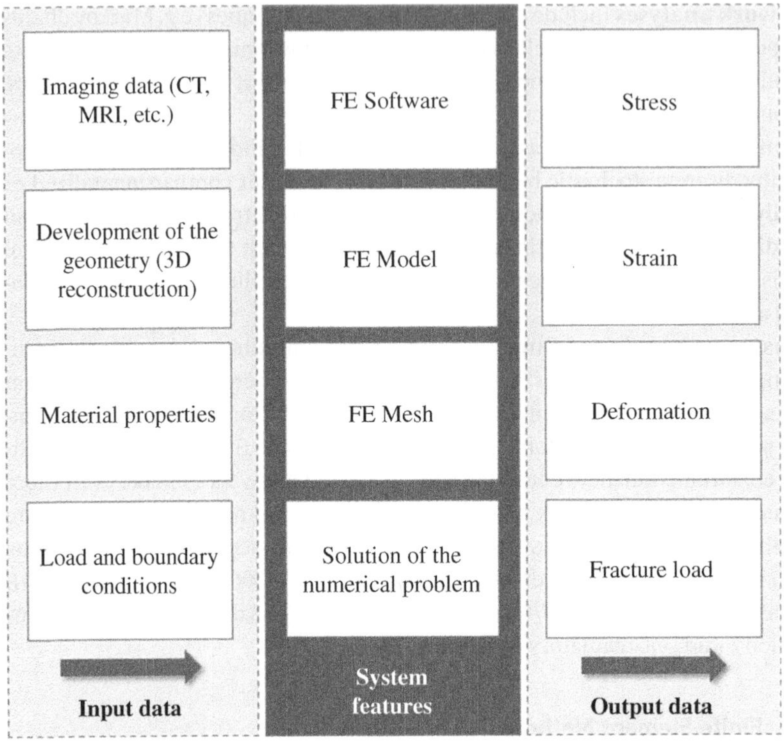


**Figure 1.9** Indicative scheme of the application of the Finite Element Method.

three-dimensional analyses). There are various types of elements formulated under each group as depicted in Figure 1.10. For example, spring, bar, and flexure elements are element types which can be categorized under line elements. Another category is general planar elements including triangular and rectangular elements. Tetrahedron, hexahedron (also known as brick), and pentahedron elements (also known as triangular prism or wedge) are examples of general solid elements.

For two-dimensional analysis, bilinear quadrilateral elements are superior compared to simple linear triangular elements in terms of meshing and accuracy. Accuracy of simple linear triangular elements can be improved by using higher-order elements, but it leads to a problem known as mesh locking, which is the main drawback of triangular elements when the analysis is done for incompressible materials [57]. It has been found that quadrilateral elements provide better accuracy compared to triangular elements [58]. As for the three-dimensional analysis, hexahedron elements may give more accurate results compared to tetrahedron elements.

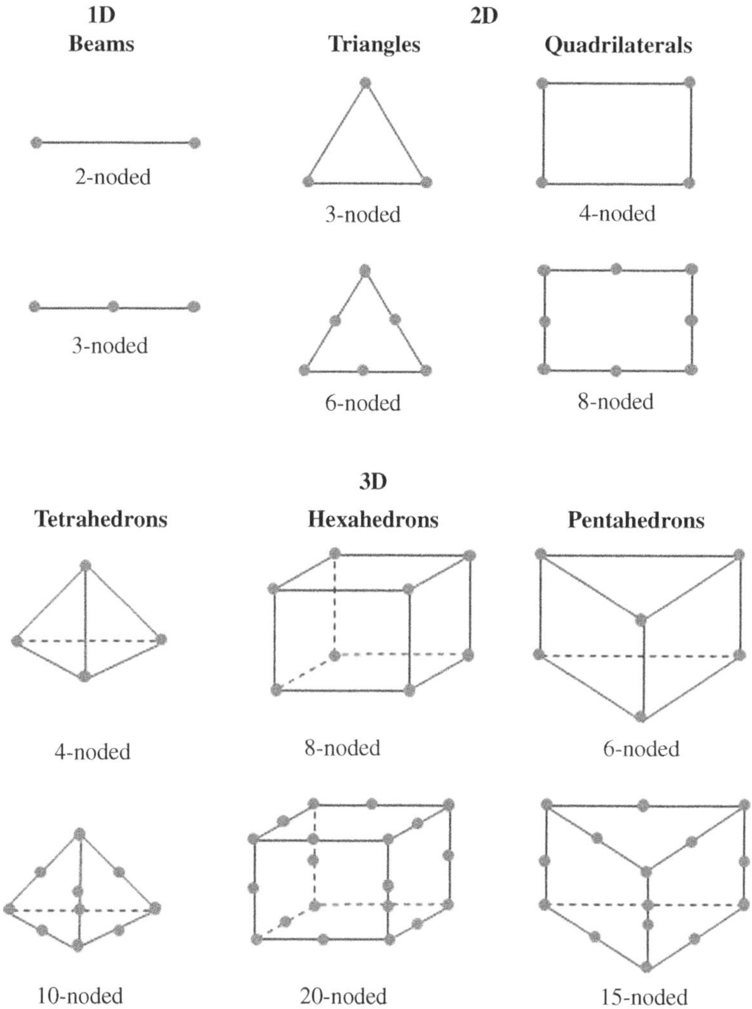


**Figure 1.10** Different types of elements for discretization using the Finite Element Method.

In biomedical applications, the first step for the development of a computational model is to acquire the necessary imaging data in order to provide an accurate visualization and representation of the studied system. According to [59–62], excellent isotropic unstructured tetrahedral grid generation algorithms exist for creating finite element or finite volume grids from imaging-derived closed triangulated manifolds. The aim of isotropic algorithms is to create tetrahedral elements with nearly equal internal angles and approximately equal edge lengths. Nevertheless, with medical imaging data, the lengths of these edges are typically more related to the resolution of the image than they are to the physical problem to be solved and do not consider the geometrical scale. Moreover, the tessellation of the volume is typically disordered. Structured grids, typically structured hexahedral grids, are computationally efficient and can be made to fit the physics of the problem, but structured grids can be laborious to construct if they can be constructed at all.

The behavior of a phenomenon can be represented using mathematical models (approximate models), which are derived based on principles and laws. There are several principles which are used to formulate FE. The principle of static equilibrium (also known as direct method) is used for phenomena which can be represented by simple governing equations. In addition, the theorem of Castigliano as well as the principle of minimum potential energy can be applied to complicated elastic structural systems. Higher mathematical principles, known as variational methods, are used to formulate FEA for phenomena governed by complex mathematical models, involving derivative terms.

Once the approximate model has been developed using principles and laws as described earlier, shape functions are then applied according to the element geometry to complete the finite element formulation. A general equation for a single finite element is represented in the following form:

$$[k]\{q\}=\{Q\}, \tag{1.1}$$

where [k] is the matrix representing characteristics of the continuum, {q} is the column matrix representing nodal values (output variable of interest), and {Q} represents the input to the continuum. For the case of stress analysis, [k] represents stiffness matrix, {q} represents vector of nodal displacements, and {Q} represents vector of nodal forces. Once individual FE are formulated, these would then be assembled to form global/assemblage equations which are represented as:

$$[K]\{r\}=\{R\}, \tag{1.2}$$

in which [K] represents assemblage property matrix, $\{r\}$ represents assemblage vector of nodal unknowns, and $\{R\}$ represents assemblage vector of nodal forcing parameters.

### 1.7.3 Boundary Element Method

Engineers who are familiar with the FEM very often ask why it is necessary to develop yet another computational technique. The answer is that FEs have been proven to be inadequate or inefficient in many engineering applications and in some cases cumbersome to use and hence difficult to implement in computer-aided engineering systems. FE analysis is still a comparatively slow process due to the need to define and redefine meshes in the piece or domain under study.

The idea of BEMs is that we can approximate the solution to a PDE by looking at the solution to the PDE on the boundary and then use that information to find the solution inside the domain. Although this idea may sound strange, it is a very powerful tool for finding solutions. BEM is used for the solution of problems for which Green's functions can be calculated. These usually involve fields in linear homogeneous media. More specifically, BEM is preferred when very large domains are simulated as in these cases, an FEM approximation would have too many elements to be practical. A typical application of BEM is presented in Figure 1.11.

Therefore, BEM has emerged as a powerful alternative to FEM, particularly especially when better accuracy is required due to problems such as stress concentration or where the domain extends to infinity. The most important feature of BEs, however, is that different to the FEM and FDM, the methodology of formulating boundary value problems as boundary integral equations describes problems only by equations with known and unknown boundary states. This implies that only the discretization of the surface is required rather than the volume, i.e. the dimension of problems is reduced by one. Consequently, the necessary discretization effort is mostly much smaller and, moreover, meshes can easily be generated and design changes do not require a complete remeshing.

BE formulations typically give rise to fully populated matrices. As a consequence, the storage requirements and computation time will tend to grow according to the square of the problem size. By contrast, FE matrices are typically banded (elements are only locally connected) and the storage requirements for the system matrices typically grow quite linearly with the problem size. Compression techniques such as adaptive cross approximation/hierarchical matrices or multipole expansions can be used to improve these problems considering the cost of added

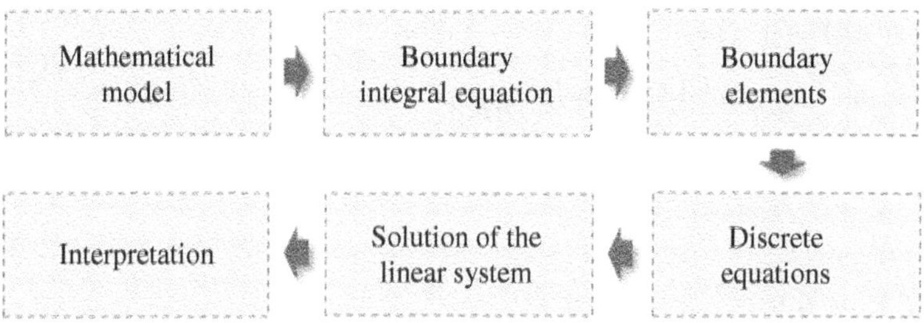


**Figure 1.11** Typical application of the Boundary Element Method.

complexity and with a success rate that depends significantly on the nature of the problem being solved and the geometry involved.

Thus, modeling an entire three-dimensional body with FE and calculating stress (or other states) at every nodal point is very inefficient because only a few of these values will be incorporated into the design analysis. Therefore, using BEs is a very effective choice of computing resources, and, furthermore, since internal points in BE solutions are optional, the user can focus on a particular interior region rather than the whole interior.

### 1.7.4 Finite Differences Method

PDEs describe many of the fundamental laws (e.g. conservation of mass), covering thus a wide range of physical phenomena [63]. Examples include Laplace's equation for steady-state heat conduction, Maxwell's equations for electromagnetic waves, Navier–Stokes equations for fluid flow, etc. Most PDEs of interest do not have analytical solutions. So, a numerical solution must be used to find an approximate solution. The necessary approximation is executed at discrete values of the independent variables and the approximation scheme is implemented via a computational tool. The use of FDM leads to the replacement of all partial derivatives and other terms in the PDE by approximations. After some manipulation, a finite difference scheme is created from which the approximate solution is obtained. More specifically, the FDM works by replacing the region over which the independent variables in the PDE are defined by a finite grid (also called a mesh) of points at which the dependent variable is approximated. The partial derivatives in the PDE at each grid point are approximated from neighboring values by using Taylor's theorem.

If we consider that the function whose derivatives are to be approximated is properly behaved, by Taylor's theorem, we can create a Taylor series expansion. Let $U(x)$ have n continuous derivatives over the interval $(a, i)$. Then for $a < x_0$, $x_0 + h < b$:

$$\begin{aligned} U(x_0 + h) = U(x_0) + hU_x(x_0) + h^2 \frac{U_{xx}(x_0)}{2!} + \ldots \\ + h^{n-1} \frac{U(n-1)(x_0)}{(n-1)!} + O(h^n), \end{aligned} \tag{1.3}$$

where:

- $U_x = \dfrac{dU}{dx}, U_{xx} = \dfrac{d^2U}{dx^2}, \ldots, U(n-1) = \dfrac{d^{n-1}U}{dx^{n-1}},$

- $U_x(x_0)$ is the derivative of $U$ with respect to $x$ evaluated at $x = x_0$,
- $O(h^n)$ is an unknown error term.

The usual interpretation of Taylor's theorem indicates that if we know the value of $U$ and the values of its derivatives at point $x_0$, then we can write down Eq. (1.3) for its value at the (nearby) point $x_0 + h$. This expression contains an unknown quantity which is written as $O(h^n)$. If we discard the term $O(h^n)$ in Eq. (1.3), we get an approximation to $U(x_0 + h)$. The error in this approximation is $O(h^n)$.

## 1.8 Challenges

The use of computer models for the simulation of biological phenomena, regardless of spatial and temporal scales, involves the associated costs in labor and computation power and storage (Figure 1.12). Initially, any modeling and simulation tool relies on well-developed means that facilitate defining models and allow simulations to be conducted in reasonable time frames. Developing robust computational tools is labor-intensive and requires significant expertise in informatics, engineering, and mathematics.

The next step refers to the development of a model, which is directly related to the research discipline. Compiling adequate model input parameters and

**Figure 1.12** Challenges in multiscale methods in systems biology. *Source:* 3DMan.eu/Adobe Stock.

representing anatomical structures (e.g. generating meshes of complex geometry) are all parts of this process. This is usually a hidden cost, not necessarily reported with the results of a study.

After the establishment of the computational model, the next step is to use the model to simulate the physical conditions of a clinical or research problem. Simulations, particularly in multiscale modeling where coupling between physical domains and scales is necessary, are computationally intensive. Solutions may be obtained, but their interpretation may also be challenging, relying on an expert (e.g. clinician and physicist), possibly spending hours to confirm the validity of results and then extracting useful information applicable to the research area.

All these general challenges can be addressed in a research setting, implying that the development cost of tools and models can be afforded. Availability of suitable modeling frameworks and simulation environments will likely be the most important factor in the wider adoption of multiscale modeling in systems biology. Most of the present modeling and simulation tools are developed for single-scale modeling and alone are not sufficient for multiscale modeling. Some will however play major roles in the new emerging frameworks, as simulation of multiscale models will probably involve assembly of many of these simulators in an integrative manner, each individual being used in its own area of specialization [64].

The aim of this book is to provide answers to questions and challenges in multiscale methods in biomedical engineering such as:

- Which tier of resolution is the most appropriate and can provide information on the underlying mechanisms of a specific, complex disease (e.g. cancer, carotid artery disease, and osteoporosis)? Is there a tier of resolution that offers the least understanding of the disease?
- Which computational method is most accurate and less time-consuming for the examined tier of resolution?
- What kind of data/input are necessary for a specific application in biomedical engineering?
- How much time is required for the preparation and interpretation of the data?

## References


1 Schmitt, F.O. and Schmitt, O.H. (1931). A vacuum tube method of temperature control. *Science* 73 (1889): 289–290.

2 Rashevsky, N. (1938). *Mathematical Biophysics: Physicomathematical Foundations of Biology*. Chicago: University of Chicago Press.

3 McCulloch, W.S. and Pitts, W. (1990). A logical calculus of the ideas immanent in nervous activity. 1943. *Bull. Math. Biol.* 52 (1, 2): 99–115; discussion 73–97.

**4** Karlebach, G. and Shamir, R. (2008). Modelling and analysis of gene regulatory networks. *Nat. Rev. Mol. Cell Biol.* 9 (10): 770–780.

**5** Fiehn, O. (2001). Combining genomics, metabolome analysis, and biochemical modelling to understand metabolic networks. *Comp. Funct. Genomics* 2 (3): 155–168.

**6** Marti-Renom, M.A., Stuart, A.C., Fiser, A. et al. (2000). Comparative protein structure modeling of genes and genomes. *Annu. Rev. Biophys. Biomol. Struct.* 29: 291–325.

**7** Telenti, A., Lippert, C., Chang, P.C., and DePristo, M. (2018). Deep learning of genomic variation and regulatory network data. *Hum. Mol. Genet.* 27 (R1): R63–R71.

**8** Ekins, S., Nikolsky, Y., and Nikolskaya, T. (2005). Techniques: application of systems biology to absorption, distribution, metabolism, excretion and toxicity. *Trends Pharmacol. Sci.* 26 (4): 202–209.

**9** Evans, W.E. and McLeod, H.L. (2003). Pharmacogenomics – drug disposition, drug targets, and side effects. *N. Engl. J. Med.* 348 (6): 538–549.

**10** Weinshilboum, R. (2003). Inheritance and drug response. *N. Engl. J. Med.* 348 (6): 529–537.

**11** Tornow, S. and Mewes, H.W. (2003). Functional modules by relating protein interaction networks and gene expression. *Nucleic Acids Res.* 31 (21): 6283–6289.

**12** Hanisch, D., Zien, A., Zimmer, R., and Lengauer, T. (2002). Co-clustering of biological networks and gene expression data. *Bioinformatics* 18 (Suppl 1): S145–S154.

**13** Ideker, T., Ozier, O., Schwikowski, B., and Siegel, A.F. (2002). Discovering regulatory and signalling circuits in molecular interaction networks. *Bioinformatics* 18 (Suppl 1): S233–S240.

**14** Segal, E., Wang, H., and Koller, D. (2003). Discovering molecular pathways from protein interaction and gene expression data. *Bioinformatics* 19 (Suppl 1): i264–i271.

**15** Spirin, V. and Mirny, L.A. (2003). Protein complexes and functional modules in molecular networks. *Proc. Natl. Acad. Sci. U.S.A.* 100 (21): 12123–12128.

**16** Knupfer, C., Beckstein, C., Dittrich, P., and Le Novere, N. (2013). Structure, function, and behaviour of computational models in systems biology. *BMC Syst. Biol.* 7: 43.

**17** Loewe, L. and Hillston, J. (2008). Computational models in systems biology. *Genome Biol.* 9 (12): 328.

**18** Thomas, M. and Schwartz, R. (2017). Quantitative computational models of molecular self-assembly in systems biology. *Phys. Biol.* 14 (3): 035003.

**19** Tourdot, R.W., Bradley, R.P., Ramakrishnan, N., and Radhakrishnan, R. (2014). Multiscale computational models in physical systems biology of intracellular trafficking. *IET Syst. Biol.* 8 (5): 198–213.

**20** Jin, J.Y., Almon, R.R., DuBois, D.C., and Jusko, W.J. (2003). Modeling of corticosteroid pharmacogenomics in rat liver using gene microarrays. *J. Pharmacol. Exp. Ther.* 307 (1): 93–109.

**21** McMullin, T.S., Hanneman, W.H., Cranmer, B.K. et al. (2007). Oral absorption and oxidative metabolism of atrazine in rats evaluated by physiological modeling approaches. *Forensic Toxicol.* 240 (1, 2): 1–14.
**22** Wang, Z., Bordas, V., and Deisboeck, T.S. (2011). Identification of critical molecular components in a multiscale cancer model based on the integration of Monte Carlo, resampling, and ANOVA. *Front. Physiol.* 2: 35.
**23** Wang, Z., Bordas, V., and Deisboeck, T.S. (2011). Discovering molecular targets in cancer with multiscale modeling. *Drug Dev. Res.* 72 (1): 45–52.
**24** Deisboeck, T.S., Wang, Z., Macklin, P., and Cristini, V. (2011). Multiscale cancer modeling. *Annu. Rev. Biomed. Eng.* 13: 127–155.
**25** Karabasov, S., Nerukh, D., Hoekstra, A. et al. (2014). Multiscale modelling: approaches and challenges. *Philos. Trans. A Math. Phys. Eng. Sci.* 372: 2021.
**26** Hoekstra, A., Chopard, B., and Coveney, P. (2014). Multiscale modelling and simulation: a position paper. *Philos. Trans. A Math. Phys. Eng. Sci.* 372: 2021.
**27** Weinan, E. (2011). *Principles of Multiscale Modeling*. Cambridge University Press.
**28** Interagency Modeling and Analysis Group (IMAG). http://www.imagwiki.nibib.nih.gov
**29** Hasenauer, J., Hasenauer, C., Hucho, T., and Theis, F.J. (2014). ODE constrained mixture modelling: a method for unraveling subpopulation structures and dynamics. *PLoS Comput. Biol.* 10 (7): e1003686.
**30** Hasenauer, J., Heinrich, J., Doszczak, M. et al. (2012). A visual analytics approach for models of heterogeneous cell populations. *EURASIP J. Bioinform. Syst. Biol.* 2012 (1): 4.
**31** Hasenauer, J., Waldherr, S., Doszczak, M. et al. (2011). Identification of models of heterogeneous cell populations from population snapshot data. *BMC Bioinf.* 12: 125.
**32** Hasenauer, J., Wolf, V., Kazeroonian, A., and Theis, F.J. (2014). Method of conditional moments (MCM) for the Chemical Master Equation: a unified framework for the method of moments and hybrid stochastic-deterministic models. *J. Math. Biol.* 69 (3): 687–735.
**33** Breitling, R. (2010). What is systems biology? *Front. Physiol.* 1: 9.
**34** Chakrabarti, A., Verbridge, S., Stroock, A.D. et al. (2012). Multiscale models of breast cancer progression. *Ann. Biomed. Eng.* 40 (11): 2488–2500.
**35** Zhang, Y., Barocas, V.H., Berceli, S.A. et al. (2016). Multi-scale modeling of the cardiovascular system: disease development, progression, and clinical intervention. *Ann. Biomed. Eng.* 44 (9): 2642–2660.
**36** Sakellarios, A.I., Rigas, G., Kigka, V. et al. (2017). SMARTool: a tool for clinical decision support for the management of patients with coronary artery disease based on modeling of atherosclerotic plaque process. *Conf. Proc. IEEE Eng. Med. Biol. Soc.* 2017: 96–99.
**37** Sakellarios, A., Bourantas, C.V., Papadopoulou, S.L. et al. (2017). Prediction of atherosclerotic disease progression using LDL transport modelling: a serial

computed tomographic coronary angiographic study. *Eur. Heart J. Cardiovasc. Imaging* 18 (1): 11–18.

**38** Sakellarios, A.I., Raber, L., Bourantas, C.V. et al. (2017). Prediction of atherosclerotic plaque development in an in vivo coronary arterial segment based on a multilevel modeling approach. *IEEE Trans. Biomed. Eng.* 64 (8): 1721–1730.

**39** Yankeelov, T.E., An, G., Saut, O. et al. (2016). Multi-scale modeling in clinical oncology: opportunities and barriers to success. *Ann. Biomed. Eng.* 44 (9): 2626–2641.

**40** Wei, S.T., Sun, Y.H., and Zong, S.H. (2017). A novel method to identify hub pathways of rheumatoid arthritis based on differential pathway networks. *Mol. Med. Rep.* 16 (3): 3187–3193.

**41** Ge, B., Tian, Y., Hu, X. et al. (2015). Construction of multi-scale consistent brain networks: methods and applications. *PLoS One* 10 (4): e0118175.

**42** Cilfone, N.A., Kirschner, D.E., and Linderman, J.J. (2015). Strategies for efficient numerical implementation of hybrid multi-scale agent-based models to describe biological systems. *Cell. Mol. Bioeng.* 8 (1): 119–136.

**43** Qin, C., Yao, D., Shi, Y., and Song, Z. (2018). Computer-aided detection in chest radiography based on artificial intelligence: a survey. *Biomed. Eng. Online* 17 (1): 113.

**44** Alcocer-Cuaron, C., Rivera, A.L., and Castano, V.M. (2014). Hierarchical structure of biological systems: a bioengineering approach. *Bioengineered* 5 (2): 73–79.

**45** Signetti, S., Bosia, F., and Pugno, N.M. (2016). Computational modeling of the mechanics of hierarchical materials. *MRS Bull.* 41 (9): 694–699.

**46** DiStefano, J. (2013). *Dynamic Systems Biology Modeling and Simulation*. Academic Press.

**47** Rao, S. (2018). *The Finite Element Method in Engineering*, 6e. Butterworth-Heinemann Elsevier.

**48** Yang, K.-H. (2017). *Basic Finite Element Method as Applied to Injury Biomechanics*, 1e. Academic Press.

**49** Katsikadelis, J.T. (2016). *The Boundary Element Method for Engineers and Scientists*. Academic Press.

**50** Aliabadi, M.H. (2002). *The Boundary Element Method, Applications in Solids and Structures*. Wiley.

**51** Langtangen, H.P. and Linge, S. (2017). *Finite Difference Computing with PDEs*. Springer.

**52** Bartocci, E., Lio, P., and Paoletti, N. (2016). *Computational Methods in Systems Biology*. Springer.

**53** Walpole, J., Papin, J.A., and Peirce, S.M. (2013). Multiscale computational models of complex biological systems. *Annu. Rev. Biomed. Eng.* 15: 137–154.

**54** Price, N.D., Reed, J.L., and Palsson, B.O. (2004). Genome-scale models of microbial cells: evaluating the consequences of constraints. *Nat. Rev. Microbiol.* 2 (11): 886–897.

**55** Haggart, C.R., Bartell, J.A., Saucerman, J.J., and Papin, J.A. (2011). Whole-genome metabolic network reconstruction and constraint-based modeling. *Methods Enzymol.* 500: 411–433.

**56** Bathe, K.-J. (2016). *Finite Element Procedures*, 2e. MIT.

**57** Hughes, T.R.J. (1987). *The Finite Element Method*. Englewood Cliffs, NJ: *Prentice Hall.*

**58** S.H. Lo, C.K. Lee., "On using meshes of mixed element types in adaptive finite element analyses," *Finite Elem. Anal. Des.*, vol. 11 pp. 307–336, 1992.

**59** Si, H. (2008). Adaptive tetrahedral mesh generation by constrained Delaunay refinement. *Int. J. Numer. Methods Eng.* 75 (7): 856–880.

**60** Sipics, M. (2006). Taking on the ITER challenge, scientists look to innovative algorithms, petascale computers. *SIAM News* 39 (7): 1–3.

**61** Kuprat, A.P. and Einstein, D.R. (2008). An anisotropic scale-invariant unstructured tetrahedral mesh generation algorithm based on local feature size. *J. Comput. Phys.* 228 (3): 619–640.

**62** Tawhai, M., Bischoff, J., Einstein, D. et al. (2009). Multiscale modeling in computational biomechanics. *IEEE Eng. Med. Biol. Mag.* 28 (3): 41–49.

**63** Causon, D.M. and Mingham, C.G. (2010). *Introductory Finite Difference Methods for PDEs*. Ventus Publishing ApS.

**64** Dada, J.O. and Mendes, P. (2011). Multi-scale modelling and simulation in systems biology. *Integr. Biol. (Camb.)* 3 (2): 86–96.

# 2

# Biomedical Imaging

## 2.1 Introduction

Visualization of the body parts, organs, tissues, or even cells is the field of medical imaging used for clinical decisions on the diagnosis, management, and treatment of a disease. For this reason, medical imaging is considered an important cornerstone in clinical practice and research. Medical imaging constitutes a sequence of steps such as image acquisition, processing, segmentation, reconstruction, classification, clustering, and visualization of the imaged objects. Disease diagnosis has been improved with the utilization of medical imaging, especially the recent years where the spatial resolution and image quality have been enhanced. Depending on the clinical field, several imaging modalities can be used to visualize different organs and tissues with different levels of accuracy. Such modalities are the X-ray imaging (radiography), the ultrasound (US), the computed tomography (CT), the magnetic resonance imaging (MRI), the positron emission tomography (PET), and single photon emission computed tomography (SPECT).

## 2.2 X-ray Radiography

X-ray radiography is the first method used for diagnostic imaging [1]. For over half a century, it has been the only method of diagnostic imaging. In its first form, X-ray imaging is the well-known classical radiography and X-ray-based projective imaging method. With the introduction of new imaging techniques in the last decades, classical radiography has lost some of its original glory. However, it still continues to be the most widely used imaging method today due to its simplicity and low cost. Based on classical X-ray projection imaging, special diagnostic

*Multiscale Modelling in Biomedical Engineering*, First Edition. Dimitrios I. Fotiadis, Antonis I. Sakellarios, and Vassiliki T. Potsika.

Published 2023 by John Wiley & Sons, Inc.

techniques with high diagnostic value have been developed in the last decades for specific diagnostic cases. Particularly important applications of X-ray imaging are the methods for tomography. X-rays are electromagnetic radiation caused at atomic level and they have energy capable of causing ionization of the matter (>30 eV). Usually, X-rays have energy from 100 eV to 100 keV or a corresponding frequency of $10^{16}$–$10^{21}$ Hz or a corresponding wavelength from $10^{-12}$ to $10^{-9}$ m.

The basic concept of X-ray imaging is the imprinting of the degree of absorption of X-rays by the tissues of the human body as this is presented in a simplified manner in Figure 2.1. The absorption of X-rays is the "removal" of photons from the beam due to their interaction with matter. Briefly, in classical radiography, the 3D anatomy is projected and recorded as a two-dimensional projection image in a detector using a 50–150 keV X-ray beam [2]. The function of X-ray imaging is based on their partial absorption or scattering by the tissues. The degree of absorption or scattering depends mainly on the energy of the beam, the atomic number of matter ($Z$), and the density of the tissues ($\rho$). The uniform beam that falls into the body will then present a weakening, proportional to the tissues it has penetrated. When the beam exits the body is no longer uniform, but intensity fluctuations are present due to the variable absorption of X-rays depending on the tissues that have penetrated. These outgoing X-rays are the signal and in classical radiography, they are directed on a radiographic film. The processing of this signal is simple and the basic principle is equivalent to processing a common photographic film [3]. Digital radiography has replaced the film use in the last years.

X-rays are artificially produced when a material is overwhelmed with a high-speed electron beam. X-ray-generating systems, also known as X-ray tubes, consist

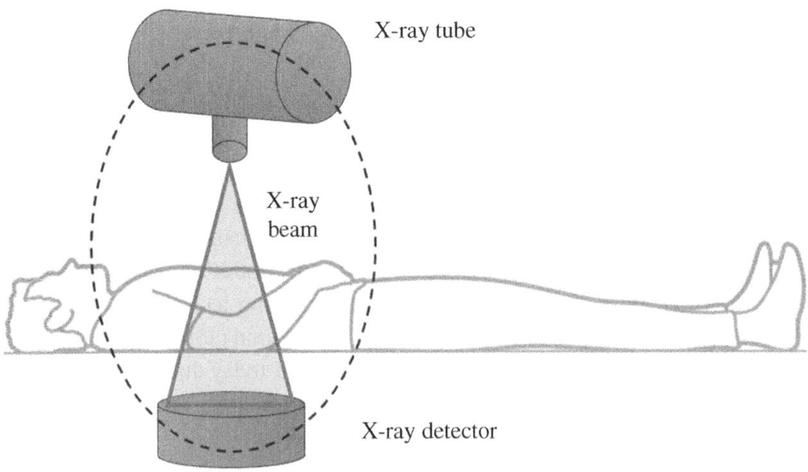


**Figure 2.1** Simplified representation of X-ray imaging.

of an anode (positively charged electrode) and a cathode (negatively charged electrode), which are in vacuum in a glass container. The cathode is essentially a wire, which is heated by the passing of electricity and the thermal energy is given to electrons emitted in the vacuum of the tube. These are accelerated due to the difference in the potential between the two electrodes and reach the anode rapidly. The ionization of the material of the anode with the moving electrons produces X-ray photons in all directions. About half of the photons created are absorbed by the material of the anode. Only those in the useful beam cone are used for the visualization. The anode (or otherwise the target of the electrons) usually consists of tungsten or tungsten and rhenium alloy (90% tungsten and 10% rhenium), which is somewhat more resistant to the high temperatures developed in the target.

### 2.2.1 X-ray Interaction with Tissues

X-rays pass through the body under examination and interact (relatively weakly) with the tissues. The main mechanisms of interaction of the X-ray photons with the tissues are the photoelectric effect and the Compton effect. Note that there are other secondary interaction mechanisms, such as nuclear photoelectric effect, nuclear resonance scattering, Rayleigh, Thomson, and Delbruck scattering [4], but they do not make a significant contribution to the range of X-rays commonly used in medical imaging. The photoelectric effect strongly depends on the atomic number $Z$ of the tissues and the probability of it occurring is proportional to $Z^4$ or $Z^5$. This means that at low energies, mainly the photoelectric effect, tissues with slight differences in $Z$ (such as muscles and other soft molecules); will have a relatively high difference in X-ray absorption resulting in good contrast to the final X-ray. Compton photons are essentially photons of scattering and energy large enough to not be fully absorbed by the surrounding tissues and reach the detector [5].

The total probability of a photon being removed from the original beam is given as the sum of the individual probabilities of the two major interaction phenomena, and they are expressed by the mass absorption coefficient ($\mu$) of the X-rays: $\mu/\rho_m$, where $\mu$ is the absorption coefficient and $\rho_m$ is the mass density. In general, the overall absorption coefficient depends on the material and energy of the photons [5]. Today tables provide the absorption coefficient for a wide range of materials and for a wide range of X-ray energies. It should be noted here that in diagnostic radiography, the following numbers should be taken into account only as an indication because, in most cases, the X-ray beam is not monoenergetic (such as the photons used to measure the following sizes), but consists of a wide spectrum of energies (continuous spectrum of X-rays).

The absorption of X-rays is related to the tissue through which the beam passes by the following equation [6]:

$$\frac{I}{I_0} = e^{(-\mu t)}, \tag{2.1}$$

where $I$ is the intensity of transmitted X-rays, $I_0$ is the intensity of incident X-rays, $\mu$ is the linear attenuation coefficient of the material, and $t$ is the thickness of material through which X-rays have traveled. Photons which are not absorbed during their pass through the tissue constitute the signal that generates the X-ray and they are often called primary radiation. Imaging of the absorption of photons by different tissues, as described earlier, is the primary image in radiography. The different tissues are reflected differently in the final X-ray because they have a different absorption coefficient for the X-rays. Soft tissues absorb lower levels of X-rays than the bones, which absorb most of the primary signal. Air cavities absorb almost no X-rays. The contrast between the various tissues can be increased by the use of contrast agents. For example, barium and iodine strongly absorb X-rays and are also absorbed by tissues that are high in these elements. They are usually introduced by injection, inhalation, or ingestion and they are selectively absorbed by specific tissues or systems and thereby selectively increase their contrast. The final X-ray image consists of the primary radiation, though noise may be produced by secondary radiation due to photons emanating from the diffusion of primary radiation into the examined body.

### 2.2.2 Medical Applications of X-rays

Radiography is used in many types of examinations and procedures where a record of a static image is desired. Some examples include dental examination, verification of correct placement of surgical markers prior to invasive procedures, mammography, orthopedic diagnostics, etc. For example, bone density scanning is based on X-ray radiography and it is also called dual-energy X-ray absorptiometry (DXA) or bone densitometry. It is an enhanced form of X-ray technology that is used to measure bone loss as the gold standard for measuring bone mineral density [7]. X-ray radiography has a significant application in cardiovascular disease in the form of catheter angiography [8]. Using this technique combined with the injection of a contrast medium, it is possible to assess the vascular disease and provide a diagnosis and treatment decision in a single procedure. It is considered an invasive technique, however, with very limited risk for the patient. Another application is the X-ray-based mammography, which enables the early detection of breast cancer before experiencing symptoms [9]. X-ray is also used to evaluate the abdomen, chest, or lower tract such as the stomach, liver, intestines, spleen, and lungs and it is the primary imaging technique used to help diagnose symptoms of unexplained pain, nausea, or vomiting. Finally, X-ray imaging has also an important role in dentistry where panoramic dental views or teeth-specific images are used for the diagnosis and proper treatment.

## 2.3 Computed Tomography

CT is another imaging modality, which is also based on X-rays. The first practical clinical application of X-ray CT was performed by Hounsfield in 1973 [10]. The mathematical principle for the production of tomography from a simple projection is usually attributed to Radon (1917) [11], who described a mathematical way to find spatial distribution of a quantity at a plane when a large number of sums of this quantity is known along axes above the plane and having different directions. Similar work on the three-dimensional case had been carried out even earlier by Lorentz in 1905. This mathematical method was first used in biomedical imaging by Cormack in 1963 for the detailed calculation of the absorbed dose from tissues during radiotherapy using emission absorption measurements. The diagnostic value of the method was sealed in 1979 by a Nobel Prize in Medicine jointly awarded to Hounsfield and Cormack. Today, it is estimated that more than 30 000 X-ray-computed tomographs are used in clinical diagnosis worldwide.

### 2.3.1 The Principle of CT Imaging

The mathematical and physical principle of the method is shown in Figure 2.2 [12]. For each irradiation, a thin X-ray beam propagates along a series of voxels. The detector records the sum of absorbance along the line. This sum is called a radial sum or line integral. The procedure is repeated for all horizontal voxel rows. The procedure is then repeated for all voxel rows in both the vertical direction and in many lateral directions. The reconstruction of the internal structure (body, tissue, or organ) is obtained by calculating the absorbance for each voxel, given the sum of the multiple projections. This mathematical problem was first solved by Radon in 1917, and since then, various mathematical methods have been proposed to solve it, which are called reconstruction techniques or algorithms. Reconstruction of the internal structure based on these methods is done in an acceptable time frame only with the use of a computer.

The philosophy of the method is described in the following steps:

1) In the first step, we take the radial summations of a direction, e.g. horizontal. For each row, we distribute the radial sum equally in all cells. Then, we calculate the sums of the cells in a different direction, e.g. perpendicular. We see that these are not the same as the values of the corresponding vertical radial sums.

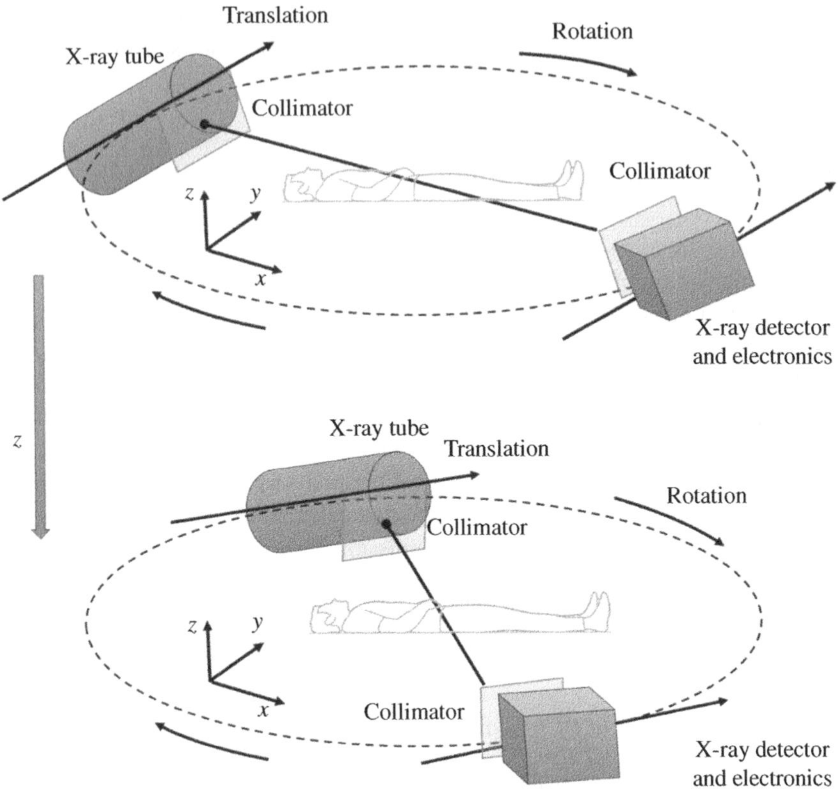


**Figure 2.2** The principles of CT imaging. *Source:* Kalender [12].

So, we calculate the difference between the sums of the cells of the first step from the real corresponding radial sums.

2) In the second step, the calculated differences from the previous step are distributed equally in the cells in the vertical direction. We calculate the new values for the cells and then calculate the sums in the third direction (diagonally). We see again that they differ from the actual radial sums in the diagonal direction, so we calculate their difference.
3) Finally, the difference in the diagonal direction is distributed again in all cells. We calculate the new values for the cells and confirm that they correspond to the actual values of the original image to complete the reconstruction.

The produced matrix of a standard medical examination is usually in the order of $512 \times 512$ cells, resulting in a much larger number of steps required to complete the reconstruction. The CT system usually takes a larger number of point views (than the minimum required) to compensate for measurement errors. In modern

CT systems for each section, approximately 600–1200 radial summations are obtained in each direction, in about 800–1500 different directions covering a 360° range. Fast and accurate image reconstruction requires the use of a high-speed computer in conjunction with the right algorithm. In the first experimental systems, reconstruction of a section took several hours. In the first professional clinical tomography, approximately 15–20 minutes were required. Today, the typical reconstruction time is in the order of 30 seconds, while there are specialized CT systems that give recording and reconstruction times of less than a second. Typical values for analyzing the reconstructed image are 512 × 512 or 1024 × 1024.

### 2.3.2 The Evolution of CT Scanners

The basic principle of CT imaging presented in Figure 2.2 has several limitations in its use in the clinical practice. The most important limitations are the increased reconstruction times and the limited resolution. For this reason, CT scanners have been evolved to provide fast and accurate image reconstruction. The four major types of CT imaging systems are represented in Figure 2.3. In the early systems (Figure 2.3a), a linear and thin X-ray beam was used to record the projections. Multiple projections are performed by scanning the body or organ (usually in 160 steps of irradiation), rotating the system at an angle of 1° and repetition of the linear translational movement to record a second view. Due to the mechanical movement of the entire system, the typical time for a plane reconstruction is ~5 minutes. In the 2nd generation (Figure 2.3b), the X-ray beam is of a low-angle circular (~10°) type opposite to a linear array of about 30 detectors. An irradiation gives a portion of a projection for a single incision, the system rotates at an angle ~10° and the motion is repeated to record a second view. The scan time is ~1 minutes to reconstruct a single plane. In the next generation (Figure 2.3c), the angle of the X-ray beam increases (~40°) opposite to a semicircular array of a large number of detectors, e.g. 300–700 detectors. This evolution leads to scan times of ~2–10 seconds for a plane. Finally, the 4th-generation CT systems have about 1000–2000 detectors in a characteristic stable circular arrangement, while the X-ray lamp rotates (Figure 2.3d). Recent developments include the helical or spiral tomography. It is a 3rd- or 4th-generation CT system, where the test table moves along the axis while receiving projections. This movement of the table during the irradiation provides a helical capture of projections, from which the final tomographic images can be reconstructed. Helical tomography enables large-volume imaging in a short period of time, e.g. <1 seconds. The typical resolution in today's clinical CT systems is about 0.5 mm, though micro CT systems have been developed for the imaging of animals that achieve a resolution of 50 μm or less.

A next generation of spiral CT scanners, which are mainly used in the current clinical practice have evolved into multirow detectors or multislice computed

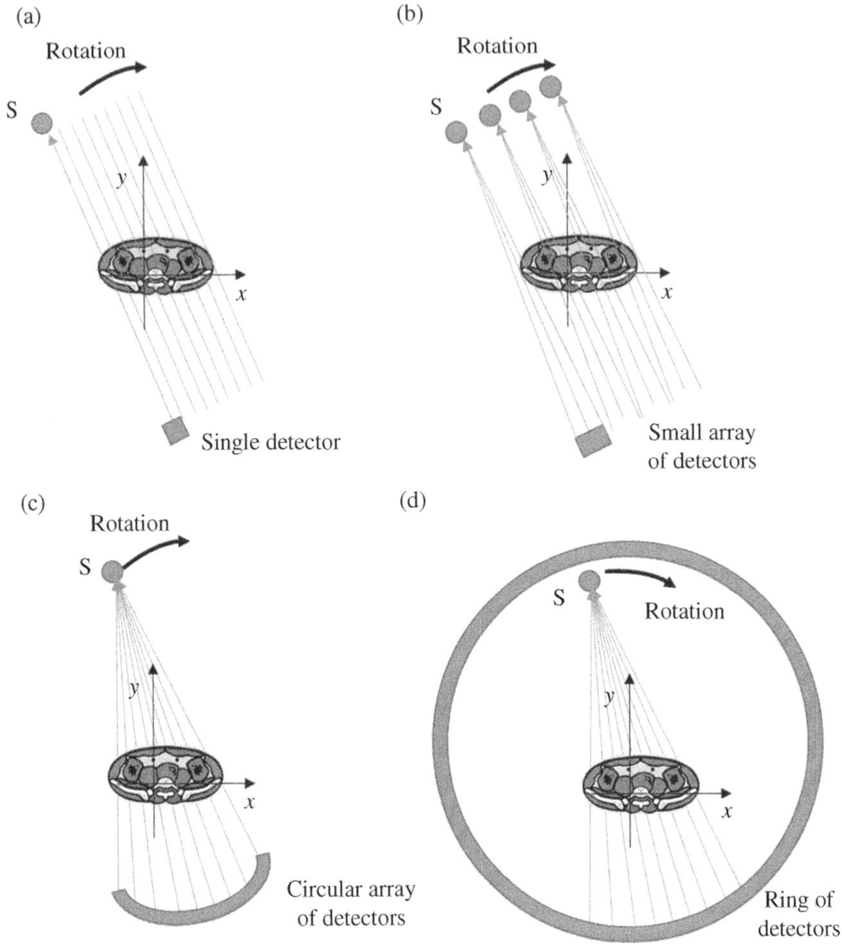


**Figure 2.3** The four historical generations of X-ray CT: (a and b) Translation–rotation CT, (a) first generation, (b) second generation. (c and d) Fan beam CT, (c) third generation, (d) fourth generation. *Source:* Peyrin and Engelke [13]/Springer Nature.

tomography scanners (MSCTs) [13]. These systems are following the same principle with the spiral systems with the main difference being that the acquisition is not performed using a single-slice detector array but a multislice detector array. The whole scan time is typically less than one minute. The main advantage is that highly accurate visualization of the whole body or organ can be achieved even with an almost isotropic spatial resolution of about 0.5 mm [13]. In this way, the artefacts are also reduced significantly. Depending on the organ and the required image resolution, the slice thickness can be chosen from 0.5 to 10 mm, enabling proper scanning of large organs (lungs and heart) or high-resolution imaging of small organs such as the coronary arteries [14]. Additionally, MSCT processing using workstations enables the utilization of algorithms and techniques for advanced visualization such as 3D volume rendering of the reconstructed images, multiplanar reformations (MPRs), and maximum intensity projections (MIPs).

### 2.3.3 Medical Applications of CT Imaging

CT imaging nowadays is considered the gold standard for diagnosis of many diseases. For example, according to the most recent guidelines in cardiology, CT imaging of the coronary arteries is considered a powerful noninvasive diagnostic tool for cardiovascular disease and it is proposed as one of the first tests to be undertaken by asymptomatic or symptomatic patients. Even though the CT imaging is based on the principle of X-ray imaging, it has a great advantage over it. In particular, CT imaging enables the imaging of the body or tissue or organ at once in 3D resolution providing also differentiation of soft tissue structures, such as liver, lung tissue, and fat. In other words, CT does not only present the region of the problem (e.g. a tumor), but more significantly, the size, its spatial location, and extent of a tumor or any other distortion depending on the examined disease and organ. For this reason, it is expected that the medical applications of CT are tremendous. Here we only present a limited glimpse of CT use in clinical practice.

#### 2.3.3.1 Application of CT Imaging in Cancer

CT imaging is used in cancer for several purposes. More specifically, CT can be used to screen the cancer and enable accurate diagnosis based on the presence, size, and type of the tumor. This also provides information about the stage of a cancer and it can be used as an indication to guide invasive diagnosis based on biopsy. Additionally, due to its noninvasive modality CT imaging can be used not only to assess the treatment response but also to guide novel treatment approaches such as radiofrequency ablation or the implantation of radioactive seeds. Concluding, CT imaging is considered an easy and accurate way of monitoring cancer progression or recurrence. For example, CT colonography can be used to screen for both large colorectal polyps and tumors. CT can be used also for lung cancer; the benefits, however, are controversial because of the higher overall rate of false-positive results.

#### 2.3.3.2 Application of CT Imaging in Lungs

CT imaging is also used to diagnose lung diseases such as the diffuse lung disease [15]. A key finding to assess effectively the characterization of diffuse lung disease is the mapping of abnormalities to the apparatus of respiration. This is achieved by using a thin-section (or high-resolution) CT with the thinnest collimation available together with a high-spatial-frequency reconstruction kernel to visualize accurately the secondary pulmonary lobule [15]. The same type of CT imaging is used to diagnose interstitial pneumonias [16]. This section of CT has also replaced the bronchography for the assessment of airway diseases. The reason for this is that high-resolution CT provides the anatomic and functional analysis of lungs in case of airway disease [17, 18]. Helical CT further improved the diagnostic capabilities of thin-section CT benefitted by the advanced visualization options such as the multiplanar reconstruction.

#### 2.3.3.3 Application of CT Imaging in Cardiovascular Disease

Another area of CT application is the heart and angiographic imaging. In particular, coronary angiography CT can be used to diagnose and assess the cardiovascular disease, as well as several pathological situations of heart function [19–22]. The spatial resolution of 0.5 mm of the CT scanners enables an accurate analysis of the coronary arteries, especially when the last-generation scanners are used where the whole heart can be visualized in less than a cardiac cycle. Moreover, novel image-processing algorithms have been implemented for the atherosclerotic plaque characterization. Furthermore, utilization of CT images combined with blood fluid dynamics can provide additional noninvasive tools for the diagnosis of the disease severity such as the virtual fractional flow reserve [23].

#### 2.3.3.4 Application of CT Imaging in Other Fields

CT imaging is not limited to the above fields but nowadays is used as a noninvasive imaging tool for almost the whole human body. Examples of using the CT imaging for diagnostic purposes are to identify liver lesions where using an intravenous contrast medium injection enables imaging of the whole liver during the hepatic arterial phase and portal venous phase in less than a minute. In combination with spiral CT, hepatocellular carcinoma and other hypervascular liver lesions can be detected with limited possible artefacts [24–26].

Injury is responsible for the highest number of years of potential life lost before 65 years of age and is one of the leading causes of death for ages less than 45 years. CT imaging has a special role in trauma since using a "panscan" (CT of the head, neck, chest, abdomen, and pelvis) provides a fast and unified diagnostic solution.

Concluding this section, CT imaging has presented great progress in the last five decades. Though the resolution still remains relatively low compared to other imaging modalities, very recently, new developments attempt to overcome some of CT's limitations. For example, dynamic computed tomography or perfusion computed tomography (PCT) is used to determine time-dependent attenuation curves of contrast agents from which functional instead of morphological parameters can be quantified [13]. The main application is the cerebral PCT to diagnose stroke [27], while recently being used for myocardium imaging, which can be used in combination with coronary CT for enhanced diagnosis [28]. Quantitative CT is another approach used especially for the quantification of tissue concentration with applications in bones or lungs. Finally, spectral CT imaging aims at the isolation of specific chemical elements e.g. iodine or calcium hydroxyapatite and at the reduction of beam-hardening artefacts or patient motion artefacts.

### 2.3.4 Radiation of CT Imaging

CT imaging has been established as a diagnostic tool with an increasing number of systems' installations. Nevertheless, CT imaging has a major limitation, which is the exposure of the patient to high levels of radiation. The level of radiation depends on the required spatial resolution, the protocol of the imaging e.g. dynamic CT has increased amounts of radiation and the size of the body, which is scanned. The Computed Tomography Dose Index (CTDI) has been established to monitor the radiation dose per protocol/scan, while many techniques have been developed to reduce the amount of radiation [13].

## 2.4 Diagnostic Ultrasound

US imaging is currently one of the most widely used diagnostic imaging methods and it has recently been expanded to the microscopic imaging of biological systems. Unlike most imaging and microscopy methods, which are based on the interaction of electromagnetic radiation with matter, USs are material waves propagated in a material medium. The rate of propagation of the material waves depends on the physical characteristics of the medium. Also, material waves that encounter during their propagation a second material medium with different physical characteristics are partly reflected. These two important properties in the propagation of material waves are exploited by US imaging to study the internal structure of the body.

Ultrasonic waves propagate in matter and are reflected by internal surfaces that separate the matter with different characteristics (e.g. blood vessel walls and blood, bones from tissue, etc.). The time from the wave emission to the reflection detection, as well as the intensity of the reflection, provide information about the internal position of the surface and the nature of the different media. The main advantages of US over other imaging modalities are:

- It is a noninvasive method without serious side effects,
- It is relatively low cost in terms of equipment,
- It is usually based on a portable equipment,
- It creates a real-time image,

- It offers good contrast for soft tissues,
- It enables static and dynamic studies as well as quantitative flow measurements.

The natural principles underlying US were introduced by Rayleigh in 1877. At about the same time, in 1880, the natural phenomenon of piezoelectricity for the practical production of US was discovered by the Curie brothers. The first medical diagnostic use of US was reported in 1942 by the Austrian brothers Dussik, who used a US beam to penetrate the patient's skull and study the anatomy of the brain. However, the first successful applications of US to medical diagnostic imaging were made after 1950 and since then, new techniques and equipment are being developed. Nowadays, US is the second most frequent diagnostic method in the world following X-ray imaging. It is used for many diagnostic applications, such as in gynecological examinations and fetal imaging, the cardiovascular system, and abdominal organs imaging. Modern developments in the field of US have allowed its application for microscopic imaging, e.g. for the study of microvasculature or in tumor biology.

### 2.4.1 The Principle of US

USs are usually defined as material waves with frequencies larger than those that can stimulate the human ear. So, USs are sound waves with frequencies above 20 kHz. Based on the frequencies, US can be used for imaging in different applications. In medical diagnostic imaging and therapeutic applications, USs are mainly applied in the frequency range from 0.8 to 15 MHz, giving a spatial resolution of a few mm. Finally, high-frequency US (12 MHz up to 1 GHz) is mainly used for microscopy with a resolution of a few μm.

Since US is a material wave, the physics of US is based on basic concepts of acoustic and wave physics. A material medium is definitely needed for sound propagation. As the molecules of matter reciprocate around their equilibrium position, they create condensations and dilutions within the material that propagate at a particular speed and in the direction of the reciprocal motion of the molecules. This type of wave where the disturbance is in the propagation direction is called a longitudinal wave. A complete path of each particle of material is called a cycle. The number of cycles per second gives the frequency of the acoustic waves, which, in the case of US, are in the order of 1 000 000 cycles/s or otherwise 1 MHz. The duration of a cycle is called period $T$ and is equal to the inverse of frequency, i.e. $T = 1/f$. An important characteristic of the different materials is the acoustic resistance ($Z$) given by the product of the density of the material ($p_0$) on the velocity of sound ($u$) in the same material. Acoustic resistance is necessary to determine the intensity of the reflected wave reaching the detector and depends only on the medium. Some indicative values for body tissues are: 138 g/sec/m$^2$ for fat, 170 g/sec/$m^2$ for muscle, 161 g/sec/$m^2$ for blood, 780 g/sec/$m^2$ for bone, etc. These differences can be used for the tissue discretization and diagnosis. The larger the difference in the acoustic resistance of two materials in contact, the larger the percentage of sound waves reflected on the surface.

The generation and detection of US are done by the same device called a transducer or transceiver. The key element of the transceiver is the piezoelectric crystal inside a special plastic cover, which conceals the entire device. An absorbent material for damping the oscillations is located at the back of the crystal. Also, the electrodes have a key role on the back and front surfaces of the piezoelectric material. The electrodes create the potential difference and polarize the crystal during US production. On the contrary, they are also used to detect the potential difference during the US detection phase.

### 2.4.2 Medical Applications of US

Two different approaches can be discriminated in US diagnostics. The first is the transmission technology, where the diagnosis is based on the different tissue absorbances of the US on distinguishing the tissues with different absorbances of US. The second approach regards the reflection technology, which is based on the reflection of the US pulse from the boundary between two different tissues with dissimilar acoustic resistance [29]. The modes which are used in US medical imaging are presented briefly below:

i) **A-mode**: A line through the body or analyzed tissue is scanned by a transducer and the echoes are captured in relation to the depth. The A-mode is also used for treatment strategies for several tumor types where A-mode allows accurate focus on the specific region of the tumor.
ii) **B-mode**: In this mode, instead of a line, a plane through the body is scanned simultaneously by an array of transducers and the results can be plotted as a 2D image.
iii) **M-mode**: In this mode, B-mode scans are produced in a rapid sequence and this enables the visualization of motion (M-mode stands for motion).
iv) **Doppler mode**: The Doppler mode enables the visualization and measurement of blood flow (velocity, pressure, etc.) and has application in cardiology and angiology.

Ultrasonography has wide use in medicine applied for diagnostic or even therapeutic purposes. Some fields of application are for visualization of abdomen organs (abdominal US: the solid organs of the abdomen such as the pancreas, aorta, inferior vena cava, liver, gall bladder, bile ducts, kidneys, and spleen are imaged), visualization of blood vessels (vascular US and intravascular US: the stenosis can be viewed, as well as the plaque type, the flow in carotid arteries, as well as the wall structure in the case of intravascular US), views of the fetus (obstetric US), visualization of the gynecological reproductive organs and male urological system (organs of the pelvic region are imaged, which includes the uterus and ovaries or urinary bladder; men are sometimes given a pelvic sonogram to check on the health of their bladder and prostate), visualization of the brain (neurosonography), assessment of heart function (echocardiography), ophthalmology, and many others [30].

US can be used for therapeutic purposes, requiring however a proper application. Usually, it aims to provide heat or agitation into the treated tissue. For this reason, the used frequency is usually significantly higher than that applied for diagnostic purposes. In [31–48], some main fields of therapeutic application of the US as approached by the Food and Drug Administration are presented in detail. Briefly, the therapy methods include unfocused or focused beam, hyperthermia, high-intensity focused US, focused US, extracorporeal lithotripsy, phacoemulsification, tissue cutting and vessel sealing, and intravascular US. The applicators can be fixed, portable, and handheld or even integrated with catheters.

## 2.5 Magnetic Resonance Imaging

In 1973, Lauterbur applied the mathematical technique of posterior projection (which had just been applied in 1972 to X-ray imaging to mark the birth of X-ray CT) and published the first MRI of a test tube with water. The back-projection method was rapidly pushed aside, and since 1975, with Ernst's contribution, the faster and more reliable method of phase and frequency coding for image reconstruction in MRI has been applied. In 1976, Mansfield presented the first MRI of a human finger, and in 1980, Edelstein's team demonstrated the applicability of the method on full-body imaging. Lauterbur and Mansfield shared the Nobel Prize in Medicine in 2003 for the establishment of the nuclear magnetic resonance (NMR) imaging. Nowadays, the method is known as "magnetic resonance imaging" or with the corresponding English abbreviation MRI (= magnetic resonance imaging) omitting the term "nuclear" as faulty related to radiation in the first years of use.

In the last 20 years, rapid technological discoveries have been made in the field of MRI and the speed of image creation is constantly increasing. For example, in 1986, we had the ability to capture images in less than five seconds, while since 1989, it has been possible to capture images at a rate suitable for filming (i.e. less than 30 ms). Spatial analysis in MRI has theoretically no limitations and is constantly being improved by continuously resolving relevant technical limitations. Today, in routine imaging in microscopy applications, the spatial resolution can reach up to 2 μm. Over the last two decades, many different imaging approaches have been developed in MRI with a variety of special applications. Since 1988, Dumoulin's team developed methods for imaging the flow and diffusion enabling the development of a significant branch of magnetic angiography. MRI techniques have also been developed that allow dynamic studies of the movement of the heart and other organs, as well as functional imaging of processes in the brain. With the technical development of MRI scanners, it is now possible to apply spectroscopy and spectroscopic imaging in a clinical routine, allowing detailed qualitative and quantitative studies of metabolic substances in the human body with noninvasive examinations and limited radiation.

The phenomenon of NMR is based on the property of the nuclei of some elements (with an odd number of protons and/or neutrons) to behave like small magnets. In a static magnetic field, the elementary magnet corresponding to each nucleus rotates around its axis with a constant frequency, characteristic of each chemical element. The way to detect a signal is to place the nuclei in a large magnetic field and irradiate them with electromagnetic radiation in the radio frequency range. Then the cores are tuned and we can get a detectable signal in the form of alternating current in the coil. The fundamental physical magnitude to which the phenomenon is due is the magnetic moment of the nuclei of the atoms of the matter. The phenomenon itself is the registration of the nuclear magnetic moment with external magnetic fields. The important feature of the NMR phenomenon is that the resonant frequency of the nuclear magnetic moment is determined unambiguously by the type of nucleus and is proportional to the intensity of the external magnetic field in the region of the nucleus.

### 2.5.1 MRI Principle

The phenomenon of NMR refers to the de-excitation of nuclear spins of nuclei with spins other than zero located in an external magnetic field of intensity $B_0$. During de-excitation, electromagnetic radiation is emitted with frequency $\omega$:

$$\omega = -\gamma B_0, \tag{2.2}$$

where $\gamma$ is the constant of the gyromagnetic ratio and has a specific, different value for each nucleus.

If we consider a sample with one type of nuclei species (of the same isotope, in the same chemical compound) and change the intensity of the external magnetic field in space, the detected frequencies depend only on the intensity of the external field, and, therefore, we can extract the distribution information of matter in space (visualization). This is the basic principle of MRI. Here we consider a sample consisting of a single type of nucleus, e.g. $^1$H in a specific chemical compound. Such a sample in a spectroscopy experiment signals a single frequency. To get an MRI tomography, we change the value of the external magnetic field in space.

As a result, the transmitted signal again consists of a set of frequencies. In this case, however, each frequency corresponds to a signal from the same core but from a different location in space, so we can distinguish the spatial origin of each signal. The intensity, and some other characteristic parameters of the signal determine the type of the tissue depicted as well as its possible pathological condition.

The basis of this technique is the application of a gradient of the external magnetic field in one direction, for example, the $x$-axis. Gradient is the gradual, linear change of the value of the external magnetic field and results in the value of the static magnetic field being different at each point in space along the $x$-axis, i.e. given by the relation:

$$B_{\text{eff}} = B_0 + B(r), \tag{2.3}$$

where $r$ is the application direction of the gradient $B(r)$, and can be any of the $x$, $y$, and $z$ axes, or any other random axis in space. The gradient of the field is implemented by applying a second static magnetic field in addition to $B_0$, with a value that varies linearly with distance and extends from negative values (i.e. decreasing the value of $B_0$) to positive values (i.e. increasing its value of $B_0$) and it is applied to the entire extent of the required field of view (FOV).

The number of frequencies we obtain are due to the nuclei of the same isotope in the same compound (e.g. $^1$H in the water molecule) but located at different positions on the $x$-axis. In this way, the spatial information in relation to a specific direction is retrieved. As the determination of the position in the 3D space requires the determination of the position with respect to at least three orthogonal directions between them, for the complete retrieval of the spatial information in MRI, the gradient of the static field in three perpendicular directions is applied. The application of the field gradient can be done during the application of the pulses $B_1$, during the reading of the signal, or even in the intervals of application of the pulses. The effect of the gradient application on the core system is proportional to its position of application during the pulse sequence. The most common types of MRI pulse series are the so-called tomographic methods where the field gradient in one direction (e.g. $z$-axis) is used to select a section to be imaged, while the field gradient in the other two directions (e.g. the $x$- and $y$-axes) are used to encode the position of the nuclei on the section plane with respect to frequency and phase.

### 2.5.2 Medical Applications of MRI

MRI has been established as a valid imaging technique which provides qualitative and quantitative information for diagnostic purposes, as well as for monitoring the treatment status of several diseased conditions. MRI can be categorized depending on the site of the body, which is imaged. In this concept, we can distinguish the head MRI, which is widely used for imaging of the brain purposing to detect tumors, aneurysms, bleedings, injuries of nerves or vessels from thrombus formations, and other abnormal conditions such as bladders and cysts. Especially, pathologic conditions of the optic and auditory nerves can be diagnosed as well. A subcategory of the head MRI is the functional (fMRI) which is used for measuring the brain activity and changes which are caused by blood flow [49]. The MRI, which is used to visualize the heart including its valves and coronary vessels, as well as the lungs, is called chest MRI. This type of MRI follows typical X-ray mammography where tumor has been diagnosed. In that case, the MRI is used for screening the tumor with increased detail [50].

A special application of magnetic resonance is the Magnetic Resonance Angiography (MRA), which is the imaging modality used for visualization of blood vessels and the blood flow through them [51–53]. It aims to identify pathologies due to atherosclerosis (such as stenosed arteries) or aneurysms. Moreover, it can be used to detect plaque vulnerability. It is common practice to use contrast agents to increase the contrast of the blood vessels.

MRI can also be used to image the abdomen and pelvic region. Using this type of MRI, tumors, bleedings, blockages, and infections can be detected in the abdomen organs such as the liver, gallbladder, pancreas, kidneys, and bladder. Also, MRI can be used to visualize the female uterus and ovaries as well as the prostate in men.

Arthritis, joint pathologies, bone marrow problems, bone tumors, cartilage problems, torn ligaments or tendons, and their infections can be depicted with the bone and joint MRI. It can also be used as an additional imaging technique when X-ray imaging cannot provide enough information for the assessment of bone fractures. Finally, spine condition in the region of the disks is usually assessed using the spine MRI to investigate the existence of spinal stenosis, disc bulges, and spinal tumors [54].

## 2.6 Positron Emission Tomography (PET)

PET is an applied method of medical imaging based on two fundamental natural phenomena: (i) the phenomenon of radioactivity $\beta^+$, and (ii) the phenomenon of positron–electron annihilation. $B^+$ radioactivity is the emission of a $\beta^+$ particle from the atomic nuclei with an insufficient number of neutrons. Annihilation is the interaction of a positron with an electron and their conversion into high-energy photons (annihilation photons). PET imaging is part of the broader field of nuclear medicine and functional and molecular imaging and uses radiopharmaceuticals that are labeled with radioactive isotope emitters of $\beta^+$ particles. The radiopharmaceutical is administered to the imaging biological system (human body, small animal, etc.) and transported to the area of diagnostic interest.

Specifically, through blood circulation, the radiopharmaceutical is transported to the examined organ. Through the recording of the emission of $\beta^+$ particles, its entire course is monitored. For each such particle, two high-energy photons emerge that travel in opposite directions. Capturing these photons requires two radiation detectors positioned in antidiameter positions. Through the recording of the emission of $\beta^+$ particles, its entire route is monitored. For each such particle, two high-energy photons are produced traveling in opposite directions. Capturing these photons requires two radiation detectors positioned in antidiameter positions. The two detectors are connected using a suitable coincident circuit which rejects any photon measurements that do not come from the same "event."

### 2.6.1 The Principle of PET

The production of radioactive $\beta^+$ isotopes is achieved by bombarding stable atomic nuclei with protons or deuterons. Intake of a proton changes the ratio of nucleons inside the nucleus (neutron deficiency). The extra proton is transformed into a neutron and at the same time, a $\beta^+$ particle is emitted as well as an electron neutrino. The typical unit for the production of radioactive isotopes, $\beta^+$ particle emissions, is based on cyclotron-type accelerators. The cyclotrons can be classified into several categories of small, medium, or large size. Respectively, the final energy given to the protons is different (e.g. 13, 18, and 32 MeV) as well as the capability of producing a wide variety of isotopes.

Small- and medium-sized machines can be installed in hospitals for on-site use. For energies of the order of 11–17 MeV, the corresponding activities are of the order of 100 GBq. The so-called negative ion technology is usually used. Suitable external sources of negative hydrogen ions ($H^-$) generating currents of 7–25 mA are used in these systems. After acceleration, the H-ions are stripped of their negative charge through thin carbon sheets. With this technology, output currents of the order of 500–600 μA and energies up to 30 MeV are achieved. Also, the acceleration efficiency of accelerated particles is very close to 100%. Stable nuclei, after being bombarded with protons, become radioactive ($\beta^+$ radioactive isotopes). The radioactive isotope is extracted from the cyclotron and purified from chemical and radioisotope impurities. The process of binding it to a pharmacological substance (biomolecule, etc.) is performed to form the radiopharmaceutical. This process takes place in the biosynthesizer unit.

The radiopharmaceutical is administered to the patient. $\beta^+$ particles are emitted from every point of the area occupied by the isotope. These travel inside the body a route in the order of mm. The length of the route depends on the energy spectrum of the $\beta^+$ particles, but also on the electronic density of the tissues. For example, in the case of the $^{18}F$, the maximum energy is 0.633 MeV and the corresponding average route in water is 1.4 mm. For the $^{15}O$, the maximum kinetic energy is

1.732 MeV and the average route in water is 2.7 mm. During its motion, the positron loses energy, through Coulomb interactions with the surrounding tissues, and slows down. At the end of the route, it interacts with an electron and an intermediate state of a hydrogen atom is created (positronium state). The positronium is vaporized and emits two photons of energy greater than 0.511 MeV (this energy corresponds to the mass of the electron or positron). The principle of conservation of momentum forces the photons to move in opposite directions. The deceleration brings the positron–electron system into a state of thermal equilibrium rather than complete immobility.

Particle behavior is governed by Fermi statistics and shows an isotropic momentum distribution in space. This results in a complete lack of linearity of the emitted photons. The photons propagate through the tissues (and air) which are located between the point of emission and the system detectors. Due to Compton scattering and photoelectric absorption, some of these photons do not reach the detectors. If the two anti-diametric detectors of the PET system record simultaneously from one photon (0.511 MeV), then these two photons come from the annihilation phenomenon. The role of the coincidence circuit is to discard measurements that are not recorded in pairs (one in each detector). This technique is called "ACD – annihilation coincidence detection." In order to record a pair of measurements, the two photons must reach the detectors in a very short period of time ("resolving time"). Due to the opposite direction of the two photons, the recorded pairs of measurements should come from a "narrow" elongated area (LOR, line of response), which forms a form of electronic collimator. An annihilation phenomenon that occurs outside the LOR does not give a pair of measurements. At least one photon will be directed away from the detectors. The absence of a metal guide significantly increases the sensitivity of the system compared to other techniques such as the SPECT imaging. It should be noted, however, that a percentage of the "coincidence measurements" comes from scattering or random coincidence of photons due to different fading phenomena. Also, the length of the route followed by the positrons (from the point of emission to the point of annihilation), as well as the deviation of the angle of their directions, are key limiting factors in the accuracy (spatial resolution) of the visualization.

### 2.6.2 Medical Applications of PET

PET succeeds in the recording of four-dimensional (spatio-temporal) distribution of molecules of a radiotracer involved in metabolic processes inside the human body. PET is characterized as the imaging method that measures and quantifies biochemical processes. It can assess local and absolute blood flow, metabolic processes of molecules, protein synthesis, gene expression, and tissue hypoxia. Also, it can assess the presence of abnormal proteins and quantify the receptors and the cellular transporter proteins that are bound to cell membranes. Over the past decade, PET has come to the forefront of oncology-related clinical practice. From a main research tool, in the recent past, it has been transformed into an imaging method of choice for diagnosis, staging, treatment monitoring and assessment of cancer recurrence. PET as an essential means of researching the biology of cancer, as well as the heart, is one of the tools of molecular imaging that helps to personalize (individualize) the treatment of diseases. PET tomography techniques are also applied to disease studies through small animal imaging (micro-PET). These techniques provide a very powerful tool for noninvasive pharmacokinetic studies, which, however, require the use of specialized high-performance systems.

PET uses isotopes of chemical elements that are a key components of biological systems. Such radioactive isotopes $\beta^+$ emitters are: $^{11}C$, $^{15}O$, and $^{13}N$. These isotopes are used to label natural, biochemical, and pharmaceutical substances. Their common feature is the very short half-life time of the order of a few minutes. However, there is a problem with their transfer from a relatively remote isotope production unit. The problem is solved by installing an accelerator, usually a cyclotron for the on-site production of isotopes.

Generally, a wide variety of radiopharmaceuticals and probes are available to help monitor and visualize the biology and physiology of cancer tumors. Today the most widely used tracer is a fluorescent analogue of glucose, FDG. FDG is generated when a deoxy-glucose substrate is labeled with fluorine-18, which is a $\beta^+$ (fluorescent deoxy-glucose) emitter. Usually, due to the presence of radioactive fluoride-18, the abbreviation $^{18}$F-FDG is used. The widespread use of $^{18}$F-FDG is due to many reasons. These include: (i) increased uptake of glucose by malignant cells, (ii) the relatively increased half-life time of $^{18}F$, which is 110 minutes, and (iii) it enables its transport from relatively remote production units when the cyclotron is out of the hospital. Also, this time is adequate for whole-body examinations in which the required duration is more than 30 minutes. It is worth noting that glucose is not a completely specific sensor (probe) of cancer tumors. However, this lack of specificity can be an advantage when it comes to detecting or staging pathology through full-body PET examinations.

## 2.7 Single Photon Emission Computed Tomography

Another medical imaging technique which is based on nuclear medicine combined with tomography is SPECT. The acquired images are similar to the ones of PET imaging. Subsequently, both imaging systems provide images related to the spatial concentration of radio-active molecules. Radionuclides are used to label the SPECT imaging emitting $\gamma$-ray photons [55]. A scintillation camera is necessary usually to capture the radiation. This type of camera permits traveling of the photons through the scintillator converting the energy of the photons to electric signals. In a post-processing step, these signals are analyzed to identify the position where the photon interacts with the scintillator's crystals. 2D images are produced by the 3D distribution of the radioactivity of the molecules caught by the tissue of the patient. The 3D representation is based on image reconstruction techniques similar to the CT by using 2D nuclear medicine images from several views around the region of interest. A major difference with PET imaging is that SPECT requires radioactive molecules which are quite common in nuclear clinics in contrast to PET imaging where generation of the photons is performed using the special equipment of cyclotrons.

### 2.7.1 The Principle of SPECT

Two main approaches are available in SPECT imaging systems for capturing the radioactivity: using one scintillation camera or using many scintillation cameras which rotate around the body. Both approaches enable the capturing of 2D images of which reconstruction can produce the 3D representation. When multiple cameras are used, the detection efficacy is increased significantly compared to the single-camera SPECT system. Both approaches are found adequate for most clinical applications. The spatial resolution of the present cameras is about 3 mm, while at 140 keV, the energy resolution is 10%. Though these values are efficient for most applications, currently, state-of-the-art SPECT imaging systems are capable to reduce the spatial resolution to less than 1–2 mm. Also, when semiconductor detectors or solid-state photon detectors are used, the spatial resolution is further reduced significantly. Such detectors are based on silicon and germanium or on cadmium zinc telluride presenting very accurate images, but their application is limited due to their high cost or the need to operate in low-temperature environments. Using these novel detectors, the limit in the spatial resolution regards the patient's allowance for the radiation dosage.

The SPECT image reconstruction in simple words is to perform a simple random transformation of the projections of the radiopharmaceutical activity. Since for each photon, attenuation and scattering are observed by the material between the detector and the source, the reconstruction of the SPECT is not the same as in CT reconstruction approaches and the photon attenuation must be considered in the process. Each photon that is found in the FOV of the detector is detected in case it is attenuated or scattered. A 3D response function is produced by combining the collimator-detector and scatter response. This function is related to the position where the photon is located in the tissue and the distance from the detector. SPECT reconstruction methodologies consist of: (i) image reconstruction algorithms analyzing the acquired projections and (ii) compensation algorithms improving image quality due to the noise produced by effects such as attenuation, scatter, and collimator-detector response [55].

### 2.7.2 Medical Applications of SPECT

SPECT imaging has nowadays a broad use especially in the field of cancer diagnosis and treatment.

| Disease | Treatment or diagnosis approach | Comment |
|---|---|---|
| Thyroid cancer | Thyroidectomy and therapy with $^{131}$I-iodide | Good results for early-stage cancer [56].<br>Worse prognosis for metastatic cancer or late stages of cancer due to reduced iodine-trapping property [57] |
| Pheochromocytomas and paragangliomas | Radioiodinated metaiodobenzylguanidine (MIBG) usually performed using the $^{123}$I labelled radiopharmaceutical | 90% accuracy of diagnosis and localization [58, 59]<br>High specificity of cancer staging<br>Improved treatment selection support using $^{131}$I-MIBG |
| Somatostatin-receptor-positive neuroendocrine tumors | $^{111}$In-octreotide scintigraphy | Accurate lesion detection [60] |
| Lymphoma | $^{67}$Ga-citrate scintigraphy | Increased diagnostic sensitivity [61] |
| Skeletal metastases | Scintigraphic imaging of bone metabolism | Cost-effective to diagnose metastases to bones and skeleton [62] from metastatic carcinomas e.g. prostate, lung, breast |
| Prostate cancer | 5 mCi $^{111}$In-ProstaScint | 76%, 44%, and 86% diagnostic accuracy, sensitivity, and specificity, respectively [63] |
| Coronary artery disease | $^{201}$Tl- and $^{99}$mTc-labelled radiopharmaceuticals | Limited accuracy due to image artefacts [64] |

## 2.8 Endoscopy

Endoscopy is a special medical imaging technique where a special equipment (i.e. endoscope) is used to examine and look inside the body. In contrast to most of the other imaging approaches, the endoscope goes precisely into the examined organ.

There is a variety of different endoscopes for each special need, while the process can be done by a specialized doctor or in some cases, by surgeons under anesthetization. The first endoscope was introduced in Mainz in 1806 in an attempt to examine the inner canals and cavities of the human body without however receiving the expected acceptance. By 1865, Dr. Cruise [65] applied endoscopy using his own endoscope in Dublin to examine the nasal and thoracic cavities. Dr. Macintyre trying to investigate the larynx developed an endoscope in 1894. The main characteristic of these first attempts of endoscope is that these were based on external light because it was impossible to have sufficient internal light. As soon as smaller bulbs were developed, endoscopes with internal lighting were performed in 1908 [66]. The terms laparoscopy and thoracoscopy were introduced by 1910 [67] and they were being used since then regularly to examine other organs as well such as the liver and the gallbladder or even for performing gynecologic laparoscopy to diagnose ectopic pregnancies or other dysfunctions.

The improvement in endoscopy was achieved after 1950 when the first fiber optics-based endoscope was produced. This leads to significant improvements in the imaging quality. The previous practice was based on using a dim red light or increased light output with the high risk however of internal burns. Further improvements in the fiber optics were related to the number of fibers which enabled the detailed and colorful visualization. Additional advancements regard the capability to steer the endoscope's tip or to remotely use the endoscope allowing only the endoscope inside the body through small holes improving in this way the wall laparoscopic process. Nevertheless, fiber optics-based endoscopies were physically limited in terms of image quality.

A more advanced approach invented by Hopkins was based on rods of glass which were used to fill the space between the lenses. This enabled to have all lenses fitted and self-aligned making it easier to operate them. Moreover, the image quality was significantly improved even for tubes of only 1-mm diameter. The production of this type of endoscope was done first by Storz [68]. Even though these endoscopes are rigid and in some cases, cannot be used when flexibility is required, the rod-lens endoscopes are still the main approach in keyhole laparoscopy. Recent developments in the field of endoscopy are: (i) the virtual reality-based simulators used for training doctors in endoscopic procedures, (ii) the disposable endoscopy which allows the use of inexpensive instruments by one patient only reducing any cross-contamination risks, (iii) the capsule endoscopy which is based on pill-sized endoscopes swallowed by the patients and used to record the visualization of the gastrointestinal tract, (iv) the endoscopic retrograde cholangiopancreatography combined with X-rays and applied to diagnose and treat pancreatic abnormalities, (v) chromoendoscopy which uses stains to identify abnormal regions, (vi) endoscopic US which combines US with the endoscope to visualize organs not regularly visible with common practices, and (vii) the endoscopic mucosal resection which is used to remove cancerous tissue.

### 2.8.1 Medical Applications of Endoscopy

There are many different types and medical applications of endoscopy as these are presented by the American Cancer Society (Table 2.1) [82]. It can be used as a diagnostic approach as well as for treatment purposes.

## 2.9 Elastography

Elastography is an imaging technique which enables the estimation of the elastic properties of the tissue as well as its stiffness. Depending on the tissue stiffness in terms of how soft or hard it is, it can be diagnosed as the state or presence of a disease. Elastography usually utilizes US or MRI for the visualization of the tissue with stiffness. Currently, there are several different approaches to elastography related to the way of operation and measurement of tissue stiffness. The common thing between all methods is that all attempt to create a distortion map in the tissue. To estimate the tissue distortion, it is necessary to identify how the tissue operates under deformation. In elastography, three major approaches exist to create distortion of the tissue: (i) to push or vibrate the surface of the tissue/organ using a tool (probe), (ii) to apply acoustic radiation based on US which applies a remote force, and (iii) to utilize any physiological distortions caused by processes such as the heartbeat.

As soon as a tissue distortion is produced, the response must be observed, and this is achieved by using another imaging technique, usually US or MRI. The outcome can also have a variety of formats including single values to lines, 2D planes, or in more advanced equipment, 3D volumes of stiffness. In general, however, the result is an image (or video) which presents the tissue with different colors which are related to the stiffness differences. At the post-processing step, the real stiffness can be estimated based on the principle that soft tissues deform more than stiff ones when the same force is applied. Moreover, it is assumed that the mechanical waves travel slower in soft tissues. The visual outcome in many cases is the elastic properties of the tissue in terms of Young's modulus and/or shear modulus.

### 2.9.1 Elastographic Techniques

As said earlier, many different techniques are available either based on US or MRI. In quasistatic elastography [83], the tissue is forced under compression and US is used to capture images pre- and post-compression. Based on the deformation map, the stiffness can be estimated between the different regions of the tissue. The acoustic radiation force impulse elastography [84] is based on the application of an acoustic force by an US beam. In that case, the tissue stiffness is

**Table 2.1** The types of endoscopy.

| Name of procedure | Name of tool | Area or organ viewed | How endoscope reaches target area |
|---|---|---|---|
| Anoscopy [69] | Anoscope | Anus and/or rectum | Inserted through the anus |
| Arthroscopy [70] | Arthroscope | Joints | Inserted through a small incision over the joint |
| Bronchoscopy [71] | Bronchoscope | Trachea, or windpipe, and the lungs | Inserted through the mouth |
| Colonoscopy [72] | Colonoscope | Entire length of the colon and large intestine | Inserted through the anus |
| Colposcopy [73] | Colposcope | Vagina and cervix | Placed at the vagina's opening after a tool called a speculum dilates the vagina. It is not inserted in the body |
| Cystoscopy [74] | Cystoscope | Inside of the bladder | Inserted through the urethra |
| Esophagoscopy [75] | Esophagoscope | Esophagus | Inserted through the mouth |
| Gastroscopy [76] | Gastroscope | Stomach and duodenum, which is the beginning of the small intestine | Inserted through the mouth |
| Laparoscopy [67] | Laparoscope | Stomach, liver, or other abdominal organs, including female reproductive organs, including the uterus, ovaries, and fallopian tubes | Inserted through a small, surgical opening in the abdomen |
| Laryngoscopy [77] | Laryngoscope | Larynx, or voice box | Inserted through the mouth |
| Neuroendoscopy [78] | Neuroendoscope | Areas of the brain | Inserted through a small incision in the skull |
| Proctoscopy [79] | Proctoscope | Rectum and sigmoid colon, which is the bottom part of the colon | Inserted through the anus |

(*Continued*)

**Table 2.1** (Continued)

| Name of procedure | Name of tool | Area or organ viewed | How endoscope reaches target area |
|---|---|---|---|
| Sigmoidoscopy [80] | Sigmoidoscope | Sigmoid colon | Inserted through the anus |
| Thoracoscopy [81] | Thoracoscope | Pleura, which are the two membranes covering the lungs and lining the chest cavity, and structures covering the heart | Inserted through a small surgical opening in chest |

*Source:* [82]/American Society of Clinical Oncology.

related to the amount that the tissue deforms along the axis of the beam. The shear wave elasticity [85] imaging elastography also uses acoustic force. In that case, the speed of wave travel is calculated and used for the calculation of the stiffness. The supersonic shear imaging [86, 87] enables the quantitative and real-time creation of a stiffness map. This technique is based on the principles of shear wave elasticity imaging. A movie is produced in this technique which presents the wave propagation in the tissue. The produced shear elasticity of the tissue is presented quantitatively using an ultrafast imaging modality and by analyzing the wave propagation outcome. The supersonic shear imaging has a great application in prostate, liver, breast, and many other organs. Finally, elastography using the principle of MRI is called magnetic resonance elastography [88] and has many medical applications especially in deep regions of the human body. A vibrator causes vibrations in the human body and the velocity of the produced waves is calculated. A 3D map is created representing the tissue stiffness while the expected MRI image is provided as well. The magnetic elastography has a major advantage over other elastographic techniques especially in regions where US cannot reach such as the deeper regions of the brain. Moreover, it is not affected by operation skills presenting high reproducibility of the results.

### 2.9.2 Elastographic Medical Applications

Using elastography, many organs can be examined in terms of their stiffness even in the deepest tissue or regions of the human body. Such organs are the liver, prostate, thyroid, lymph nodes, skin, cervix, and brain [89]. A common application of elastography is in cancer to differentiate the benign from the malignant lesions. In this way, tumor classification, lymph node metastasis, and vascular invasion are possible to be identified and quantified providing a useful prognostic tool [90, 91]. Elastography is also being used for the assessment of liver stiffness to diagnose fibrosis and cirrhosis [92, 93]. It has high discrimination power for fibrosis, but an average one for cirrhosis especially between advanced to moderate disease. Elastography can also be used to predict preterm birth by estimating the softening process across the gestation [94].

## 2.10 Conclusions and Future Trends

Medical imaging is a powerful tool in clinical practice, and it can be used for diagnostic, prognostic, and therapeutic purposes. As presented earlier, several types of medical imaging exist, capable to cover different areas of application and visualization of organs, tissues, or even cells. A comparison of the main medical imaging modalities is presented in Table 2.2 [95].

It could be argued that medical imaging and especially its applications have limitless boundaries. In particular, a major novel application in biomedical engineering and bioengineering is the 3D printing using or not using bioinks. In such applications, medical imaging is used to provide the 3D reconstruction of the organ geometries. This step enables the employment of patient-specific experimental designs or even provides the computer-aided diagnosis by examining the 3D-printed object.

Technical improvements are expected also in the medical equipment. This especially regards the state-of-the-art modalities of CT and MRI imaging. For example, CT imaging systems are provided with even less radiation or integrating deep learning approaches with real-time analysis. MRI systems, on the other hand, are provided with even larger magnetic fields $>3T$, which can boost the signal-to-noise ratio enabling fast imaging with high resolution. Additionally, the 3D imaging concept of the CT and MRI imaging enables cinematic rendering. Currently, volume rendering is provided by most workstation providers. However, the future trend is to enhance the visualization of the segmented tissues/organs with cinematic approaches improving the way that clinicians can make a diagnosis.

Another area with high novelty is the employment of deep learning and artificial intelligence algorithms for the analysis of medical images. This has increased significantly the segmentation or detection results. As, for example, it could be mentioned the deep learning analysis of intravascular US and optical coherence tomography (OCT) images for the segmentation and atherosclerotic plaque classification [96, 97] or the detection of breast cancer using mammography [98]. Another trend in medical imaging is the use of Enterprise Imaging. More specifically, a movement from the local picture archiving and communications system (PACS) to enterprise imaging and data management systems is observed. The concept of enterprise imaging is to use the imaging data accompanied by other patient's information (reports, exams, and waveforms) from one place so any doctor can access and assess them.

**Table 2.2** Comparison of the main medical imaging modalities.

| Medical imaging modality | X-ray | US | CT | MRI | PET | SPECT |
|---|---|---|---|---|---|---|
| Penetration level | Organ-tissue | Organ-tissue | Organ-tissue | Organ-tissue | Tissue-cellular-molecular | Tissue-cellular |
| Spatial resolution | ~0.1 mm | 50–500 μm | 50–200 μm | 25–100 μm | 1–2 mm | 1–2 mm |
| Cost | Low | Low | Moderate | High | Very high | High |
| Advantages | Fast<br>High applicability | High spatial and temporal resolution<br>Low cost<br>Real time | High spatial resolution<br>Anatomic imaging | Highest spatial resolution<br>Soft tissue contrast<br>Functional imaging | High sensitivity<br>Unlimited depth penetration | Unlimited depth penetration<br>Imaging multiple probes simultaneously |
| Disadvantages | Limits number of images<br>Radiation | Operator dependency | No target-specific imaging<br>Radiation<br>Poor soft tissue contrast | Cost<br>Low sensitivity<br>Long scan and processing time | Cost<br>Cyclotron is needed<br>Low spatial resolution<br>Radiation | Limited spatial resolution<br>Radiation |

*Source:* Adapted from Jang et al. [95].

## References


1 Howell, J.D. (2016). Early clinical use of the X-ray. *Trans. Am. Clin. Climatol. Assoc.* 127: 341–349.

2 Thüring, T., Abis, M., Wang, Z. et al. (2014). X-ray phase-contrast imaging at 100 keV on a conventional source. *Sci. Rep.* 4 (1) Art. no. 1: https://doi.org/10.1038/srep05198.

3 Berger, M., Yang, Q., and Maier, A. (2018). Chapter 3 – X-ray imaging. In: *Medical Imaging Systems: An Introductory Guide* (ed. A. Maier, S. Steidl, V. Christlein and J. Hornegger), 37–52. Cham: Springer.

4 Bernstein, A.M. and Mann, A.K. (1958). Scattering of gamma rays by a static electric field. *Physiol. Rev.* 110 (4): 805–814. https://doi.org/10.1103/PhysRev.110.805.

5 Carlsson, C.A. and Carlsson, G.A. Basic physics of X-ray imaging. 31.

6 Hawkes, D.J. and Jackson, D.F. (1980). An accurate parametrisation of the X-ray attenuation coefficient. *Phys. Med. Biol.* 25 (6): 1167–1171. https://doi.org/10.1088/0031-9155/25/6/014.

7 Dhainaut, A., Hoff, M., Syversen, U., and Haugeberg, G. (2016). Technologies for assessment of bone reflecting bone strength and bone mineral density in elderly women: an update. *Womens Health (Lond.)* 12 (2): 209–216. https://doi.org/10.2217/whe.15.94.

8 de Roos, A. and Higgins, C.B. (2014). Cardiac radiology: centenary review. *Radiology* 273 (2S): S142–S159. https://doi.org/10.1148/radiol.14140432.

9 Pauwels, E.K.J., Foray, N., and Bourguignon, M.H. (2016). Breast cancer induced by X-ray mammography screening? A review based on recent understanding of low-dose radiobiology. *Med. Princ. Pract.* 25 (2): 101–109. https://doi.org/10.1159/000442442.

10 G. N. Hounsfield, "Computerized transverse axial scanning (tomography). 1. Description of system," *Br. J. Radiol.*, vol. 46, no. 552, pp. 1016–1022, Dec. 1973, https://doi.org/10.1259/0007-1285-46-552-1016.

11 Radon, J. (1986). On the determination of functions from their integral values along certain manifolds. *IEEE Trans. Med. Imaging* 5 (4): 170–176. https://doi.org/10.1109/TMI.1986.4307775.

12 Kalender, W.A. (2006). *Computertomographie: Grundlagen, Gerätetechnologie, Bildqualität, Anwendungen*, 2e. Erlangen: Publicis.

13 Peyrin, F. and Engelke, K. (2012). CT imaging: basics and new trends. In: *Handbook of Particle Detection and Imaging* (ed. C. Grupen and I. Buvat), 883–915. Berlin/Heidelberg: Springer.

14 Willi A. Kalender (2011). Computed Tomography: Fundamentals, System Technology, Image Quality, Applications, 3, Wiley. https://www.wiley.com/en-us/Computed+Tomography%3A+Fundamentals%2C+System+Technology%2C+Image+Quality%2C+Applications%2C+3rd+Edition-p-9783895786440 (accessed 09 December, 2019).

**15** Rubin, G.D. (2014). Computed tomography: revolutionizing the practice of medicine for 40 years. *Radiology* 273 (2S): S45–S74. https://doi.org/10.1148/radiol.14141356.

**16** Johkoh, T., Müller, N., Cartier, Y. et al. (1999). Idiopathic interstitial pneumonias: diagnostic accuracy of thin-section CT in 129 patients. *Radiology* 211 (2): 555–560. https://doi.org/10.1148/radiology.211.2.r99ma01555.

**17** Grenier, P., Maurice, F., Musset, D. et al. (1986). Bronchiectasis: assessment by thin-section CT. *Radiology* 161 (1): 95–99. https://doi.org/10.1148/radiology.161.1.3763889.

**18** Lynch, D.A., Brasch, R.C., Hardy, K.A., and Webb, W.R. (1990). Pediatric pulmonary disease: assessment with high-resolution ultrafast CT. *Radiology* 176 (1): 243–248. https://doi.org/10.1148/radiology.176.1.2353097.

**19** Leber, A.W., Becker, A., Knez, A. et al. (2006). Accuracy of 64-slice computed tomography to classify and quantify plaque volumes in the proximal coronary system: a comparative study using intravascular ultrasound. *J. Am. Coll. Cardiol.* 47 (3): 672–677. https://doi.org/10.1016/j.jacc.2005.10.058.

**20** Chow, B.J., Abraham, A., Wells, G.A. et al. (2009). Diagnostic accuracy and impact of computed tomographic coronary angiography on utilization of invasive coronary angiography. *Circ. Cardiovasc. Imaging* 2 (1): 16–23. https://doi.org/10.1161/CIRCIMAGING.108.792572.

**21** Klass, O., Kleinhans, S., Walker, M.J. et al. (2010). Coronary plaque imaging with 256-slice multidetector computed tomography: interobserver variability of volumetric lesion parameters with semiautomatic plaque analysis software. *Int. J. Cardiovasc. Imaging* 26 (6): 711–720. https://doi.org/10.1007/s10554-010-9614-3.

**22** Takahashi, S., Kawasaki, M., Miyata, S. et al. (2016). Feasibility of tissue characterization of coronary plaques using 320-detector row computed tomography: comparison with integrated backscatter intravascular ultrasound. *Heart Vessels* 31 (1): 29–37. https://doi.org/10.1007/s00380-014-0577-5.

**23** Siogkas, P.K., Anagnostopoulos, C.D., Liga, R. et al. (2018). Noninvasive CT-based hemodynamic assessment of coronary lesions derived from fast computational analysis: a comparison against fractional flow reserve. *Eur. Radiol.* https://doi.org/10.1007/s00330-018-5781-8.

**24** Bonaldi, V.M., Bret, P.M., Reinhold, C., and Atri, M. (1995). Helical CT of the liver: value of an early hepatic arterial phase. *Radiology* 197 (2): 357–363. https://doi.org/10.1148/radiology.197.2.7480677.

**25** Oliver, J.H. and Baron, R.L. (1996). Helical biphasic contrast-enhanced CT of the liver: technique, indications, interpretation, and pitfalls. *Radiology* 201 (1): 1–14. https://doi.org/10.1148/radiology.201.1.8816509.

**26** Kamel, I.R., Choti, M.A., Horton, K.M. et al. (2003). Surgically staged focal liver lesions: accuracy and reproducibility of dual-phase helical CT for detection and characterization. *Radiology* 227 (3): 752–757. https://doi.org/10.1148/radiol.2273011768.

**27** Hoeffner, E.G., Case, I., Jain, R. et al. (2004). Cerebral perfusion CT: technique and clinical applications. *Radiology* 231 (3): 632–644. https://doi.org/10.1148/radiol.2313021488.

**28** Andreini, D., Mushtaq, S., Pontone, G. et al. (2019). CT perfusion versus coronary CT angiography in patients with suspected in-stent restenosis or CAD progression. *J. Am. Coll. Cardiol. Imaging* https://doi.org/10.1016/j.jcmg.2019.05.031.

**29** Carovac, A., Smajlovic, F., and Junuzovic, D. (2011). Application of ultrasound in medicine. *Acta Inform. Med.* 19 (3): 168–171. https://doi.org/10.5455/aim.2011.19.168-171.

**30** Venkata Ranganayakulu, S., Nathani, R., and Gahane, L. (2016). Ultrasound applications in medical sciences. *ijnter* 287–297.

**31** Miller, D.L., Smith, N.B., Bailey, M.R. et al. (2012). Overview of therapeutic ultrasound applications and safety considerations. *J. Ultrasound Med.* 31 (4): 623–634. https://doi.org/10.7863/jum.2012.31.4.623.

**32** Robertson, V.J. and Baker, K.G. (2001). A review of therapeutic ultrasound: effectiveness studies. *Phys. Ther.* 81 (7): 1339–1350.

**33** Samulski, T.V., Grant, W.J., Oleson, J.R. et al. (1990). Clinical experience with a multi-element ultrasonic hyperthermia system: analysis of treatment temperatures. *Int. J. Hyperth.* 6 (5): 909–922. https://doi.org/10.3109/02656739009140972.

**34** Tempany, C.M.C., Stewart, E.A., McDannold, N. et al. (2003). MR imaging-guided focused ultrasound surgery of uterine leiomyomas: a feasibility study. *Radiology* 226 (3): 897–905. https://doi.org/10.1148/radiol.2271020395.

**35** Burgess, S.E., Silverman, R.H., Coleman, D.J. et al. (1986). Treatment of glaucoma with high-intensity focused ultrasound. *Ophthalmology* 93 (6): 831–838. https://doi.org/10.1016/s0161-6420(86)33672-8.

**36** Klingler, H.C., Susani, M., Seip, R. et al. (2008). A novel approach to energy ablative therapy of small renal tumours: laparoscopic high-intensity focused ultrasound. *Eur. Urol.* 53 (4): 810–816; discussion 817, 818,. https://doi.org/10.1016/j.eururo.2007.11.020.

**37** Ninet, J., Roques, X., Seitelberger, R. et al. (2005). Surgical ablation of atrial fibrillation with off-pump, epicardial, high-intensity focused ultrasound: results of a multicenter trial. *J. Thorac. Cardiovasc. Surg.* 130 (3): 803–809. https://doi.org/10.1016/j.jtcvs.2005.05.014.

**38** Alam, M., White, L.E., Martin, N. et al. (2010). Ultrasound tightening of facial and neck skin: a rater-blinded prospective cohort study. *J. Am. Acad. Dermatol.* 62 (2): 262–269. https://doi.org/10.1016/j.jaad.2009.06.039.

39 Weizer, A.Z., Zhong, P., and Preminger, G.M. (2007). New concepts in shock wave lithotripsy. *Urol. Clin. North Am.* 34 (3): 375–382. https://doi.org/10.1016/j.ucl.2007.07.002.

40 Lowe, G. and Knudsen, B.E. (2009). Ultrasonic, pneumatic and combination intracorporeal lithotripsy for percutaneous nephrolithotomy. *J. Endourol.* 23 (10): 1663–1668. https://doi.org/10.1089/end.2009.1533.

41 Haake, M., Buch, M., Schoellner, C. et al. (2003). Extracorporeal shock wave therapy for plantar fasciitis: randomised controlled multicentre trial. *BMJ* 327 (7406): 75. https://doi.org/10.1136/bmj.327.7406.75.

42 Packer, M., Fishkind, W.J., Fine, I.H. et al. (2005). The physics of phaco: a review. *J. Cataract. Refract. Surg.* 31 (2): 424–431. https://doi.org/10.1016/j.jcrs.2004.11.027.

43 Mann, M.W., Palm, M.D., and Sengelmann, R.D. (2008). New advances in liposuction technology. *Semin. Cutan. Med. Surg.* 27 (1): 72–82. https://doi.org/10.1016/j.sder.2008.01.005.

44 Koch, C., Borys, M., Fedtke, T. et al. (2002). Determination of the acoustic output of a harmonic scalpel. *IEEE Trans. Ultrason. Ferroelectr. Freq. Control* 49 (11): 1522–1529. https://doi.org/10.1109/tuffc.2002.1049734.

45 Parikh, S., Motarjeme, A., McNamara, T. et al. (2008). Ultrasound-accelerated thrombolysis for the treatment of deep vein thrombosis: initial clinical experience. *J. Vasc. Interv. Radiol.* 19 (4): 521–528. https://doi.org/10.1016/j.jvir.2007.11.023.

46 Smith, N.B. (2008). Applications of ultrasonic skin permeation in transdermal drug delivery. *Expert Opin. Drug Delivery* 5 (10): 1107–1120. https://doi.org/10.1517/17425247.5.10.1107.

47 Potsika, V.T., Protopappas, V.C., Grivas, K.N. et al. (2017). Numerical evaluation of the backward propagating acoustic field in healing long bones. *J. Acoust. Soc. Am.* 142 (2): 962. https://doi.org/10.1121/1.4998722.

48 Vavva, M.G., Grivas, K.N., Carlier, A. et al. (2018). Effect of ultrasound on bone fracture healing: a computational bioregulatory model. *Comput. Biol. Med.* 100: 74–85. https://doi.org/10.1016/j.compbiomed.2018.06.024.

49 Watanabe, T., Sasaki, Y., Shibata, K., and Kawato, M. (2017). Advances in fMRI real-time neurofeedback. *Trends Cogn. Sci. (Regul. Ed.)* 21 (12): 997–1010. https://doi.org/10.1016/j.tics.2017.09.010.

50 Leithner, D., Wengert, G.J., Helbich, T.H. et al. (2018). Clinical role of breast MRI now and going forward. *Clin. Radiol.* 73 (8): 700–714. https://doi.org/10.1016/j.crad.2017.10.021.

51 Russo, V., Renzulli, M., Buttazzi, K., and Fattori, R. (2006). Acquired diseases of the thoracic aorta: role of MRI and MRA. *Eur. Radiol.* 16 (4): 852–865. https://doi.org/10.1007/s00330-005-0028-x.

52 Dellegrottaglie, S., Sanz, J., and Rajagopalan, S. (2006). Technology insight: clinical role of magnetic resonance angiography in the diagnosis and

management of renal artery stenosis. *Nat. Clin. Pract. Cardiovasc. Med.* 3 (6): 329–338. https://doi.org/10.1038/ncpcardio0556.

**53** Okudera, T. (1995). The current role of cerebral angiography in diagnosis of cerebrovascular disease. *Rinsho Shinkeigaku* 35 (12): 1569–1571.

**54** Waldt, S., Gersing, A., and Brügel, M. (2014). Measurements and classifications in spine imaging. *Semin. Musculoskelet. Radiol.* 18 (3): 219–227. https://doi.org/10.1055/s-0034-1375565.

**55** National Research Council (US) and Institute of Medicine (US) Committee on the Mathematics and Physics of Emerging Dynamic Biomedical Imaging. Washington (DC) (1996). *Single Photon Emission Computed Tomography.* National Academies Press (US).

**56** "Thyroid Cancer in Clinical Practice | I. Ross McDougall | Springer." https://www.springer.com/gp/book/9781846285448 (accessed 22 November 2022).

**57** Edge, S., Byrd, D.R., Compton, C.C. et al. (ed.) (2010). *AJCC Cancer Staging Handbook: From the AJCC Cancer Staging Manual*, 7e. New York: Springer-Verlag.

**58** Kaltsas, G.A., Besser, G.M., and Grossman, A.B. (2004). The diagnosis and medical management of advanced neuroendocrine tumors. *Endocr. Rev.* 25 (3): 458–511. https://doi.org/10.1210/er.2003-0014.

**59** Rufini, V., Calcagni, M.L., and Baum, R.P. (2006). Imaging of neuroendocrine tumors. *Semin. Nucl. Med.* 36 (3): 228–247. https://doi.org/10.1053/j.semnuclmed.2006.03.007.

**60** Lodish, M., Dagalakis, U., Chen, C.C. et al. (2012). $^{111}$In-octreotide scintigraphy for identification of metastatic medullary thyroid carcinoma in children and adolescents. *J. Clin. Endocrinol. Metab.* 97 (2): E207–E212. https://doi.org/10.1210/jc.2011-2766.

**61** Palumbo, B., Sivolella, S., Palumbo, I. et al. (2005). $^{67}$Ga-SPECT/CT with a hybrid system in the clinical management of lymphoma. *Eur. J. Nucl. Med. Mol. Imaging* 32 (9): 1011–1017. https://doi.org/10.1007/s00259-005-1788-x.

**62** Even-Sapir, E. (2005). Imaging of malignant bone involvement by morphologic, scintigraphic, and hybrid modalities. *J. Nucl. Med.* 46 (8): 1356–1367.

**63** Fareedy, S., Deh, K., Koney, N. et al. (2013). Ex vivo SPECT/CT imaging of prostate cancer via $^{111}$In-J591 monoclonal antibody. *J. Nucl. Med.* 54 (supplement 2): 282–282.

**64** Hasbek, Z., Ertürk, S.A., Çakmakçılar, A. et al. (2019). Evaluation of myocardial perfusion imaging SPECT parameters and pharmacologic stress test with adenosine versus coronary angiography findings: are they diagnostically concordant? *Mol. Imaging Radionucl. Ther.* 28 (2): 53–61. https://doi.org/10.4274/mirt.galenos.2019.47450.

**65** Cruise, F.R. (1865). The endoscope as an aid to the diagnosis and treatment of disease. *Br. Med. J.* 1 (223): 345–347.

**66** Shawki, O., Deshmukh, S., and Pacheco, L.A. (2017). *Mastering the Techniques in Hysteroscopy*. JP Medical Ltd.

**67** Litynski, G.S. (1997). Laparoscopy – the early attempts: spotlighting Georg Kelling and Hans Christian Jacobaeus. *JSLS* 1 (1): 83–85. PMID: 9876654; PMCID: PMC3015224.

**68** Cockett WS, Cockett AT. The Hopkins rod-lens system and the Storz cold light illumination system. *Urology*. 1998;51(5A Suppl):1–2. doi:https://doi.org/10.1016/s0090-4295(98)00060-0

**69** London, S., Hoilat, G.J., and Tichauer, M.B. (2020). Anoscopy. In: *StatPearls*. Treasure Island, FL: StatPearls Publishing.

**70** Villar, R.N. (1994). Arthroscopy. *BMJ* 308 (6920): 51–53. doi: 10.1136/bmj.308.6920.51. PMID: 8298357; PMCID: PMC2539138.

**71** Mahmoud, N., Vashisht, R., Sanghavi, D., and Kalanjeri, S. (2022). Bronchoscopy. In: *StatPearls. Treasure Island (FL)*. StatPearls Publishing.

**72** Seward, E. (2019). Recent advances in colonoscopy. *F1000Res.* 8: https://doi.org/10.12688/f1000research.18503.1.

**73** Landers, E.E., Erickson, B.K., Bae, S., and Huh, W.K. (2016). Trends in colposcopy volume: where do we go from here? *J. Low. Genit. Tract Dis.* 20 (4): 292–295. https://doi.org/10.1097/LGT.0000000000000258.

**74** Assmus, M.A., McLarty, R., Senthilselvan, A., and De, S.K. (2020). Direct to cystoscopy: a prospective quality assessment of patient preferences. *Can. Urol. Assoc. J.* 14 (4): 118–121. https://doi.org/10.5489/cuaj.6013.

**75** Sabirin, J., Abd Rahman, M., and Rajan, P. (2013). Changing trends in oesophageal endoscopy: a systematic review of transnasal oesophagoscopy. *ISRN Otolaryngol.* 2013: https://doi.org/10.1155/2013/586973.

**76** Chen, L, Liang, X, and Tan, X, et al. (2019). Safety and efficacy of combined use of propofol and etomidate for sedation during gastroscopy: Systematic review and meta-analysis. *Medicine (Baltimore)* 98 (20): e15712.

**77** Pieters, B.M.A., Maas, E.H.A., Knape, J.T.A., and van Zundert, A.A.J. (2017). Videolaryngoscopy vs. direct laryngoscopy use by experienced anaesthetists in patients with known difficult airways: a systematic review and meta-analysis. *Anaesthesia* 72 (12): 1532–1541. https://doi.org/10.1111/anae.14057.

**78** Shim, K.W., Park, E.K., Kim, D.-S., and Choi, J.-U. (2017). Neuroendoscopy: current and future perspectives. *Journal of Korean Neurosurgical Society* 60 (3): 322–326.

**79** Sørensen, H.T., Ejlersen, E., Møller-Petersen, J. et al. (1992). Overall use of proctoscopy in general practice and possible relation to the stage of rectal cancer. *Family Practice* 9 (2): 145–148.

**80** Niedermaier, T., Weigl, K., Hoffmeister, M., and Brenner, H. (2017). Diagnostic performance of flexible sigmoidoscopy combined with fecal immunochemical test in colorectal cancer screening: meta-analysis and modeling. *European Journal of Epidemiology* 32 (6): 481–493.

**81** Murthy, V. and Bessich, J.L. (2017). Medical thoracoscopy and its evolving role in the diagnosis and treatment of pleural disease. *J. Thorac. Dis.* 9 (Suppl 10): S1011–S1021. https://doi.org/10.21037/jtd.2017.06.37.

**82** "Types of endoscopy," *Cancer. Net.*, 2011. https://www.cancer.net/navigating-cancer-care/diagnosing-cancer/tests-and-procedures/types-endoscopy (accessed 22 November 2022).

**83** Ophir, J., Céspedes, I., Ponnekanti, H. et al. (1991). Elastography: a quantitative method for imaging the elasticity of biological tissues. *Ultrason. Imaging* 13 (2): 111–134. https://doi.org/10.1016/0161-7346(91)90079-W.

**84** Nightingale, K.R., Palmeri, M.L., Nightingale, R.W., and Trahey, G.E. (2001). On the feasibility of remote palpation using acoustic radiation force. *J. Acoust. Soc. Am.* 110 (1): 625–634. https://doi.org/10.1121/1.1378344.

**85** Sarvazyan, A.P., Rudenko, O.V., Swanson, S.D. et al. (1998). Shear wave elasticity imaging: a new ultrasonic technology of medical diagnostics. *Ultrasound Med. Biol.* 24 (9): 1419–1435. https://doi.org/10.1016/S0301-5629(98)00110-0.

**86** Bercoff, J., Tanter, M., and Fink, M. (2004). Supersonic shear imaging: a new technique for soft tissue elasticity mapping. *IEEE Trans. Ultrason. Ferroelectr. Freq. Control* 51 (4): 396–409. https://doi.org/10.1109/TUFFC.2004.1295425.

**87** Gennisson, J.-L., Rénier, M., Catheline, S. et al. (2007). Acoustoelasticity in soft solids: assessment of the nonlinear shear modulus with the acoustic radiation force. *J. Acoust. Soc. Am.* 122 (6): 3211–3219. https://doi.org/10.1121/1.2793605.

**88** Sarvazyan, A.P., Skovoroda, A.R., Emelianov, S.Y. et al. (1995). Biophysical bases of elasticity imaging. In: *Acoustical Imaging* (ed. J.P. Jones), 223–240. Boston, MA: Springer US.

**89** Lu, J., Chen, M., Chen, Q.-H. et al. (2019). Elastogram: physics, clinical applications, and risks. *Maternal Fetal Med.* 1 (2): 113–122. https://doi.org/10.1097/FM9.0000000000000024.

**90** Evans, A., Whelehan, P., Thomson, K. et al. (2012). Invasive breast cancer: relationship between shear-wave elastographic findings and histologic prognostic factors. *Radiology* 263 (3): 673–677. https://doi.org/10.1148/radiol.12111317.

**91** Athanasiou, A., Tardivon, A., Tanter, M. et al. (2010). Breast lesions: quantitative elastography with supersonic shear imaging-preliminary results. *Radiology* 256 (1): 297–303. https://doi.org/10.1148/radiol.10090385.

**92** L. Sandrin B. Fourquet, J.-M. Hasquenoph *et al.*, "Transient elastography: a new noninvasive method for assessment of hepatic fibrosis," *Ultrasound Med. Biol.*, vol. 29, no. 12, pp. 1705–1713, Dec. 2003, https://doi.org/10.1016/j.ultrasmedbio.2003.07.001.

**93** Sánchez-Conde, M., Montes-Ramırez, M.L., Miralles, P. et al. (2010). Comparison of transient elastography and liver biopsy for the assessment of liver fibrosis in HIV/hepatitis C virus-coinfected patients and correlation with noninvasive serum markers. *J. Viral Hepat.* 17 (4): 280–286. https://doi.org/10.1111/j.1365-2893.2009.01180.x.

**94** Peralta, L., Molina, F., Melchor, J. et al. (2017). Transient elastography to assess the cervical ripening during pregnancy: a preliminary study. *Ultraschall Med.* 38 (4): 395–402. https://doi.org/10.1055/s-0035-1553325.

**95** Jang, H.-Y., Kim, H.-R., Kang, M.-S. et al. (2014). The demand for quantitative techniques in biomedical image informatics. *Biomed. Eng. Lett.* 4 (4): 319–327. https://doi.org/10.1007/s13534-014-0169-4.

**96** Olender, M.L., Athanasiou, L.S., Michalis, L.K. et al. (2020). A domain enriched deep learning approach to classify atherosclerosis using intravascular ultrasound imaging. *IEEE J. Sel. Top. Signal Process.* 14 (6): 1210–1220. https://doi.org/10.1109/JSTSP.2020.3002385.

**97** Athanasiou, L.S., Olender, M.L., Hernandez, J.M.d.l.T. et al. (2019). *A Deep Learning Approach to Classify Atherosclerosis Using Intracoronary Optical Coherence Tomography*, vol. 10950. SPIE.

**98** Shen, L., Margolies, L.R., Rothstein, J.H. et al. (2019). Deep learning to improve breast cancer detection on screening mammography. *Sci. Rep.* 9 (1), Art. no. 1: https://doi.org/10.1038/s41598-019-48995-4.

# 3

# Computational Modeling at Molecular Level

## 3.1 Introduction

Modeling at molecular level deals with the science and research of investigating molecular structure and function using computational means. The research domain of biomolecular modeling is considered relatively new, having started in the 1950s, and having gained significant attention since the mid-1980s following the technology revolution, which enabled computer technology to be employed in biomedical engineering. Nowadays, this research domain is developing presenting tremendous results. The establishment of a computational model can be as simple as plastic templates or metal rods, or as sophisticated as interactive, animated color stereographics, and laser-made wooden sculptures [1]. The field of molecular modeling includes both theoretical and computational methods, which can be applied to model or mimic the behavior of molecules based on quantum mechanics, molecular dynamics, molecular mechanics, structure/activity relationships, chemical/biochemical information and databases, as well as other established procedures. Computational chemistry, biomedical engineering, pharmacogenetics and pharmacogenomics, computational biology, and materials science are only some of the research areas which study molecular systems ranging from small chemical systems to large biological molecules and material assemblies. While simple calculations can be performed by hand, when complex numerical systems are investigated at the molecular level, the use of computers is inevitable in order to model and simulate any reasonably sized system.

Molecular mechanics is considered the main aspect of the science of molecular modeling, as it encompasses the use of classical mechanics (Newtonian mechanics) to describe the physical and fundamental bases for the development of numerical models. A traditional description of molecular models refers to atoms

*Multiscale Modelling in Biomedical Engineering*, First Edition. Dimitrios I. Fotiadis, Antonis I. Sakellarios, and Vassiliki T. Potsika.

Published 2023 by John Wiley & Sons, Inc.

(nucleus and electrons collectively) as point charges with an associated mass. The interactions between neighboring atoms can be described by spring-like interactions representing chemical bonds and Van der Waals forces. The electrostatic interactions are computed employing Coulomb's law. The Cartesian space or internal coordinates are used to assign coordinates to atoms. The assignment of velocities is based on dynamic simulations. The atomic velocities can be related to the system's temperature, which is a macroscopic quantity. The collective mathematical expression depends on the system's internal energy, a thermodynamic quantity which is equal to the sum of potential and kinetic energies.

Molecular modeling techniques have grown exponentially in recent years and are becoming essential components of many chemical, physical, and biological studies. Two primary reasons for this remarkable evolution exist, which are described in what follows [2–5].

First, molecular imaging, the visualization, characterization, and measurement of biological processes at the cellular, subcellular level, or even molecular level in living subjects, has rapidly gained importance in the dawning era of personalized medicine [2]. Molecular imaging has opened new horizons in the diagnosis and management of several diseases including cancer and neurological and cardiovascular diseases. With the accomplishment of human genome project, rapid advancements in molecular and cell biology have been achieved, and ever-increasing molecular targets and biomarkers have been identified, especially those aberrantly expressed in tumor malignancy, invasiveness, and metastasis [2, 3]. Structural data from proteins are derived from X-ray crystallography, nuclear magnetic resonance (NMR) spectroscopy, and electron microscopy studies. More specifically, X-ray crystallography is a technique which allows the determination of the three-dimensional (3D) structure of molecules, including complex biological macromolecules such as proteins and nucleic acids. This technique is considered a powerful tool which can give an insight into the 3D structure of a molecule at atomic resolution. The electron density can be reconstructed according to the diffraction pattern obtained from X-ray scattering of the periodic assembly of molecules or atoms in the crystal [5]. NMR is a spectroscopic technique which enables the observation of local magnetic fields around atomic nuclei and provides detailed information about the structure, dynamics, reaction state, and chemical environment of molecules. NMR is largely used by biochemists to investigate proteins and other complex molecules. According to this technique, the sample is placed in a magnetic field and the NMR signal is generated through the excitation of the nuclei sample with radio waves into NMR. The intramolecular magnetic field around an atom in a molecule alternates the resonance frequency and reveals details on the electronic structure of a molecule and its individual functional units.

Second, over the last few decades, accompanying this evolution in the area of structural biology, advancements in computer and software technologies have reached an impressive level. As computers continue to increase in speed and capacity, the scientific community is able to tackle larger and more complex systems. This remarkable technical development has enhanced the evolution of "big data," which is a field that aims to analyze, systematically extract information from, and deal with datasets that are too large or complex to be examined with traditional data-processing application software.

## 3.2 Introduction to Molecular Mechanics

The aim of this section is to present fundamental principles in molecular mechanics such as chemical formulas, molecular structure, and polarity.

### 3.2.1 Chemical Formulas

A molecule can be defined as a group of two or more atoms held together by chemical bonds and is electrically neutral. Each cell is considered a "bustling factory" with a wide range of prolific arrays of molecular machinery. This complex molecular machinery self-organizes to accomplish the complex biochemical activities of living. The functions of a biomolecule are prescribed by its structure implying that its shape and chemical properties facilitate interactions with other molecules and determine its role in the cell. The function of a protein depends on the precise folding of encoded linear stretches of amino acids into 3D shapes, while protein misfolding may cause various diseases [6, 7]. For example, according to [8], many neurodegenerative diseases involve the misfolding and aggregation of specific proteins into abnormal, toxic species and therapeutic options target different steps in the production and processing of proteins including synthesis, chaperone-assisted folding and trafficking, as well as degradation *via* the proteasome and autophagy pathways.

In Figure 3.1a, the levels of structural organization in the human body are presented focusing on the molecular level. More specifically, a molecular formula is a representation using chemical symbols to give the total number of atoms of each element in each molecule of a substance. In particular, Figure 3.1b presents the composition of the molecule of sulfur. It can be depicted as a structural formula, a ball-and-stick model, or a space-filling model. In addition, molecular formulas are useful as abbreviations for the names of compounds. For example, $H_2$ is a molecular formula which represents a diatomic molecule of hydrogen, consisting of two atoms of the element that are chemically bonded together. The expression 2H, on the other hand, indicates two separate hydrogen atoms that are not combined as a unit. The expression $2H_2$ represents two molecules of diatomic hydrogen. The symbols H, 2H, $H_2$, and $2H_2$ represent very different entities (Figure 3.1c) [7].

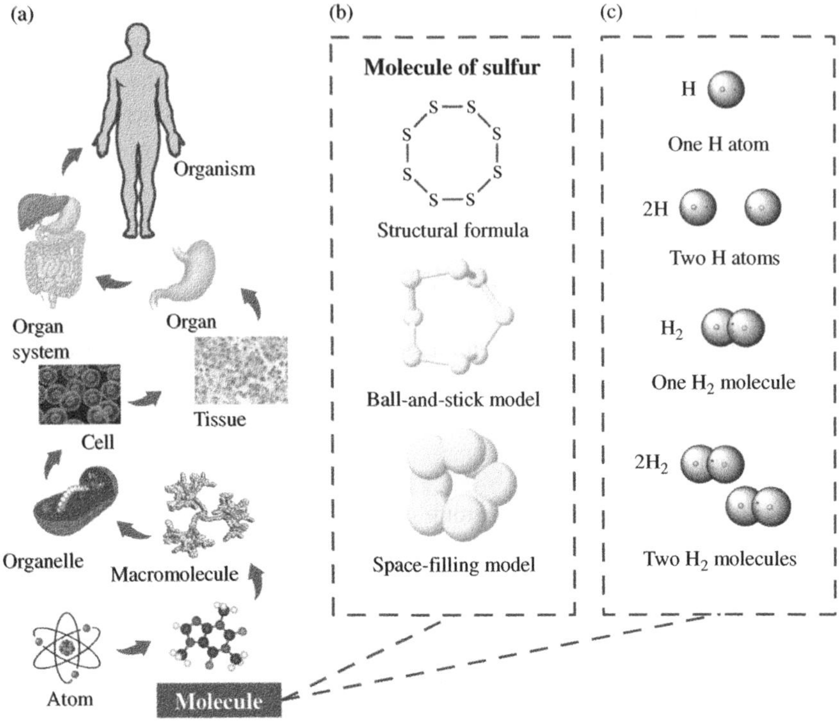


**Figure 3.1** (a) Levels of structural organization in the human body, (b) a molecule of sulfur is composed of eight sulfur atoms and is therefore written as $S_8$, (c) the symbols H, 2H, $H_2$, and $2H_2$ represent different entities. *Source:* SciePro/Adobe Stock.

A compound is formed when two or more elements combine chemically, leading to the formation of bonds. For instance, hydrogen and oxygen can react to form water, and sodium and chlorine can react to form table salt. In some cases, the composition of these compounds is described with empirical formulas indicating the types of atoms present and the simplest whole-number ratio of the number of atoms (or ions) in the compound. For example, titanium dioxide has the empirical formula $TiO_2$, which identifies the elements titanium (Ti) and oxygen (O) as the constituents, and indicates the presence of twice as many atoms of the element oxygen as atoms of the element titanium [7].

### 3.2.2 Molecular Structure and Polarity

This section deals with the structures of small molecules based on valence shell electron pair repulsion (VSEPR) theory and focuses on the term molecular polarity. The investigation of the molecular 3D structure is essential for the thorough

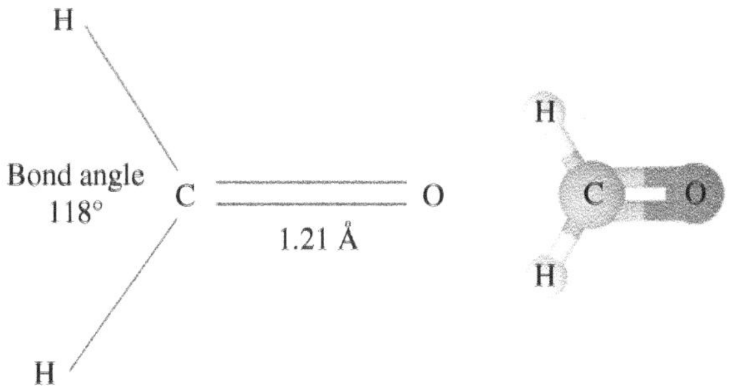


**Figure 3.2** The bond distance and angle for the formaldehyde molecule ($H_2CO$).

comprehension of the details of its function and can give valuable insights for the establishment of effective strategies for experiments [8].

Considering the fact that the molecular structure is 3D, molecular bonds can be analyzed in terms of their distances, relative position in space and angles (Figure 3.2). More specifically, a bond angle is defined as the angle between any two bonds that include a common atom. Bond angles are usually measured in degrees. A bond distance, also known as bond length, is the distance between the nuclei of two bonded atoms along the straight line joining the nuclei. The unit of measurement bond distances is Ångstrom ($1\,Å = 10^{-10}\,m$) or picometer ($1\,pm = 10^{-12}\,m$, $100\,pm = 1\,Å$) [7].

Lewis structures, which are also called Lewis dot diagrams, electron dot structures, or Lewis electron dot structures, represent atom and their positions in the molecule structure with its chemical symbol. Atoms which are bonded to one another are connected with lines, or alternatively, pairs of dots can be used instead of lines. VSEPR is a theory which considers electron-pair repulsions and is applied to predict the arrangement of electron pairs around each central atom and the arrangement of atoms in a molecule.

Polarity establishment, the spontaneous generation of asymmetric molecular distributions, is a crucial component of many cellular functions [8]. Molecular polarity characterizes the separation of electric charge leading to a molecule or its chemical groups having an electric dipole moment, with a negatively charged end and a positively charged end. In polar molecules, polar bonds are contained. The polarity of a bond is determined by the relative electronegativities of the bonded atoms. Electronegativity is the ability of an atom in a molecule or an ion to attract electrons. Thus, there is a direct correlation between electronegativity and bond polarity. A bond is nonpolar if the bonded atoms have equal electronegativities. If the electronegativities of the bonded atoms are not equal, however, the bond is polarized toward the most electronegative atom. Thus, a polar molecule with two or more polar bonds must have a geometry, which is asymmetric in at least one direction so that the bond dipoles do not cancel each other. Polar molecules interact through hydrogen bonds and dipole–dipole intermolecular forces.

Molecular polarity determinants localize to specific domains of the plasma membrane and then act to polarize the action of other cellular systems. These polarity determinants can orient a whole host of cellular functions, such as cell shape, cell adhesion, cell migration, cell division, cell fate determination, and the uptake and release of molecules [9]. Molecular polarity has an impact on a number of physical properties such as surface tension, solubility, as well as melting and boiling points.

#### 3.2.2.1 Mathematical Modeling of Polarizing Biochemical Systems

The use of mathematical models to investigate the spontaneous pattern formation by biochemical systems started in 1950s [8, 10, 11]. In these models, diffusion-driven instabilities are used to generate symmetry breaking without relying on mechanisms such as diffusional barriers, directed transport, and molecular cross-linking. More specifically, these mathematical models require: (i) positive feedback to amplify local fluctuations, (ii) chemical species that diffuse at different rates, and (iii) a mechanism to limit the growth of the polarity site [8]. Models in which patterning can be induced by an arbitrarily weak perturbation (e.g. molecular-level fluctuations) are called turing-type [8, 11]. Turing-type models are attractive theoretically as they can account for spontaneous polarization, achieve a high degree of amplification, and maintain the polar pattern after the signal is removed [12]. Other, non-turing-type models of polarity require perturbations of finite strength to induce pattern formation [13].

According to the literature, the spatiotemporal dynamics of polarizing biochemical systems can be modeled using reaction-diffusion equations (RDEs) in the form of nonlinear partial differential equations (PDEs) [8]. RDEs are deterministic and ignore stochastic effects intrinsic to chemical reactions and thermal diffusion. In some systems, stochastic effects have been shown to expand the parameter space that leads to patterning and accelerate pattern formation [11, 14]. Many modeling approaches have been used to study stochastic effects in biological signaling networks, including stochastic differential equations, such as the chemical Langevin equations [8, 15].

In a recent study [16], numerical methods were employed to investigate planar polarity referring to any polarized protein or structure that breaks cellular symmetry in the plane of the tissue, occurring via multiple independent pathways. A key consideration in these computational studies is the timescale over which a tissue can establish or remodel the asymmetric distribution of planar polarity components within a cell, versus the timescale over which mechanical changes occur. Models of polarity establishment typically assume that the dynamics of protein localization occur on a much faster timescale than cell shape changes, and thus consider a static cell packing geometry. Studying dynamic cell shape changes requires the coupling of models of planar polarity with tissue mechanics. To this end, several "cell-based" models have been developed, which allow for the incorporation of cell signalling and feedback [17]. These include vertex [18] and cellular Potts [19] models, which approximate each cell's apical surface by a polygon whose vertices move according to a force balance equation, or a set of pixels that change stochastically to minimize an energy function, respectively.

In addition, the rate of planar polarization is likely to be dependent on mechanisms such as directed vesicular transport and recycling of planar polarity components. Nevertheless, these parameters have not been considered so far in the literature. Furthermore, the significance of stochasticity and variability in polarity protein interactions and signal interpretation remain to be addressed, even though *in vivo* these are likely to contribute a significant degree of noise [16]. Moreover, while two-dimensional computational models of patterned epithelial have established themselves as important tools, 3D models remain limited and are typically restricted to imposed, static anisotropies in mechanical properties [16, 20]. The extension of such models to allow for the dynamic simulation of planar polarity signalling remains to be tackled. For example, an intriguing link between core pathway planar polarity and 3D tissue deformations was shown in [21], which indicated that planar polarity-dependent polarization of the recycling endosome marker Rab11 is required for apical constriction and subsequent epithelial folding in the Xenopus (a genus of highly aquatic frogs native to sub-Saharan Africa) neural plate.

## 3.3 Molecular Bioengineering in Areas Critical to Human Health

Cellular and molecular bioengineering (CMBE) is considered a scientific domain, which plays a key role in the analysis of human diseases. Human diseases usually occur as a consequence of abnormal molecular and cellular processes, such as abnormalities in migration, adhesion, cell division, or transport, which could potentially be detectable with advanced bioengineering tools. Therefore, the analysis of cell and molecular effects based on bioengineering principles and techniques is a powerful tool in order to provide insights into the mechanisms that molecular-level interactions give rise to cell, tissue, and organ behaviors. Figure 3.3 shows that CMBE studies drive technological advances that are pushing the boundaries of our knowledge of both basic and translational sciences, including biology, preventive medicine, therapeutic medicine, and diagnostic medicine [22].

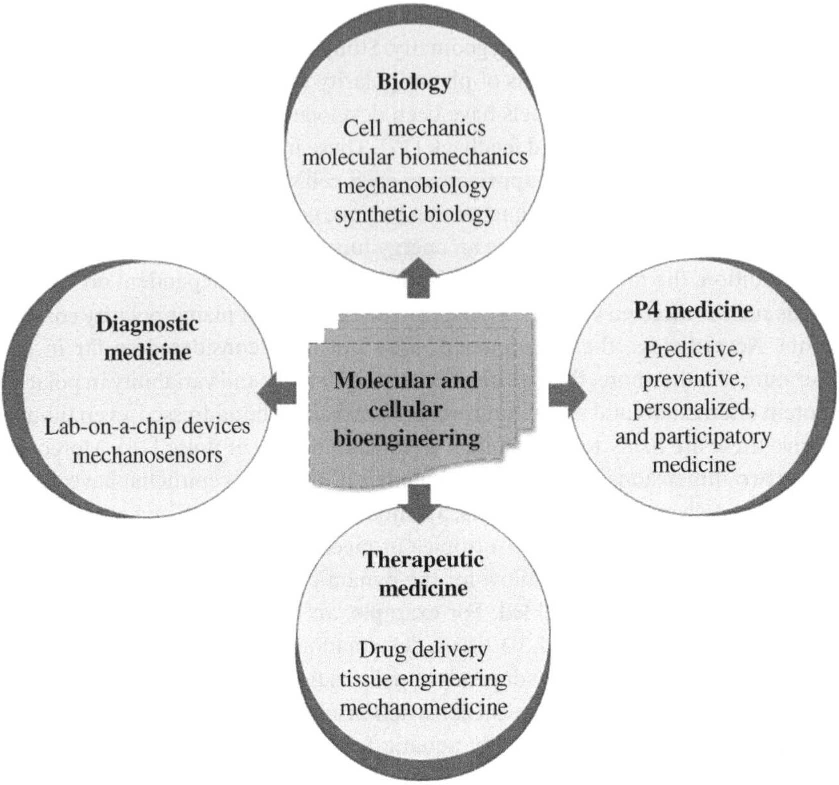


**Figure 3.3** Cellular and molecular bioengineering role in areas critical to human health.

### 3.3.1 Cell Biology

The domain of **biomechanics** has made significant progress in the areas of organ and tissue mechanics allowing a deeper understanding of tissue behavior under physiologic and pathologic conditions and leading to the development of successful medical devices such as artificial joints, heart valves, and stents. On the other hand, the role of mechanics in biology has been studied to a lesser extent despite the fact that mechanical signals are fundamental regulators of cell behavior. Therefore, the manner in which mechanical cues are sensed and transduced at the molecular level to regulate gene expression has long remained enigmatic [23, 24]. It should be emphasized that the knowledge of how cells sense and respond to mechanical signals has a crucial impact on the evolution of various diseases such as atherosclerosis, cancer, osteoporosis, and osteoarthritis.

**Mechanobiology** is an emerging field of science which combines biological and engineering principles focusing on the way that physical forces and changes in cells or tissue mechanics contribute to physiologic and pathologic conditions. The renaissance of modern mechanobiology started with seminal discoveries on the mechanisms of mechanoperception at the level of cell–cell and cell–extracellular matrix adhesion sites, on the role of cytoskeletal proteins in these events, and by key demonstrations that cell mechanics and cell shape control cell proliferation, death, and stem cell differentiation [24, 25]. Nowadays, mechanobiology can be related to biological functions and consequences of mechanical loading, with detailed, real-time, and 3D characterization and quantification of how mechanical signals are transduced into quantitative biochemical responses in a cell. A major challenge in the field is to understand **mechanotransduction**, the process of translating mechanical signals into a cellular response [26]. For example in [27], mechanotransduction in bone was examined to follow mechanical loading from the macro scale down to the molecular level, to outline potential candidates for molecular transduction of mechanical to biochemical signaling, and then trace the biological response upward as it is integrated into the organ level.

Cell biology aims to address fundamental queries such as how the expression of specific genes is regulated inside the nucleus of a living cell covering physiological processes from embryogenesis, development, and pattern formation to differentiation and several other biological processes [28]. A most recent finding on the sensitive responses from a subnuclear organelle protein interactome to a local surface force of physiologic magnitudes highlights the vital roles of force and mechanics in subnuclear functions [29]. In this direction, **genomics biomechanics** or **mechanoepigenomics** are relatively new research domains which could play a significant role in the investigation of the mechanisms which control the "command center" of the cell [22].

In addition, as smaller hierarchical scales are considered in areas critical to human health, biomechanical observations can also be made in the regions surrounding individual cells [22]. In these pericellular regions, physical signals evolve which stimulate cells' sense and response. Thus, **pericellular mechanics** could provide further information for the comprehension of what physical signals cells experience and how they are transduced [27]. Nevertheless, our understanding of pericellular mechanics remains limited [27]. This lack of understanding of pericellular and cellular mechanics is considered a roadblock to further insights into mechanotransduction mechanisms in nonexcitable, nonspecialized cells [27].

#### 3.3.1.1 Biology of Growth Factor Systems

Growth factors are secreted proteins which enable cellular growth, proliferation, healing, and cellular differentiation by activating chemical signal transduction pathways through binding of specific cell-surface receptors [30]. Several physiological processes depend on molecular regulation by growth factors. According to [31] genetic, epigenetic, and somatic changes deregulate the expression of growth factor receptors (GFRs), leading to cancer initiation and progression. Computational and numerical techniques could provide insights into molecular complexity and multiscale nature involved in modeling growth factor system interactions and effects. For example, in [32], epidermal GFR mutations were found to predict differential drug sensitivity profiles toward kinase inhibitors based on computational analysis.

One particular growth factor system involved in the *in vivo* regulation of blood vessel growth is called the vascular endothelial growth factor (VEGF) system. Angiogenesis, the growth of new blood vessels from the existing vasculature, is critical for the maintenance of health and response to injury. The vascular endothelial growth factor and its receptor (VEGFR) have been shown to play an essential role not only in physiological but also in most pathological cases of angiogenesis, such as cancer. In the ischemic disease, the process is impaired, but therapies targeting VEGF, have failed to provide an effect in clinical practice. This indicates the need for deeper understanding of the complex regulation underlying angiogenic signaling [30]. In recent years, there has been tremendous progress in elucidating different VEGF/VEGFR signaling functions in both the blood and lymphatic vascular systems [33].

More specifically, several numerical studies have been presented in recent years to simulate VEGF functions [30, 34–36]. In [30], a computational model was presented integrating detailed regulation of VEGF and the related placental growth factor distribution and binding to VEGF receptor 1 and VEGF receptor 2 by sR1 (secreted by endothelial cells into the tissue interstitial space), the extracellular matrix, and coreceptor NRP1 into a multiscale pharmacokinetic/pharmacodynamic (PK/PD) framework. The resulting model predicted that all of these features interact and contribute to regulation of tissue-level VEGF family signaling. In [34], a computational bioregulatory model was presented on the effect of ultrasound on bone fracture healing which accounts for the ultrasound effect on angiogenesis and more specifically on the transport of the VEGF. The results made clear that ultrasound enhances angiogenesis mechanisms during bone healing. The proposed model could be regarded as a step toward monitoring the effect of ultrasound on bone regeneration. In [35], outline approaches were presented to multiscale molecular-detailed computational modeling of the VEGF family, with particular emphasis on PK/PD modeling of current or potential therapeutic interventions in cancer. At the molecular and cellular scales, a biophysically accurate kinetic model of ligand–receptor interactions and/or intracellular signaling was presented. Finally, the paper of [36] introduced a variety of *in silico* models that seek to recapitulate different aspects of VEGF system biology at various spatial and temporal scales: molecular-level kinetic models focus on VEGF ligand-receptor interactions at and near the endothelial cell surface; mesoscale single-tissue 3D models can simulate the effects of multicellular tissue architecture on the spatial variation in VEGF ligand production and receptor activation; compartmental modeling allows efficient prediction of average interstitial VEGF concentrations and cell-surface VEGF signaling intensities across multiple large tissue volumes, permitting the investigation of whole-body inter-tissue transport (e.g. vascular permeability and lymphatic drainage). The aforementioned studies are considered indicative examples demonstrating the significance of computational modeling in enhancing both basic science and clinical research on VEGF systems biology, as well as, in general, the domain of growth factor systems [36].

### 3.3.2 Diagnostic Medicine

This section presents the role of computational modeling at the molecular level in diagnostic medicine describing applications on lab-on-a-chip devices and biosensors.

#### 3.3.2.1 Lab-on-a-Chip Devices

The term "microfluidics" refers to the behavior, precise control, and manipulation of fluids that are geometrically constrained to a small, typically submillimeter, scale at which capillary penetration governs mass transport. In the first half of the 21st century, applications in microfluidics may be compared with microelectronics of the second half of the 20th century. Microfluidics systems work by using a pump and a chip which is a device that enables a tiny amount of liquid to be processed or visualized. The development of the integrated circuit led to the reduction of the size of electrical devices and, at the same time, to the increase in their speed. Similarly, it is expected that many of the large, expensive chemical and biological analyses that are currently being performed can be replaced by integrated microfluidic devices [37, 38], often called labs-on-chip, resulting in a similar revolution [39]. Microfluidic systems have several advantages such as increased automation, faster analysis, easy-to-process, less power consumption, compatibility with mass manufacturing, and integrating multiplexing analysis, which provide the necessary conditions to simulate *in vivo* micro-environments in an *in vitro* system for both fundamental research and point-of-care (POC) applications [40].

A lab-on-a-chip (LOC) is a device of several square millimeters to centimeters which integrates into a single chip one or several analyses that are usually done in a laboratory environment (e.g. DNA sequencing and biochemical detection). Microfluidics principles are applied in LOCs and analytical processes are miniaturized to enhance mobility, efficiency, and fulfill the requirements for a POC diagnostic device (e.g. fast turnaround time for analysis and low consumption of sample and reagents) [41].

Nowadays, the evolution of LOCs is based on several aspects such as the miniaturization and integration of diagnostic tests, the improvement of the robustness and reliability without compromising sensitivity, the measurement of multiple parameters in one test and the use of nanotechnology which could make devices smaller and more sensitive when measuring analytes. In addition, the development of LOCs could potentially leverage next-generation diagnostics for personalized medicine. Recently, the unique capability of 3D printing strategy for patterning micro-sized features with complex 3D architectures in a single process has caught increasing interest in the development of LOC devices [40].

Therefore, as microfluidic devices become increasingly complex, optimal fluidic and transport designs become more and more difficult to be performed experimentally, there is an increasing demand in the field for the integration of simulation tools which allow users to "numerical prototype" highly integrated devices. An overview of indicative techniques, applications, and tools of computational simulation for integrated microfluidics devices was presented in [39]. Numerical studies aim to address not only fundamental microfluidics problems such as species mixing and sample dispersion but also unique approaches which integrate a more global overview of the entire system. In addition, recent numerical studies have focused on the research domain of microscale transport such as thermal analysis and chemical reactivity and specificity. In general, the most comprehensive and globally applicable models adapt to the following steps: (i) calculate the flow field using a finite element or similar technique, (ii) compute the various hydrodynamic and other important forces (e.g. buoyancy, electrostatic, and magnetic) acting on the particle, (iii) update its position using Newton's law, (iv) remesh and project the previous timesteps velocity onto the new mesh [39].

Finally, it should be mentioned that one of the great strengths of microfluidic systems is that they have provided researchers with an unprecedented ability to sort, manipulate, analyze, and store individual microscale particles and macromolecules enabling single-cell analysis via such techniques as capillary electrophoresis [42], sorting, and cytometry and dielectrophoresis [43], for such applications as drug development, tissue engineering [44], and biosensors [45].

#### 3.3.2.2 Biosensors

Nanotechnology can be defined as the branch of the science of engineering involved in the design, synthesis, characterization, and application of materials and devices whose smallest functional organization in at least one dimension is on the nanometer size. In applications to medicine and physiology, these materials and devices can interact with cells and tissues at the molecular (i.e. subcellular) level with high functional specificity, allowing a degree of integration between technology and biological systems which was not attained previously [46].

A major breakthrough in nanotechnology applications includes the industry of nanosensors. The term nanosensor refers to a sensor developed on the atomic scale which conducts measurements of nanometers with several applications in the domains of medicine, integrated circuits, aerospace, national security, etc. Nanosensors are used in order to obtain data on the atomic scale and transfer this type of information into data that can be easily analyzed. These sensors are extremely sensitive and are able to detect single virus particles or even very low concentrations of a substance that could be potentially harmful [47].

One of the most well-known areas of research in nanosensors with significant financial profit is biosensors, which can be used to detect specific types of DNA. The significant development of this domain can be explained by the wide range of possibilities that this technology could lead to the early detection of cancer or other diseases. The sensors are consisted of synthetic polymers called dendrimers and are created layer-by-layer into spheres with a diameter of less than 5 nm. Their small size allows their positioning transdermally [46]. Figure 3.4 presents the configuration of a biosensor showing biorecognition, interface, and transduction elements [48].

Several research groups have used computational modeling to simulate different biosensor functions and processes. More specifically, the area of mathematical modeling of biosensors investigates processes that occur in the sensors' layers and at their interface and provides numerical and analytical methods to solve diffusion and enzymatic kinetic equations [49]. The action of single enzyme and poly-enzyme biosensors has been examined using numerical means, as well as the modeling of biosensors which contain perforated membranes and multipart mass transport profiles. Also, several studies have investigated how signals can be biochemically amplified, how signals are processed via a chemometric approach and artificial neuronal networks, and how cascades of enzymatic substrate conversion are triggered. In [50], single-walled carbon nanotubes were used as biomedical sensors where a numerical approach to the influence of biomolecules on

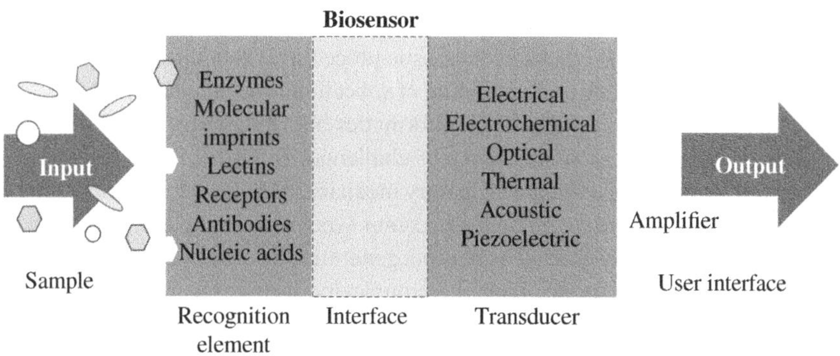


**Figure 3.4** Configuration of a biosensor showing biorecognition, interface, and transduction elements.

nanotube electronic properties through nanotube-biomolecule interactions was established, furnishing the ability to test nanotube electronic properties in realistic biological environments. Molecular modeling was applied to evaluate scaffold proteins, redesign apolipoproteins found in mammalian plasma that hold the discoidal membranes in the proper shape, and predict the assembly and final structure of the nanodiscs.

### 3.3.3 Preventive Medicine

In the 21st century, due to the significant advancement of high-throughput technologies and diagnostic techniques, molecular underpinnings of many disorders have been identified. Nevertheless, the incorporation of patient-specific molecular mechanisms into tailored clinical applications is considered a challenging task requiring the integration of multidimensional molecular and clinical data into patient-centric models. To this end, the terms preventive medicine, personalized medicine, precise medicine, and genomic medicine have been used frequently in recent years [51]. Personalized medicine aims to: (i) ensure that patients receive the right treatment at the right dose, at the right time, with minimum adverse consequences and maximum efficacy, and (ii) enhance clinical decisions according to a patient's molecular profile [52].

Personalized medicine advances depend significantly on the evolution of molecular imaging [52]. For instance, molecular imaging with a tissue biomarker could assist in the discovery of novel drugs and also in predictive biopsies [53, 54]. Numerous crucial markers of diseases have already been accepted into clinical practice [54]. In addition, the response to therapy has a multifaceted link to genotype, dysregulation of signalling conduits using gene expression, protein activity, protein–protein interactions, and disease phenotypic traits. These data can be collected using molecular imaging as well as other established techniques, but imaging analysis is in several cases problematic considering the thousands of low-specific markers. Progress in diagnostic procedures, genomics, and proteomics could provide a better understanding of subcellular mechanisms [52].

In this direction, translational bioinformatics and integrative modeling have opened new avenues to address current challenges in P4 medicine (predictive, preventive, personalized and participatory medicine) [55]. The application of bioinformatics and integrative modeling becomes even more challenging when dealing with complex diseases such as neurodegenerative disorders. The main goal of systems approaches is to decipher the complexity of diseases by integrating all types of heterogeneous biological information into predictive and actionable models (Figure 3.5). Following a more holistic approach to integrating information of the interacting parts and subentities into a single mathematical representation or model, could potentially offer better clues as to the causal chain of events

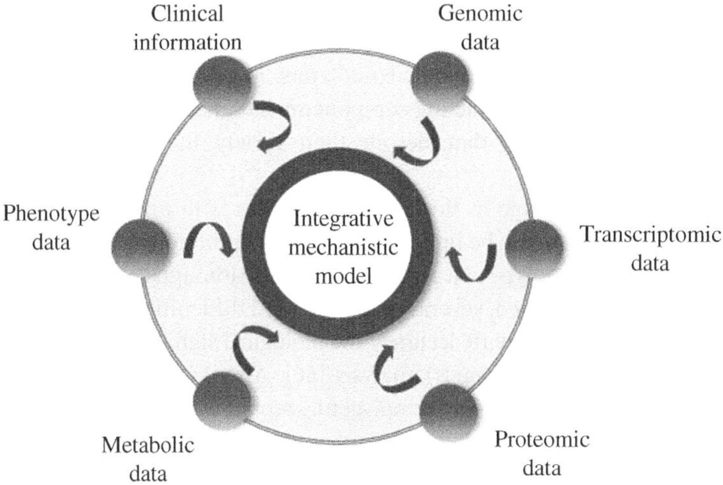


**Figure 3.5** Disease-specific models at the center of data integration. Different data types across multiple biological scales can be aggregated and integrated into an integrative mechanistic model.

that leads to the apparent phenotypic manifestations and how to remedy the situation [56]. Such models are considered crucial to provide new knowledge to the premises underlying P4 medicine as they could assist researchers to bridge the "translational gap" between basic and clinical research more efficiently.

Integrative, also known as holistic, disease modeling is at the vanguard of translational research which supports the shift from descriptive to causal mechanistic diagnosis and treatment of complex diseases by exploiting the power of computational methods to collect, integrate, model, store, and interpret accumulated disease information across different biological scales (from molecules to phenotypes) [55]. The contribution of P4 medicine depends on the ability to manage and integrate heterogeneous data types across multiple scales ranging from molecular datasets to medical history and clinical information of patients. The main challenge to be addressed for the proper exploitation of this complex and multidimensional information by translational bioinformatics is the problem of linking omics molecular data and affected biological pathways to clinical readouts and epidemiological backgrounds of patients with complex diseases. Despite the fact that one set of methods in this sense-making process comes from bioinformatics, the key objective is to use mathematical modeling to integrate the data in what is called "holistic multiscale models" [57]. These models decipher biological complexity by showing how all elements in biological systems interact with each other to produce health and disease states. In [58], the term "network theory" is used to describe the set of mathematical and computer simulation models and tools that have been developed to study network architectures and dynamics. Network theory ranges from dynamical systems theory to network topology, from random boolean network models to coupled oscillators. The evolution of networks with strongly and recurrently interacting components has enabled researchers to study holistic systems, showing that, despite their variety, they share specific generic properties [57, 58].

Finally, the greatest challenge in the post-genomic era is to understand how heterogeneous "-omics" data can be integrated into a coherent holistic model that can explain the disease phenotype in a personalized way. Although computational tools are being constantly refined, several challenges and difficulties remain such as the major difficulty separate with accuracy the biological signal from technical noise in differential expression (omics) analysis [56]. Also, considering the fact that risk factors of complex diseases are polygenic, technical heterogeneity and batch effects are difficult to be distinguished from gene, protein, and epigenetic perturbations causing a specific disease. To ameliorate this problem, computational methods integrate enormous amounts of data into models or network representations. Nevertheless, each bioinformatic approach has different advantages and drawbacks implying that the lack of standard biostatistical and algorithmic procedures may lead to uncertainty, as far as the validity of the results is concerned, and, in some cases, cannot fully guarantee reproducibility. This indicates the significance of experimental validation and replication of the results.

### 3.3.4 Therapeutic Medicine

This section deals with the role of computational modeling in therapeutic medicine focusing on applications of drug delivery and tissue engineering.

#### 3.3.4.1 Drug Delivery

Mathematical modeling of drug delivery and predictability of drug release is a research domain with particular scientific and industrial interest as the rapid growth of information technology enhances the *in silico* optimization of novel drug delivery systems and the further development of pharmaceutical technology. It is only a matter of time for computational tools to be routinely used in daily clinical practice to optimize the design of novel dosage forms. Considering the desired type of administration, drug dose to be incorporated and targeted drug release profile, mathematical predictions could improve the required composition, dimensions, preparation procedure, and geometry of the respective dosage forms [59].

In the literature, several mathematical theories have been presented, but most of them still have limited accuracy and/or easiness of application [59–61]. The beginning of the quantitative treatment of drug release from pharmaceutical dosage forms was made in 1961 with the first mathematical model of drug delivery presented in [62, 63]. More specifically, these first studies presented an equation allowing for a surprisingly simple description of drug release from an ointment base exhibiting a considerable initial excess of non-dissolved drug within an inert matrix with film geometry [62, 63]. Since then, various computational models have been proposed, which can be divided into two main categories including empirical/semiempirical, as well as mechanistic mathematical theories [59].

Concerning the empirical/semiempirical approaches, the mathematical approximation is mainly descriptive and not based on real physical, chemical, and/or biological phenomena implying that limited insight into the underlying drug release mechanisms can be provided. Another main drawback of these kinds of computational models is that their predictive power is, in some cases, limited. It should be noted that approaches based on artificial neural networks are considered an exception, as they have shown good predictive power.

On the other hand, mechanistic mathematical theories are based on real phenomena, such as diffusion, swelling, dissolution, erosion, precipitation, and degradation [64, 65]. This kind of model enables the determination of system-specific parameters that can offer deeper insight into the underlying drug release mechanisms. For example, the relative importance of several processes involved can be evaluated such as drug diffusion and polymer swelling. The dosage form is not treated as a "black box," but as a real drug delivery system, the mechanisms of which can be understood [59]. Therefore, the use of these models has several advantages especially during product development as they take into account the quantitative prediction of the effects of formulation and processing parameters (e.g. the initial tablet height and radius) on the resulting drug release kinetics. This indicates that the required shape, size, composition, and preparation procedure of a novel dosage form with desired properties could become theoretically predictable. Also, potential challenges during production are much easier to be regulated when having a thorough view of how the system works.

In conclusion, the proper application of mathematical theories to model drug delivery depends on the following factors: (i) the accuracy of a mathematical theory generally increases with increasing model complexity, while caution must be paid as too complex models are cumbersome to be used, (ii) computational results should always be compared with experimental findings. A model describing drug delivery is a simplification of the real system based on specific assumptions and its suitability may be restricted to certain cases, (iii) there is no general mathematical theory which can be used to simulate all the different kinds of drug delivery systems. Some theories may be appropriate for a limited number of drug delivery systems, while others may have a much wider application spectrum.

#### 3.3.4.2 Tissue Engineering

Tissue engineering and regenerative medicine are domains of life science which employ both biological and engineering principles to create new tissues and organs and promote the regeneration of damaged or diseased tissues and organs [66]. More specifically, it refers to the *in vitro* regeneration or growth of a tissue or an organ for the purpose of biomedical studies in areas such as toxicology, tissue/organ repair, and medical device development [67]. The different tissue engineering applications may vary from blood vessels over trachea to cartilage and bone.

The main challenge in tissue engineering is the translation of biological knowledge on complex cell and tissue behavior into a predictive and robust engineering process. In this direction, computational modeling could contribute significantly in: (i) the optimization and quantification of the tissue engineering process (e.g. by adapting bioreactor design to improve quality and quantity of the final product) and (ii) the optimization and quantification of the tissue engineering product (e.g. by adapting scaffold design to optimize microenvironmental signals), (iii) the evaluation of the influence of the *in vivo* environment on the behavior of the tissue engineering product (e.g. by investigating vascular ingrowth) [68].

Computational tools have been traditionally used to assist in engineering design and modeling from the aspect of representation, analysis, and manufacturing. Advances in computer-aided design (CAD) software and biomedicine have been a major driving force for the evolution of research in tissue engineering. Figure 3.6 shows that CAD tissue engineering provides critical information on geometrical, morphological, biological, and biophysical properties for the modeling, design, and fabrication of complex tissue substitutes [69]. The common name used for novel techniques which allow an initial computational control of fabrication using CAD is rapid prototyping (RP). RP leads to better pore reproducibility of regular shape as well as to better interconnection than the conventional methods.

The exploitation of computational modeling and simulation in the study of optimization of *in vitro* tissue growth environments could lead to the decrease of experimental costs and allow easy variable manipulation to determine effect [67]. Numerical modeling is currently limited to activities which are capable of predicting or characterizing the processes which occur within a given cell culture system, while the combination of well-validated models describing the molecular metabolism, mass transport, scaffold decomposition, and many other phenomena on multiple levels requires further investigation.

In particular, in [67], modeling approaches were presented focusing on biology, the complex interactions between biology and the microenvironmental signals and the ways in which modeling might be able to assist in investigating and mastering this complexity. The main focus was on the optimization of mechanical signals, mass transport, and fluid flow encountered by the cells in scaffolds and

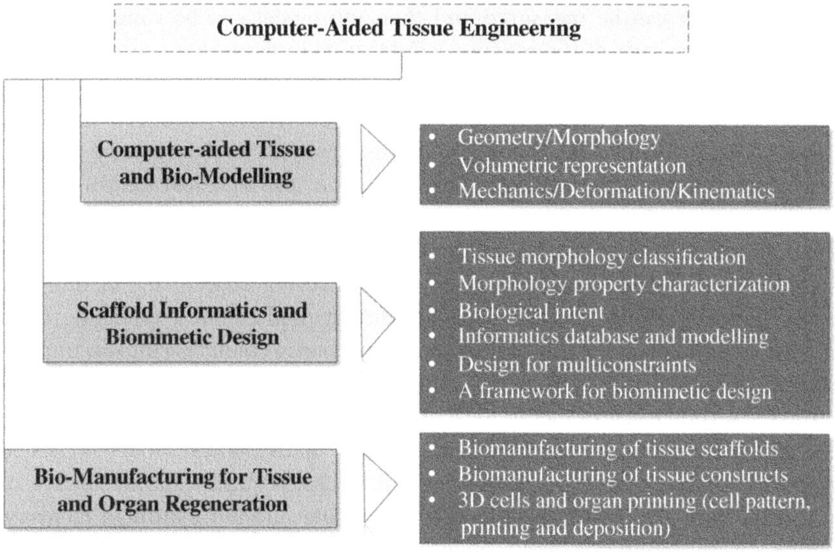


**Figure 3.6** Computer-aided tissue engineering.

bioreactors as well as on the optimization of the cell population itself. In another study [70], computational modeling and novel experimental methods were presented to develop tissue engineering scaffolds as delivery devices for exogenous and endogenous cues, including biochemical and mechanical signals, to drive the fate of mesenchymal stem cells seeded within. Computational fluid dynamics (CFD) was applied coupled with finite element analysis (FEA) modeling to predict flow regimes within the scaffolds and optimize flow rates to simulate seeded cells. The proper application of CFD and FEA aims to provide a more powerful tool to predict cell-organ scale system behavior, seamlessly, over multiple time scales that span the life of the patient or test subject.

Despite the fact that most of the studies in the literature on computational modeling in tissue engineering applications focus on the cell level and will be discussed in further detail in Chapter 4, many studies have investigated the effects of environment (i.e. fluid velocity), porosity, and scaffold material on degradation rates by surface and bulk erosion, as well as by chemical reactions [67, 71, 72]. For instance, in [73], polymer degradation by hydrolysis was modeled based on equations depending on the mole concentration of ester bonds of polymer chains, the mole concentration of monomers, the rate constant of the reaction which was independent of any catalyst, the rate constant of a reaction catalyzed by the previously formed products, and the disassociation power of the acid end group. These reactions are then related to the molecular weight of the polymer present in the scaffold, which can be measured and then the model can be validated. Also, in [67], the properties of the porous scaffolds used in tissue engineering, such as biocompatible material, porosity, and tensile strength were examined using numerical means. In another recent study on multiscale mathematical modeling in dental tissue engineering, a new model was presented for determining the stiffness of a hydroxyapatite-based granular biomaterial, through upscaling of compositional and morphological information [74]. More specifically, it was shown for the first time how an advanced multiscale mathematical modeling approach may support the design of a successful tissue engineering concept for mandibular bone. For this reason, the latest state-of-the-art mathematical modeling techniques in the field of multiscale micromechanics were combined into an unprecedented suite of highly efficient, semianalytically defined computation steps resolving several levels of hierarchical organization, from the millimeter- down to the nanometer-scale. It was found that the new mathematical model suggested that early stiffness recovery, within several weeks, requires total avoidance of microcracks in the hydroxyapatite scaffolds, while mid-term stiffness recovery, within several months, is additionally promoted by provision of small granule sizes, in combination with high bone formation and low scaffold resorption rates. Finally, in [75], a molecular modeling approach was presented for bone morphogenetic protein (BMP) for tissue engineering applications. More specifically, molecular dynamics simulations were used to study the dynamic behavior of BMP-2 on a flat gold substrate which is typically used in medical devices. Protein adsorption was analyzed for three different temperatures. The results indicated that BMP-2 was resistant to high temperatures at the nanoscale level since no denaturation occurred.

In conclusion, future research on tissue engineering and biomaterials should focus on a system's multilevel perspective and computational modeling could unravel new knowledge by addressing several questions which remain unanswered. Biochemical responses need to be evaluated, possibly including functional genomics, proteomics, metabolomics, or other system-wide "-omic" responses [76]. The chemical and physical properties of biomaterials would strongly influence cellular response and these parameters depend on both the underlying biomaterial as well as adsorbed proteins. On top of this are the biochemical and biophysical cues from the culture environment and the host microenvironment [76]. Therefore, the use of multiscale modeling approaches, which expand from (i) the molecular interaction of the scaffold coating with the molecules present in the surrounding medium, to (ii) the cellular interaction with the scaffold surface, and to (iii) the formation of extracellular matrix tissue, will provide insights into the interactions between material scaffolds and biological entities within the next decades [77].

## References


1 Schlick, T. (2010). *Molecular Modeling and Simulation: An Interdisciplinary Guide*, 2e. Springer Science+Business Media, LLC.

2 Chen, K. and Chen, X. (2010). Design and development of molecular imaging probes. *Curr. Top. Med. Chem.* 10 (12): 1227–1236.

3 Hartwell, L., Mankoff, D., Paulovich, A. et al. (2006). Cancer biomarkers: a systems approach. *Nat. Biotechnol.* 24 (8): 905–908.

4 Saxena, A., Wong, D., Diraviyam, K., and Sept, D. (2009). The basic concepts of molecular modeling. *Methods Enzymol.* 467: 307–334.

5 Brenner, S. and Miller, J. (2013). *Brenner's Encyclopedia of Genetics*, 2e. Academic Press.

6 Barber, N. and Stark, L. (2014). Engaging with molecular form to understand function. *CBE Life Sci. Educ.* 13 (1): 21–24.

7 Flowers, P., Theopold, K., Langley, R., and Robinson, W. (2015). *Chemistry*, 1e. OpenStax College, Rice University.

8 Pablo, M. and Ramirez, S. (2018). Particle-based simulations of polarity establishment reveal stochastic promotion of Turing pattern formation. *PLoS Comput. Biol.* 14 (3): 1–25.

9 Thompson, B. (2013). Cell polarity: models and mechanisms from yeast, worms and flies. *Development* 140: 13–21.

10 Turing, A. (1952). The chemical basis of morphogenesis. *Philos. Trans. R. Soc. London Ser. B* 237 (641): 37–72.

11 Dziekan, P., Lemarchand, A., and Nowakowski, B. (2012). Particle dynamics simulations of Turing patterns. *J. Chem. Phys.* 137 (7): 074107.

12 Jilkine, A. and Edelstein-Keshet, L. (2011). A comparison of mathematical models for polarization of single eukaryotic cells in response to guided cues. *PLoS Comput. Biol.* 7 (4). e1001121. https://doi.org/10.1371/journal.pcbi.1001121.

13 Mori, Y., Jilkine, A., and Edelstein-Keshet, L. (2008). Wave-pinning and cell polarity from a bistable reaction-diffusion system. *Biophys. J.* 94 (9): 3684–3697.

14 Lemarchand, A. and Nowakowski, B. (2011). Do the internal fluctuations blur or enhance axial segmentation? *Europhys. Lett.* 94 (4): 48004.

15 McKane, A., Biancalani, T., and Rogers, T. (2014). Stochastic pattern formation and spontaneous polarisation: the linear noise approximation and beyond. *Bull. Math. Biol.* 76 (4): 895–921.

16 Fisher, K.H., Strutt, D., and Fletcher, A.G. (2017). Integrating planar polarity and tissue mechanics in computational models of epithelial morphogenesis. *Curr. Opin. Syst. Biol.* 5: 41–49.

**17** Fletcher, A., Cooper, F., and Baker, R. (2017). Mechanocellular models of epithelial morphogenesis. *Philos. Trans. R. Soc. London Ser. B* 372 (1720), pii: 20150519.

**18** Fletcher, A., Osterfield, M., Baker, R., and Shvartsman, S. (2014). Vertex models of epithelial morphogenesis. *Biophys. J.* 106: 2291–2304.

**19** Marée, A., Grieneisen, V., and Hogeweg, P. (2007). The Cellular Potts Model and biophysical properties of cells, tissues and morphogenesis. In: *Single-Cell-Based Models in Biology and Medicine. Mathematics and Biosciences in Interaction* (ed. A.R.A. Anderson, M.A.J. Chaplain and K.A. Rejniak), 107–136. Basel: Birkhäuser. https://doi.org/10.1007/978-3-7643-8123-3_5.

**20** Bielmeier, C., Alt, S., Weichselberger, V. et al. (2016). Interface contractility between differently fated cells drives cell elimination and cyst formation. *Curr. Biol.* 26: 563–574.

**21** Ossipova, O., Kim, K., Lake, B. et al. (2014). Role of Rab11 in planar cell polarity and apical constriction during vertebrate neural tube closure. *Nat Commun.* 5: 3734. https://doi.org/10.1038/ncomms4734.

**22** Brown, G., Butler, P., Chang, D. et al. (2012). Cellular and molecular bioengineering: a tipping point. *Cell. Mol. Bioeng.* 5 (3): 239–253.

**23** Eyckmans, J., Boudou, T., Yu, X., and Chen, C. (2001). A hitchhiker's guide to mechanobiology. *Dev. Cell.* 21 (1): 35–47.

**24** Panciera, T., Azzolin, L., Cordenonsi, M., and Piccolo, S. (2017). Mechanobiology of YAP and TAZ in physiology and disease. *Nat. Rev. Mol. Cell Biol.* 18: 758–770.

**25** Vogel, V. and Sheetz, M. (2006). Local force and geometry sensing regulate cell functions. *Nat. Rev. Mol. Cell Biol.* 7: 265–275.

**26** K. Jansena, D. Donatob, H. Balcioglue, T. Schmidt, E. Danenc, and G. Koenderink, A guide to mechanobiology: where biology and physics meet Biochim. Biophys. Acta, Mol. Cell. Res., vol. 1853, no. 11 Part B, pp. 3043–3052, 2015.1

**27** Jacobs, C., Temiyasathit, S., and Castillo, A. (2010). Osteocyte mechanobiology and pericellular mechanics. *Annu. Rev. Biomed. Eng.* 12: 369–400.

**28** Dahl, K., Kalinowski, A., and Pekkan, K. (2010). Mechanobiology and the microcirculation: cellular, nuclear and fluid mechanics. *Microcirculation* 17 (3): 179–191.

**29** Poh, Y., Shevtsov, S., Chowdhury, F. et al. (2012). Dynamic force-induced direct dissociation of protein complexes in a nuclear body in living cells. *Nat. Commun.* 3 (866) https://doi.org/10.1038/ncomms1873.

**30** Clegg, E. and Gabhann, F. (2017). A computational analysis of in vivo VEGFR activation by multiple co-expressed ligands. *PLoS Comput. Biol.* 13 (3): e1005445.

**31** Tiash, S. and Chowdhury, E. (2015). Growth factor receptors: promising drug targets in cancer. *J. Cancer Metastasis Treat* 1: 190–200.

**32** Akula, S., Kamasani, S., Sivan, S. et al. (2018). Computational analysis of epidermal growth factor receptor mutations predicts differential drug sensitivity profiles toward kinase inhibitors. *J. Thoracic Oncol.* 13 (5): 721–726.

**33** Karaman, S., Leppänen, V.-M., and Alitalo, K. (2018). Vascular endothelial growth factor signaling in development and disease. *Development* 145, https://doi.org/10.1242/dev.151019.

**34** Vavva, M., Grivas, K., Carlier, A. et al. (2018). Effect of ultrasound on bone fracture healing: a computational bioregulatory model. *Comput. Biol. Med.* 1: 74–85.

**35** Gabhann, F.M., Stefanini, M.O., and Popel, A.S. (2012). Simulating therapeutics using multiscale models of the VEGF receptor system in cancer. In: *Modeling Tumor Vasculature* (ed. T. Jackson). New York, NY: Springer. https://doi.org/10.1007/978-1-4614-0052-3_2.

**36** Wu, F., Stefanini, M., Gabhann, F. et al. (2010). A systems biology perspective on sVEGFR1: its biological function, pathogenic role and therapeutic use. *J. Cell. Mol. Med.* 14 (3): 528–552.

**37** Vilkner, T., Janasek, D., and Manz, A. (2004). Micro total analysis systems. Recent developments. *Anal. Chem.* 76: 3373–3385.

**38** Paegel, B., Blazej, R., and Mathies, R. (2003). Microfluidic devices for DNA sequencing: sample preparation and electrophoretic analysis. *Curr. Opin. Biotech.* 14: 42–50.

**39** Erickson, D. (2005). Towards numerical prototyping of labs-on-chip: modeling for integrated microfluidic devices. *Microfluid. Nanofluid.* 1: 301–318.

**40** Zhang, Y., Ge, S., and Yu, J. (2016). Chemical and biochemical analysis on lab-on-a-chip devices fabricated using three-dimensional printing. *TrAC, Trends Anal. Chem.* 85 (part C): 166–180.

**41** Hermsen, S., Roszek, B., van Drongelen, A., and Geertsma, R. Lab-on-a-chip devices for clinical diagnostics Measuring into a new dimension. RIVM report 080116001/2013.

**42** Stuart, J. and Sweedler, J. (2003). Single-cell analysis by capillary electrophoresis. *Anal. Bioanal. Chem.* 375: 28–29.

**43** Cui, L., Holmes, D., and Morgan, H. (2001). The dielectrophoretic levitation and separation of latex beads in microchips. *Electrophoresis* 22: 3893–3901.

**44** Saltzman, W. and Olbricht, W. (2002). Building drug delivery into tissue engineering. *Nat. Rev. Drug Discovery* 1 (3): 177–186.

**45** Pancrazio, J., Whelan, J., Borkholder, D. et al. (1999). Development and application of cell-based biosensors. *Ann. Biomed. Eng.* 27: 697–711.

**46** Agrawal, S. and Prajapati, R. (2012). Nanosensors and their pharmaceutical applications: a review. *Int. J. Pharm. Sci. Nanotechnol.* 4 (4): 1526–1535.

**47** Riu, J., Maroto, A., and Rius, F. (2006). Nanosensors in environmental analysis. *Talanta* 69: 288–301.

**48** Chambers, J., Arulanandam, B., Matta, L. et al. (2008). Biosensor recognition elements. *Curr. Issues Mol. Biol.* 10: 1–12.

**49** Baronas, R., Ivanauskas, F., and Kulys, J. (2010). *Mathematical Modeling of Biosensors. An Introduction for Chemists and Mathematicians*, Springer Series on Chemical Sensors and Biosensors. Springer. https://www.amazon.com/Mathematical-Modeling-Biosensors-Introduction-Mathematicians/dp/9048132428.

**50** Lu, D., Aksimentiev, A., Shih, A. et al. (2006). The role of molecular modeling in bionanotechnology. *Phys. Biol.* 3 (1): S40–S53.

**51** Thrall, J. (2004). Personalized medicine. *Radiology* 231: 613–616.

**52** Weissleder, R., Ross, B., Rehemtulla, A., and Gambhir, S. (2010). *Molecular Imaging: Principles and Practice*. Shelton, CT: People's Medical Publishing House.

**53** Figueiras, R., Padhani, A., Goh, V. et al. (2011). Novel oncologic drugs: what they do and how they affect images. *Radiographics* 11: 2059–2091.

**54** Abi-Jaoudeh, N., Duffy, A., Greten, T. et al. (2013). Personalized oncology in interventional radiology. *J. Vasc. Interv. Radiol.* 24: 1083–1092.

**55** Younesi, E. and Hofmann-Apitius, M. (2013). From integrative disease modeling to predictive, preventive, personalized and participatory (P4) medicine. *EPMA J.* 4 (1).

**56** Noell, G., Faner, R., and Agustí, A. (2018). From systems biology to P4 medicine: applications in respiratory medicine. *Eur. Res. Rev.* 27 (147), pii: 170110. https://doi.org/10.1183/16000617.

**57** Vogt, H., Hofmann, B., and Getz, L. (2016). The new holism: P4 systems medicine and the medicalization of health and life itself. *Med. Health Care Philos.* 19 (2): 307–323.

**58** Moreno, A., Ruiz-Mirazo, K., and Barandiaran, X. (2011). The impact of the paradigm of complexity on the foundational frameworks of biology and cognitive science. In: *Philosophy of Complex Systems* (ed. C. Hooker), 311–333. Amsterdam: Elsevier.

**59** Siepmann, J. and Siepmann, F. (2008). Mathematical modeling of drug delivery. *Int. J. Pharm.* 364: 328–343.

**60** Arifin, D., Lee, L., and Wang, C. (2006). Mathematical modeling and simulation of drug release from microspheres: implications to drug delivery systems. *Adv. Drug Deliv. Rev.* 58: 1274–1325.

**61** Lin, C. and Metters, A. (2006). Hydrogels in controlled release formulations: network design and mathematical modeling. *Adv. Drug Deliv. Rev.* 58: 1379–1408.

**62** Higuchi, T. (1961). Physical chemical analysis of percutaneous absorption process from creams and ointments. *J. Soc. Cosmet. Chem.* 11: 85–97.

**63** Higuchi, T. (1961). Rate of release of medicaments from ointment bases containing drugs in suspensions. *J. Pharm. Sci.* 50: 874–875.

**64** Frenning, G. and Stromme, M. (2003). Drug release modeled by dissolution, diffusion, and immobilization. *Int. J. Pharm.* 250: 137–145.

**65** Raman, C., Berkland, C., Kim, K., and Pack, D. (2005). Modeling small-molecule release from PLG microspheres: effects of polymer degradation and nonuniform drug distribution. *J. Control. Release* 103: 149–158.

**66** Dzobo, K., Thomford, N., Senthebane, D. et al. (2018). Advances in regenerative medicine and tissue engineering: innovation and transformation of medicine. *Stem Cells Int.*, https://doi.org/10.1155/2018/2495848.

**67** German, C. and Madihally, S. (2016). Applications of computational modelling and simulation of porous medium in tissue engineering. *Computation* 4 (1) https://doi.org/10.3390/computation4010007.

**68** Liesbet, G. (2013). Computational modeling in tissue engineering. In: *Studies in Mechanobiology, Tissue Engineering and Biomaterials* (ed. L. Geris). Springer https://link.springer.com/book/10.1007/978-3-642-32563-2.

**69** Sun, W., Starly, B., Nam, J., and Darling, A. (2005). Bio-CAD modeling and its applications in computer-aided tissue engineering. *Comput. Aided Des.* 37: 1097–1114.

**70** Song, M., Dean, D., and Knothe Tate, M. (2013). Computational modeling of tissue engineering scaffolds as delivery devices for mechanical and mechanically modulated signals. *Stud. Mechanobiol. Tissue Eng. Biomater.* 10: 127–143.

**71** Chen, Y., Zhou, S., and Li, Q. (2011). Mathematical modeling of degradation for bulk-erosive polymers: applications in tissue engineering scaffolds and drug delivery systems. *Acta Biomater.* 7: 1140–1149.

**72** Shirazi, R., Ronan, W., Rochev, Y., and McHugh, P. (2016). Modelling the degradation and elastic properties of poly(lactic-*co*-glycolic acid) films and regular open-cell tissue engineering scaffolds. *J. Mech. Behav. Biomed. Mater.* 54: 48–59.

**73** Wang, Y., Pan, J., Han, X. et al. (2008). A phenomenological model for the degradation of biodegradable polymers. *Biomaterials* 29: 3393–3401.

**74** Scheiner, S., Komlev, V., Gurin, A., and Hellmich, C. (2016). Multiscale mathematical modeling in dental tissue engineering: toward computer-aided design of a regenerative system based on hydroxyapatite granules, focussing on early and mid-term stiffness recovery. *Front. Physiol.* 7 (383) https://doi.org/10.3389/fphys.2016.00383.

**75** Marquetti, I. and Desai, S. (2018). Molecular modeling of bone morphogenetic protein for tissue engineering applications. In: *Proceedings of the 2018 IISE*

*Annual Conference* (ed. K. Barker, D. Berry and C. Rainwater). at Florida: https://par.nsf.gov/servlets/purl/10075098.

**76** A. Ananthanarayanan, L. Tucker-Kellogg, B. Narmada, L. Venkatraman, N. Abdul Rahim, Y. Wang, C. Kang, and H. Yu, H. (2011). "Chapter: 5.513 – Systems biology in biomaterials and tissue engineering", Comprehensive Biomaterials, pp. 177–188. https://doi.org/10.1016/b978-0-08-055294-1.00196-3, Elsevier Ducheyne, P.

**77** Olivares, A. and Lacroix, D. Computational methods in modeling of scaffolds for tissue engineering. In: *Computational Modeling in Tissue Engineering. Studies in Mechanobiology, Tissue Engineering and Biomaterials*, vol. 10 (ed. L. Geris). Berlin, Heidelberg: Springer. https://doi.org/10.1007/8415_2012_136.

# 4

# Computational Modeling at Cell Level

## 4.1 Introduction

Cells are entities that can generate and sustain mechanical forces within their environment as part of their physiology. Cells are able to identify mechanical stimulation through the activation of mechanosensitive signaling pathways and respond to physical cues through cytoskeletal re-organization and force generation. The cytoskeletal systems of different organisms consist of similar proteins. In eukaryotes, the cytoskeletal matrix is a dynamic structure composed of three main proteins, which are able to grow rapidly or disassemble depending on the cell's requirements. Pathogens and genetic mutations which disrupt the cytoskeletal architecture can lead to changes in the mechanical properties of cells such as elasticity, viscosity, and adhesiveness. Nowadays, several experimental techniques and theoretical models have been presented based on soft matter physics and mechanical engineering which aim to characterize cell mechanical properties [1]. The interaction of cell mechanical characterization techniques with cell biology now paves the way for enhancing our fundamental understanding of cell mechanics and its role in development, physiology, and disease.

Within the last decades, computational models play a crucial role in the advancement of science. Undoubtedly, it is significant that researchers recognize the implications of the mathematical and algorithmic steps of modeling various constructions and comprehend the wide applicability of computational modeling in biomedical engineering. Models can contribute to research through several paths including hypothesis testing, deepening understanding, suggesting and interpreting experiments, sensitivity analyses, integrating knowledge, and inspiring new approaches in experimental and clinical research [2]. Although computational studies cannot replace experiments or clinical trials, they can demonstrate

*Multiscale Modelling in Biomedical Engineering*, First Edition. Dimitrios I. Fotiadis, Antonis I. Sakellarios, and Vassiliki T. Potsika.

Published 2023 by John Wiley & Sons, Inc.

whether or not a proposed mechanism is sufficient to produce an observed phenomenon and be used for validation means.

This chapter deals with computational modeling at the cell level. The mechanical deformation characteristics of living cells have an important effect not only on their biological and chemical behavior but also on the occurrence, progression, and consequences of human diseases.

According to the literature, several studies have been presented that focus on the numerical modeling and mechanical behavior of a specific cell type (e.g. red blood cells, leukocytes, etc.). Through the numerical evaluation of individual cells, biological observations can be directly translated into simulation rules. For example, in [3], a detailed discussion is presented on the mechanics of large deformation of the human red blood cell using computational simulations. The authors presented a systematic parametric analysis to investigate the cell membrane shear and bending rigidity, the cell interior volume, the variations in cell diameter, conditions of contact between the cell membrane and the beads trapped by the laser on the large deformation characteristics of the cell. Computational simulations were performed in the context of a variety of available constitutive descriptions of the cell membrane. As a second example, we can refer to numerical models of cancer stem cells which offer valuable insights into the driving forces of cancer biology. In a more recent study [4], the role of adipose-derived stromal cells was examined in promoting vascular regeneration which is believed to have high potential in vascular regeneration.

Other research groups have presented numerical models at the cell level focusing on a specific type of disease. For instance, in [5], the dynamics of cancer were intimately linked to a dysregulation of the cell cycle and signalling pathways. More specifically, the cancer cell cycle was simulated as a means for rational drug development and a systems pharmacology approach to cyclotherapy was presented. The impact of drug on multiple cell populations was simulated, facilitating simultaneous prediction of effects on normal and transformed cells. To summarize, several mathematical models at the cell level have been presented in the literature related to different cancer types aiming to provide insight into: (i) the interaction of the tumor with its microenvironment (e.g. cancer cells with the extracellular matrix). In some cases, the transport of drugs, oxygen, and growth factors is also considered which allows the association of cancer development with the microenvironmental conditions, (ii) the effect of chemotherapy on cancer through simulations representing cancer response for a continuous drug delivery, (iii) the population-level dynamics resulting from different stem cell kinetics, (iv) mechanical models of tumor invasion that are influenced by chemical factors and physical forces necessary to degrade the host tissue [6–9]. These numerical models can be used to study complicated situations or even contribute to alternative therapeutic strategies such as immunotherapy.

Except for the case of cancer, several mathematical models have been presented at the cellular level to investigate musculoskeletal and cardiovascular pathologies. For instance, in [10], a computational approach was presented on the basis of coupling a pharmacokinetics model of denosumab with a pharmacodynamics model to investigate the impact of denosumab on bone remodeling. The pharmacodynamics model was developed based on two different approaches including: (i) an integrated systems biology-continuum micromechanics approach based on a bone cell population model, considering the governing biochemical factors of bone remodeling, (ii) a multiscale micromechanics-based bone mechanics model considering the mechanobiology of bone remodeling. Moreover, in [11], computational models of post-traumatic osteoarthritis were found to reveal cellular responses to contact stress that damage articular cartilage and present targets for intervention. It was also emphasized that by incorporating information from data derived from animal assays into parameters for mathematical models, mathematics can translate experimental data into clinically relevant knowledge. In [12], a mathematical model was presented incorporating intraplaque neovascularization and hemodynamic calculation with plaque destabilization for the quantitative evaluation of the role of neoangiogenesis and intraplaque hemorrhage in the vulnerable atherosclerotic plaque formation. More specifically, an angiogenic microvasculature was established by two-dimensional nine-point discretization of endothelial cell proliferation and migration from the vasa vasorum. The plaque progression model included three key cells (endothelial cells, macrophages, and smooth muscle cells) and three key chemicals (extracellular matrix, vascular endothelial growth factors, and matrix metalloproteinase). It was indicated that intraplaque hemorrhage can significantly speed up plaque vulnerability by increasing necrotic core and thinning fibrous cap.

## 4.2 Introduction to Cell Mechanics

Cell mechanics is considered a research domain which plays a significant role in cellular and tissue biology, from tissue and organ development to wound healing and cancer cell metastasis and migration [13]. To study the mechanical behavior of cells, computational models have been presented that approach cell mechanics as a whole of biochemical and biomechanical processes. These models aim to explain experimental observations by providing a framework of underlying cellular mechanisms. They have also enhanced predictive *in silico* studies, which would otherwise be difficult or impossible to be conducted according to current experimental approaches [14].

### 4.2.1 Cell Material Properties

The response of a medium to mechanical stimuli is determined by a group of characteristics referred to broadly as its "mechanical properties" [1]. This term refers to how a medium deforms in response to an applied stress, and how this deformation evolves over time. Young's modulus, often referred to as the material's elasticity, is a mechanical property that measures the stiffness of a solid material. It determines the relationship between strain (defined as proportional deformation) and stress (defined as force per unit area) in a material in the linear elasticity regime of a uniaxial deformation. In contrast to elastic solids, fluids flow under the application of stress and are unable to store elastic energy. The rate at which a fluid flows under a defined load is quantified by its viscosity. However, several materials exhibit both elastic and viscous properties and are known as viscoelastic materials.

Under physiologically relevant timescales, cells are considered intrinsically viscoelastic media, displaying a combination of both elastic and time-dependent responses to deformation. A viscoelastic material undergoing deformation simultaneously stores and dissipates mechanical energy and thus mechanical stress relaxes and deformation increases over time. In order to investigate the interactions between stress, strain, and the rate of flow, rheological measurements have been conducted to reveal how cells flow rather than deform purely elastically in response to an applied force [1]. In addition, early cell mechanical measurements have shown: (i) links between local increases in cellular elasticity and subcellular structures such as stress fibers [15–16] and (ii) changes to cellular elastic and viscous properties under various treatments. Nevertheless, the applicability of these terms in a strict engineering sense is demanding and a range of complex rheological characteristics for cellular systems have been observed.

Cells are complex heterogeneous media consisting of a range of proteins, filaments, subcellular structures, and organelles, which can have various effects on cell elasticity and viscosity. For instance, the role of the nucleus in defining whole cell elasticity is further discussed as follows. More specifically, the nucleus is stiffer than the cytoplasmic portion of the cell [17]. During cell compression experiments, the cell cortex and membrane are deformed within the first 200 nm of compression [1, 17]. By increasing the compression, a larger contribution from the underlying stiff nucleus is observed, implying that the stress–strain relationship cannot be characterized by a simple linear relationship [1].

In comparison to typical materials such as plastics, metals, or glass, cells are considered "small" and "soft" media. Typical values for the eukaryotic cell elastic modulus range are depicted in Figure 4.1 in comparison with other entities at different hierarchical scales. The development of advanced technologies has enabled the measurement of the mechanical properties of the cell on spatially accurate

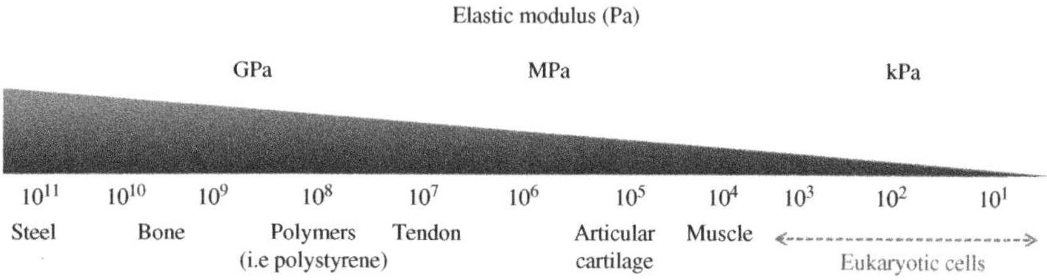


**Figure 4.1** Elastic modulus of eukaryotic cells and comparison with other entities.

length scales. The mechanical measurement techniques can be categorized into two main groups: (i) methods that can be used to apply controlled deformations and forces on part of, or on the entire cell (such as magnetic bead cytometry [1, 18], optical tweezers [1, 19], cell stretchers [20], flow rheometry [21], and atomic force microscopy (AFM) [22]), (ii) techniques which monitor the ability of a cell to generate forces and deform its environment (such as traction force microscopy (TFM) [23] and micropillar arrays [1]). Also, the extraction of local cell mechanical properties can be achieved by tracking the motion of endogenous cellular structures, such as the movement of actin filaments, microtubules, mitochondria, or embedded particles of various sizes that are excited thermally or driven by an external force [24].

### 4.2.2 Cell Composition and Structure

The living cell is the fundamental unit of all structures in biology. Regardless of an organism is multicellular or single-celled, it must be able to perform the same basic functions in order to survive. These functions include the acquisition of nutrients and energy sources, disposal of unusable and toxic materials, reproduction, locomotion, and interaction with components in the environment [25]. Also, the aforementioned functions need the proper coordination of short-range activities (e.g. sensation) and long-range activities (e.g. growth).

Concerning cell composition and structure, all cells are surrounded by a plasma membrane which encapsulates the cytoplasm and creates internal compartments where cell functions evolve (Figure 4.2). The cellular membrane consists of a thin lipid bilayer with thickness from 5 to 10 nm and bending stiffness from $10^{-19}$ to $10^{-20}$ N·m.

This section describes the two basic types of cells which are prokaryotes and eukaryotes and their differences are summarized in Table 4.1. Eukaryotic cells have a nucleus enclosed within membranes, while prokaryotes (Bacteria and Archaea) have no membrane-bound organelles. Also, eukaryotic cells contain other membrane-bound organelles such as mitochondria and the Golgi

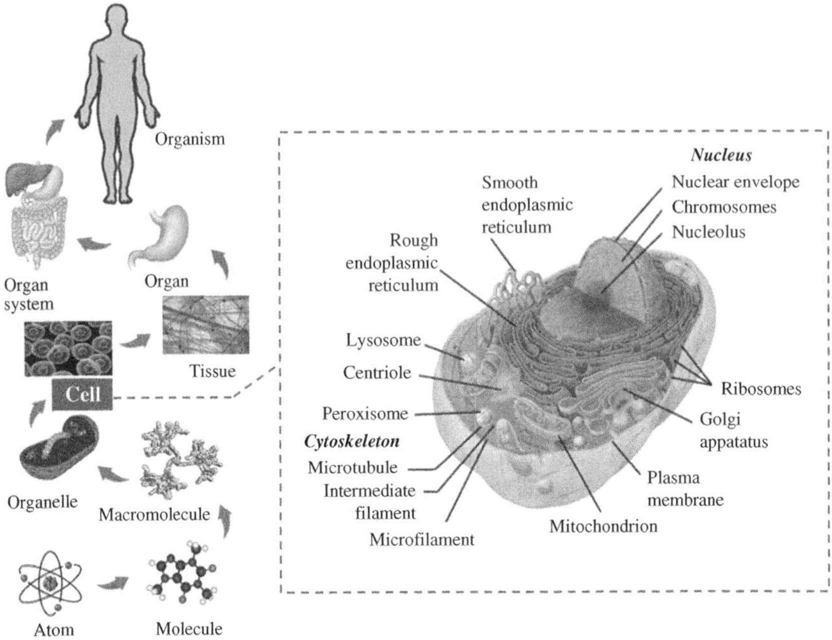


**Figure 4.2** Typical structure of eukaryotic cells. *Source:* SciePro/Adobe Stock.

**Table 4.1** Difference between prokaryotic and eukaryotic cells.

| | Prokaryotes | Eukaryotes |
|---|---|---|
| Genetic information | DNA is circular (free-floating in cytoplasm) | DNA is linear (in nucleus) |
| Organelles | Without nucleus or membrane-bound organelles | With nucleus and membrane-bound organelles (i.e. mitochondria, chloroplasts, Golgi body) |
| Size | Small (1–5 μm) | Larger (10–100 μm) |
| Organisms | Bacteria/archaea | Animals, plants, fungi, and protists |
| Cell structure | Unicellular | Unicellular or multicellular |

apparatus. Some cells of algae and plants contain chloroplasts. Compared to bacteria and unicellular archaea, eukaryotes may also be multicellular and include organisms consisting of many cell types composing various types of tissue. Animals and plants are the most popular eukaryotes.

As far as prokaryotic cells are concerned, they are characteristically small and possess minimal intracellular structures. Distinct membrane-bound organelles are absent in most cases. In contrast, eukaryotic cells are larger and contain numerous intracellular organelles of various functions and structures, each of which is specialized for a specific function such as respiration, digestion, biosynthesis, or secretion. Both bacterial and eukaryotic cells must carry out the same essential processes to live and reproduce. Therefore, the differences in cell structure do not refer to divergent modes of life, but to the presence or absence of compartments specifically designed to fulfill significant functions. In addition, in eukaryotes, all processes are conducted in a spatially isolated domain, while these processes operate largely within a single compartment for the case of prokaryotes. Although the prokaryotic cell may appear less complex compared to the cells of higher organisms, the functional dissection of the latter may be simpler.

In addition, the energy generation in eukaryotic and prokaryotic cells differs. In eukaryotic organelles, the mitochondrion is the organelle of respiration. Mitochondria possess outer and inner membranes of entirely different compositions. The mitochondrion can be considered a multicompartmental structure as its four domains include the inner and outer membranes, the matrix between these two membranes, and the inner matrix. In the outer membrane, the enzyme monoamine oxidase can be found, whereas the protein components of the electron transport chain and the oxidative phosphorylation system can be identified in the inner membrane. Adenylate kinase can be detected in the outer matrix, whereas the enzymes which oxidize fatty acids, amino acids, and di- and tricarboxylic acids appear exclusively in the inner matrix. The inner matrix also includes deoxyribonucleic acid (DNA), as well as all the necessary means to translate and transcribe the nucleotide sequences of nucleic acids into proteins.

The interior of a single cell has a fluid nature containing macromolecules, organelles, and structures that fulfill different functions. A network of subcellular filaments, known as cytoskeleton, forms meshes and bundles that enhance individual cells with the ability to sustain external mechanical forces of higher order. The three cytoskeletal filaments which are of significant importance for cell mechanical properties are microtubules, actin microfilaments, and intermediate filaments. Actin is considered one of the most abundant proteins in eukaryotic cells forming polarized filaments interacting with an array of ancillary proteins. From a mechanical point of view, actin filaments are semiflexible on the length scale of the cell, having a persistence length on the order of the cellular length $\sim$20 μm. Moreover, actin filaments are highly dynamic and rapidly reorganize to allow cells to migrate and alter shapes. Actin is known to be vital for the cell mechanical function. For example, the actin cytoskeleton forms a 200 nm-thick mesh below the apical plasma membrane which enhances the cell's mechanical integrity, basal fibers anchor the cell to the extracellular matrix, and linear bundles coordinate and sustain forces along intercellular junctions.

## 4.3 Cellular Bioengineering in Areas Critical to Human Health

Cellular and molecular bioengineering (CMBE) is a scientific domain which plays a key role in the analysis of human diseases. The analysis of cell effects based on bioengineering principles and techniques is a powerful tool in order to provide new knowledge to cell, tissue, and organ behaviors. Figure 4.3 shows that CMBE studies drive technological advances that are pushing the boundaries of our knowledge on both basic and translational sciences, including biology, diagnostic medicine, therapeutic medicine, and P4 medicine [26]. The contribution of computational modeling methods to unwind the cellular mechanical behavior and interactions in a variety of CMBE studies is discussed in the following subsections.

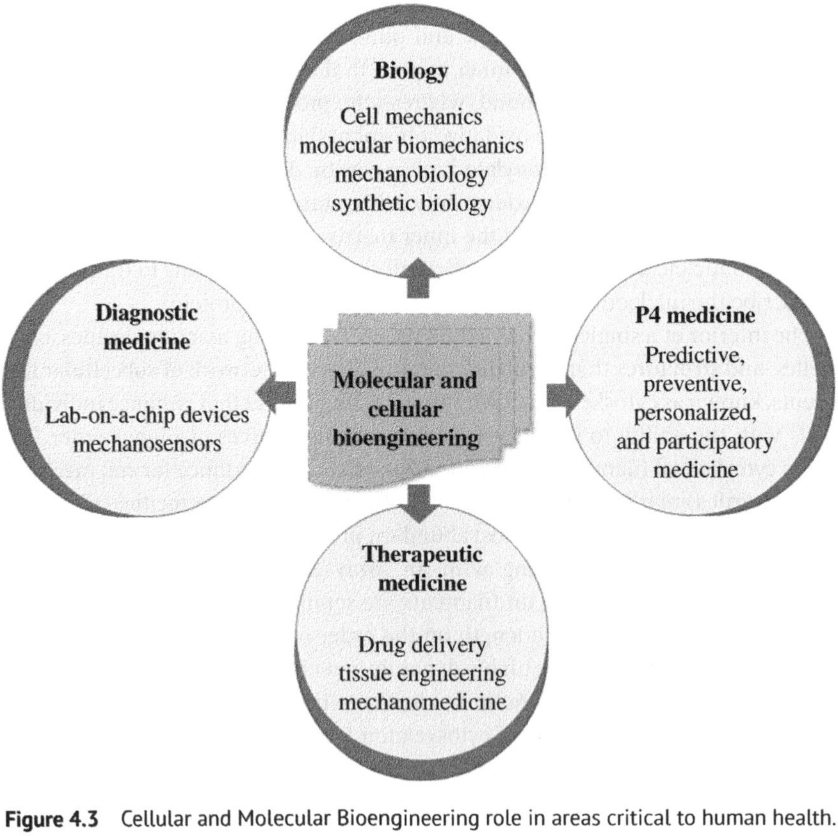


**Figure 4.3** Cellular and Molecular Bioengineering role in areas critical to human health.

### 4.3.1 Biology

Several pathological diseases including asthma, osteoporosis, atherosclerosis, cancer, osteoarthritis, glaucoma, and muscular dystrophy may occur due to irregular cellular functions. Nevertheless, the specific biological mechanisms which lead to these disease states, as well as how they can be remedied, are still unknown.

The mechanical properties of the cells, the forces produced internally as well as forces applied to them externally are factors which play an important role in their normal function. To study the mechanics of cells, computational models have been presented that approach cell mechanics as a whole of biochemical and biomechanical processes. These models have been advantageous in explaining experimental observations by providing a framework of underlying cellular mechanisms [14]. To better understand cell mechanics, it is important to closely translate results from *in vitro* experiments into mathematical relationships for *in silico* models.

Cellular cytoskeletal mechanics is widely applied in many aspects of human health from organ development to wound healing, tissue homeostasis and cancer metastasis. Several studies have been presented: (i) aiming to model cellular stiffness and mechanics and the cytoskeletal components and factors that regulate them, (ii) focusing on key experiments that have assisted model parameterization and compare the advantages of different models that have been used to recapitulate these experiments, (iii) dealing with feed-forward mechanisms from signaling to cytoskeleton remodeling. A thorough understanding of the molecular mechanisms that affect cell and tissue mechanics will underpin innovations in medical device technologies of the future [13].

To this end, computational models provide a tool which can contribute significantly to better comprehending the physical properties of cells. Depending on the level of sophistication of the model, this interpretation can be simple (e.g. describe the elastic stiffness of the cytoplasm) or it can be more complex (e.g. estimate the rate constants for cytoskeletal remodeling and force generation). Two main categories can be distinguished including active and passive models. In order for a model to simulate cell behavior, it must be able to predict the mechanotransduction response of a cell to its physical environment. In addition, in order to be truly predictive, an active model must consider the key mechanisms which determine how a cell senses and responds to its physical environment. While active models incorporate key biomechanical processes that underlie cell behavior, passive models have also been presented which assume that a cell will respond passively to mechanical stimuli. By extension, active models possess the ability to simulate the remodeling of cells when exposed to physical stimuli [14]. The next paragraphs present the most popular active and passive models in the literature.

One of the most important categories of passive models includes the Newtonian liquid drop model, compound Newtonian liquid drop model, shear-thinning liquid drop model, and the Maxwell liquid drop model. These models assume that the cells are composed of one or more layers of cytoplasmic fluids, surrounded by prestressed cortical shells. These formulations have been used to computationally represent leukocytes and account for the rheological measurements of these cells using micropipette aspiration [14, 27].

In particular, as far as the Newtonian liquid drop model is concerned, the interior of the cell is assumed to be a homogeneous, viscous, Newtonian liquid. The cell cortex is modeled as a layer of isotropic, viscous fluid with static tension and no bending resistance. Next, a compound Newtonian liquid drop model was developed aiming to account for some of the discrepancies between experimental and numerical results found when applying the Newtonian liquid drop model. In this model, cells are defined as being composed of two different layers, the nucleus and the cytoplasm. Concerning the Maxwell liquid drop model, it has been applied to cells in order to capture their rapid entry into a micropipette at the beginning of an aspiration test. This model is based on the Newtonian liquid drop model, but it differs in the assumption that the cytoplasm acts as a Maxwell element (elastic spring in series with a dashpot) [14]. The Maxwell model is able to predict the rapid entry and recovery behavior of leukocytes subject to small strains from micropipette aspiration. Nevertheless, when this same model was used for the finite element simulations of large deformations by micropipette aspiration, it was unable to replicate the experimental results. This indicates the need for the establishment of active models that incorporate dynamic remodeling of the cell in response to mechanical stimuli.

In solid elastic models, the cell models consist of one or more homogeneous material layers. There are two different types of linear elastic solid models, the linear-elastic and the linear-viscoelastic. In the linear-elastic solid model, a cell is considered solid with homogeneous, elastic properties. Alternatively, if the cell is modeled as a linear-viscoelastic solid, then this is equivalent to having a Maxwell model in parallel with a spring element (a standard linear solid).

The power-law rheology model deals with the responses of adherent cells that are subjected to time-varying forces in their environments. It is based on empirical observations claiming that a large number of "soft" materials exhibit similar rheological properties. It is considered that these materials have microscopic constituents which are unable to equilibrate thermodynamically (e.g. glass) and for this reason, they are also known as soft-glassy materials. Numerical models of cells applicable to such materials can also describe the frequency-dependent rheology of cells which follows a power-law trend.

Finally, the biphasic model considers both the liquid and solid compositions of the cytoplasm assuming that it consists of a linear-elastic solid phase and a fluid phase having no or negligible viscosity. In this model, the liquid phase is able to diffuse through the solid phase. Momentum exchange takes place through friction between the two different phases.

### 4.3.2 Diagnostic Medicine

Computational modeling in diagnostic medicine aims to combine all the existing knowledge in order to develop improved therapies specialized to the needs of individuals. Computational modeling can provide insights into the function and structure of living systems in disease, and the knowledge obtained from modeling approaches can, in turn, be exploited to establish improved methods for disease diagnosis and treatment. This emerging approach is also known in the literature as "computational medicine" [28].

The appearance of diseased cells is due to perturbations in biological networks owing to the net effect of interactions among multiple molecular agents, including inherited and somatic DNA variants, alterations in mRNA and microRNA, protein expression, and epigenetic factors (e.g. DNA methylation) [28]. A significant amount of data for these disturbances can be derived by next-generation sequencing and microarray experiments of large patient cohorts, making it feasible to investigate the driving differences in the abundance and activity of key biomolecules.

In order to optimize disease and treatment options, the development of models spanning different levels of biological organization is required. For instance, cancer begins at the molecular level as a result of mutations of oncogenes, tumor suppressor genes, miRNA, and changes in DNA methylation. This leads to dysregulation of many different signaling pathways involved in cell growth, angiogenesis, and apoptosis. Tumor cells also interact with their local environment (extracellular matrix and cells) by secreting signaling molecules. Changes to this signaling can ultimately affect cell adhesion and migration, tissue invasion, and metastasis. Cancer is clearly a multiscale disease, and mechanistic models spanning these scales are being developed and applied to predict the progression of tumors [29–31]. Nevertheless, the model-building process should start with a minimal set of assumptions regarding the numbers of states and parameters, and then test the extent to which this "minimal model" reproduces as well as predicts experimental data [32].

The following subsections present two main domains of diagnostic medicine, namely organ chip technology and mechanosensors.

#### 4.3.2.1 Organ Chip Technology

Organ chips are microfluidic cell culture devices consisting of optically clear plastic, glass, or flexible polymers, such as polydimethylsiloxane (PDMS). They contain perfused hollow microchannels populated by living cells which recapitulate *in vivo* organ-level physiology and pathophysiology by recreating tissue-level and organ-level structures and functions *in vitro* [33]. They are called "chips" because they were initially fabricated using micromanufacturing methods adapted from computer microchip fabrication [34]. Microfluidic devices can be used to create "tissue chips" with a single channel lined by cells from one type of tissue or more complex organ chips that combine two or more tissue types that can be interfaced directly across a porous extracellular matrix (ECM)-coated membrane or separated by a gel that fills one or more microchannels [35].

Using the technology of organ chips, one can start studying the effects of a single microenvironmental component or drug on cancer cells, cultured either alone or in various combinations with normal epithelial cells, fibroblasts, endothelial cells, immune cells, numerous mechanical cues, and even microbiota, while visualizing cellular responses in real time at high resolution [36]. The chips can be developed based on multiomics analysis and the use of primary cells or induced pluripotent stem cells isolated from patients. These features show the unprecedented ability to identify key molecular, chemical, cellular, and biophysical contributors to human cancer formation and progression under controlled conditions *in vitro*. The implementation of this synthetic biology approach with organ chips has enabled the discovery of new mechanisms for the evaluation of cancer progression and led to new insights into the control of tumor invasion and growth, which could not have been discovered using traditional *in vitro* methods. The incorporation of human organ chips into clinical care and drug development pipelines could be very promising as soon as challenges related to simplifying their fabrication and use, as well as increasing their robustness have been resolved.

To fill the gap between *in vivo* and *in vitro* conditions, researchers working on chip technology and other tissue engineering research fields have applied microfluidic devices to cell culture applications since 2000 [37]. Figure 4.4 presents a

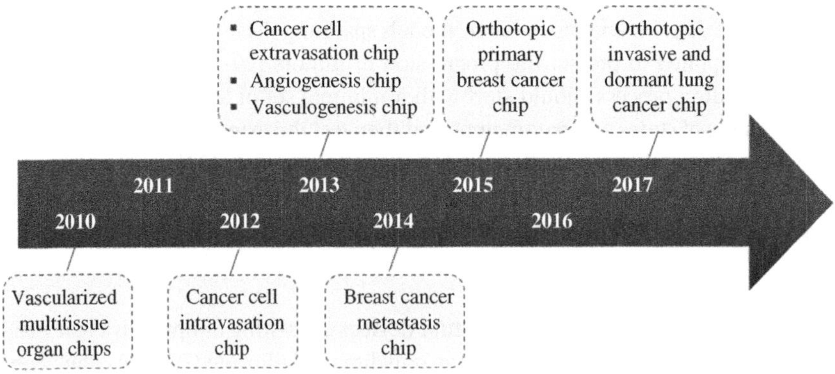


**Figure 4.4** A timeline showing the development of different cancer organ chips within the last decade.

timeline showing the development of the technology of cancer organ chips within the last decade [35].

Recent studies have used computational approaches for the characterization of microfluidic devices [38, 39] and simulations have been carried out to characterize the flow properties of the microfluidic device. More specifically, the influence of the Reynolds number, the chamber dimensions, and the outlet channels aspect ratio on the stream tracer distribution are some of the main parameters of interest. In order to support the simulation results, some preliminary experimental validations can be conducted to ensure that the model can accurately characterize the flow in the geometry under discussion.

The outcome of both experimental and numerical approaches can be very useful to advance the design of continuous-flow particle sorting lab-on-a-chip (LOC) devices. Computational modeling has many advantages for LOC research, including robust device design and the ability to simulate complex and coupled physics rapidly and at a low cost. The LOC community uses computational tools in a myriad of ways, from basic research to device design [39]. For example, level-set methods can be used to simulate bubble formation, coalescence, and biomolecular sorting in microfluidic devices [40]. Spectral methods are very useful tools for the simulation of nonlinear electrokinetic flow in simple microchannel geometries [41–42]. Model order reduction methods can be very important for rapid LOC prototyping [39].

#### 4.3.2.2 Mechanosensors

Cells reside within complex microenvironments consisting of both biophysical and biochemical cues. The dynamic interactions of cells with their microenvironment depend on biophysical cues that control critical cellular behaviors. A significant effort has been made in order to identify and mathematically model these fundamental mechanobiological mechanisms. The focus of this section is the use of mathematical models to obtain new knowledge on the way that cellular mechanosensitivity arises from the range of biophysical sensing modes inside the cells.

Several mathematical models have been established investigating the responses of cells to the biophysical cues in their microenvironments, including dynamic strain, osmotic shock, fluid shear stress, mechanical force, matrix rigidity, porosity, and matrix shape [43]. For example, cardiac fibroblasts and cardiomyocytes experience periodic deformations of the heart wall, endothelial cells experience pulsatile shear flows and pressures, and cancer cells can bear compression deformation in a tumor [44]. Several studies investigate how and why the cells could sense and respond to external dynamic strain with the ultimate aim to find novel ways to cure diseases of mechanotransduction. Furthermore, emerging models try to explain how cellular mechanosensing integrates dynamic interactions between biomolecules across wide temporal and spatial scales in the cell microenvironment [45–46]. Mathematical models based upon hypotheses of the mechanics and kinetics of biochemical and biophysical processes are needed to further unwind the biophysics of cellular mechanosensing. The main mathematical models which have been presented in the literature can be categorized as follows: (i) one-dimensional stress fiber model [47], (ii) two-dimensional stress fiber model [43, 48], (iii) Rho-regulated mechanochemical stress fiber model [49, 50], and (iv) minimum free energy stress fiber model [51].

In addition, cells can sense cycles of osmotic shock (e.g. cycles in which cells are subjected to a hypotonic shock followed by a hypertonic shock). These responses are induced through the activation of mechanosensitive ion channels at the cell membrane-cortex surface. These mechanosensitive ion channels are controlled by the mechanical force balance at the cell membrane-cortex surface, and their action influences the homeostatic values of cell volume and membrane tension. To understand the pathways underlying this volumetric regulation, a chemomechanical model was proposed in [52], which modeled activation of the Rho signaling pathway by the opening of mechanosensitive ion channels. The ion channel chemomechanical model consists of three main parts: force balance at the cell edge, membrane tension-mediated myosin activation, and cell volume change that depend upon local traction. There are two main assumptions in this model. The first is that the geometry of the model accounts for a spherical or cylindrical cell so as to simplify both cell geometry and cell adhesion. The second is that the cortical actomyosin layer coupled to membrane is modeled as an active viscoelastic gel-like fluid.

Concerning endothelial cells, they are subjected to two kinds of mechanical cues from their vascular microenvironment: shear stress from blood flow and tensile stretch from the cardiac pressure cycle [53]. The shearing may affect signal activation, gene expression, and cytoskeletal realignment. A mathematical model was proposed in [54] to study cellular mechanosensing in response to fluid shear stress in which a cell's actin cytoskeleton network transferred shear stress to cell adhesions on the basal surface of the cell, leading to changes in adhesion dynamics. An assumption of this model is that fluid shear stress is not explicitly introduced as an input as the model applies a transient decrease of the Rho concentration to represent shear stress loading on the cell.

### 4.3.3 Therapeutic Medicine

Computational modeling can serve to better comprehend the complex structure and function of living systems in disease, and insights gained from modeling approaches can be used to establish improved methods for disease treatment. In order to fulfill the needs of therapeutic medicine, advanced models have been developed integrating cellular networks and physiological processes as well as modeling anatomic shapes layered with physiological function. Although modeling approaches in therapeutic medicine vary significantly, the common goal is the proper exploitation of quantitative means to understand altered structure and function in disease.

#### 4.3.3.1 Drug Delivery

Drug delivery is considered the main research area of therapeutic medicine. During the last decade, several numerical studies have been presented at the cell level, which mainly deal with tumors and cancer [55–57].

For instance, the ability to predict how far a drug will penetrate the tumor microenvironment within its pharmacokinetic (PK) lifespan would provide valuable information about therapeutic response. As the PK profile is directly related to the route and schedule of drug administration, an *in silico* tool that can predict the drug administration schedule that results in optimal drug delivery to tumors could enhance the proper clinical trial design. In this direction, in [55], the application of mathematical and computational modeling techniques was studied as a means to improve our knowledge of the fundamental mechanisms underlying drug delivery and compare the performance of a simple model with more complex approaches. To this end, the binding model was established considering that the interaction of the chemotherapeutic agent with the microenvironment is restricted to drug binding only. Figure 4.5a shows that the interaction of the chemotherapeutic agent with the microenvironment of cells can be described by a three-compartment model including both the extracellular space $V_1$ with a concentration $C_1$ of free drug and the intracellular space $V_2$ with concentrations $C_2$ and $C_3$ corresponding to free and bound drug, respectively. It is noted that for this model, the term bound refers to both DNA-intercalated drug and drug bound to the cell in other ways. The binding is described by a simple kinetic model as drug binds reversibly to sites within the cell. Applying the principle of mass action

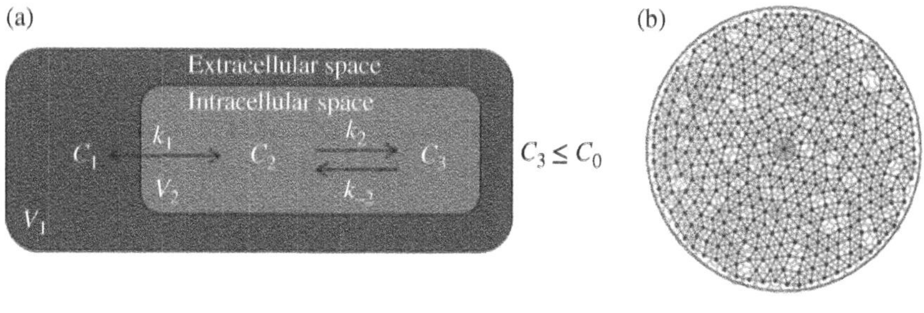


**Figure 4.5** (a) A three-compartment model of drug distribution, (b) a representative tumor cord geometry for the two-dimensional cell-center model in which the computational nodes (cell centers) surround a central vessel.

leads to three coupled ordinary differential equations which describe the system of Figure 4.5a [55]:

$$V_1 \frac{dC_1}{dt} = ak_1(C_2 - C_1), \tag{4.1}$$

$$V_2 \frac{dC_2}{dt} = ak_1(C_1 - C_2) - V_2 k_2 C_2 (C_0 - C_3) + V_2 k_{-2} C_3, \tag{4.2}$$

$$V_2 \frac{dC_3}{dt} = V_2 k_2 C_2 (C_0 - C_3) - V_2 k_{-2} C_3, \tag{4.3}$$

where $k_1$ represents the rate constant for the transmembrane transport of drug, $a$ is the area of the interface between the extracellular and intracellular spaces (the surface area of the cell), $k_2$ is the drug association rate, and $C_0$ is the concentration of binding sites within the cell. It is assumed that mixing in each compartment is instantaneous implying that intracellular and extracellular diffusion evolves fast on the scale of an individual cell.

Another more general modeling approach is illustrated in Figure 4.5b showing a multidimensional model in which each biological cell in the cord is depicted as a computational node (cell center) characterized by geometrical and biological information such as position, cell radius, binding rates, and concentration of binding sites. Each of these nodes in isolation acts in accordance with the three-compartment-binding model illustrated in Figure 4.5a, providing the potential to insert a bespoke model of the microenvironment for each node/cell. Therefore, in this way, the binding model is augmented by a description of the spatial transport of the drug, described locally by defining transport terms between nodes and their nearest neighbors. In addition, the binding model can be further extended with a spatial component by exploiting the shell-like nature of tumor cords and the structural property that cells are arranged in concentric circles around a central blood vessel. In this case, it can be assumed that the rate of transport of drugs between neighboring shells is proportional to the shared interface area and the difference in concentration across the interface.

In a more recent study [57], an *in silico* model of drug transport in a tumor cord was presented to investigate the effect of different drug regimens over a 72-hours period and how alterations in PK parameters affect tumor exposure to the cytotoxic drug doxorubicin. This numerical approach described the axial and radial distribution of the drug in the tumor cord as a function of changes in the transport rate across the cell membrane, flow rate, blood vessel, and intercellular permeability, as well as the binding and unbinding ratio of drug within the cancer cells. This study examined how changes in these parameters may affect cellular exposure to the drug. The model explored and provided information on: (i) the extent to which the distance from the supplying vessel influences drug levels, (ii) the effect of dosing schedule in relation to saturation of drug-binding sites, and (iii) the impact on drug distribution of the abnormal vasculature seen within tumors. The values of the kinetic rate constants for the binding process were derived from the three-compartment model of drug distribution in Figure 4.5a [55].

Finally, intraperitoneal administration of chemotherapy is considered an alternative treatment for peritoneal carcinomatosis, allowing for higher intratumor concentrations of the cytotoxic agent in comparison with intravenous administration. In [57], a three-dimensional computational fluid dynamics (CFD) model was developed for drug penetration in a tumor nodule. The impact of a number of key parameters on the drug transport and penetration depth during intraperitoneal chemotherapy was examined. It was found that smaller tumors show better penetration than larger ones, which could be attributed to the lower interstitial fluid pressure in smaller tumors. Nevertheless, drug penetration depths are still limited to a few millimeters and further computational research could lead to better comprehend the limiting factors behind this poor penetration in order to improve intraperitoneal chemotherapy delivery.

#### 4.3.3.2 Tissue Engineering

Tissue engineering studies living cells in a variety of ways aiming to create functional tissues and organs which are able to replace or repair those which have become defective due to trauma, disease, or aging. Tissues respond to both chemical and mechanical cues and carefully specified and controlled biochemical and biomechanical environments are required to create a new tissue. The growth of biological tissue is a complex process and a major factor impeding progress in the field is an incomplete understanding of the various processes that contribute to the production of a functioning organ or tissue. Advances in the understanding of tissue growth processes can improve significantly the viability and suitability of the resulting tissue constructs. In this direction, the contribution of computational modeling can contribute significantly as they are powerful tools which can lead to new knowledge on how tissues and organs develop and are maintained and repaired [58].

Undoubtedly, it is challenging to model a tissue culture system which consists of the bioreactor, culture medium, and developing construct. Therefore, a novel and robust modeling approach is required to provide a solution in an acceptable time frame with routine computing resources. This kind of problem can be addressed by a multiscale modeling approach in which single-scale models operating at different time and length scales interact to give a representation of the system as a whole. For example, in [58], the authors aim to address this by developing a multiscale framework based on a hierarchical aggregation of coupled cellular automata and agent-based models.

Computational models can simulate and predict the role of mechanical forces in cell differentiation, adhesion, proliferation, motility, and secretion of extracellular matrix proteins regarding tissue engineering scaffold environments. For specific tissue engineering scaffolds, computational approaches can be used to evaluate mechanical stimuli to cells seeded within aiming to achieve the targeted tissue manufacturing and integration [59]. Moreover, computational models can be used to investigate cell deformations in near real time under shear flow or even predict deformation based on a given geometry and mechanical or chemical properties under external forces (passive behavior) [60]. Nevertheless, active morphological changes in cells can occur through biological changes such as gene expression and reorientation of cytoskeletal filaments [59, 61]. This active behavior can also be analyzed by computational models based on given results under forces, which is an important approach to elucidate mechanisms of mechanoadaptation [62].

In addition, CFD can be applied to predict flow regimes around cells in known flow chamber geometries and flow fields, to deliver controlled mechanical signals to cells seeded within. *In situ* microscopy allows for concomitant imaging of live cell mechanoadaptation while tracking the cell's dynamic mechanical environment [59, 63]. Specifically, microscale particle image velocimetry enables: (i) the validation of CFD predictions at the length scale of the coverslip (2D) and scaffolds (3D) onto which cells are seeded for mechanotransduction studies and (ii) the observation of cell seeding effects on 3D flow field in the vicinity of cells.

In a more recent study [64], numerical approaches were applied to study bone tissue engineering in bioreactor systems. In particular, the effect of the bioreactor environment on the cell culture in bones was examined by identifying a beneficial range of operational settings to stimulate tissue production. Several parameters were examined numerically such as scaffold properties, flow velocities, and nutrient and growth factor concentrations. However, further numerical research is required to: (i) optimize the bioreactor operating conditions and establish reproducible protocols leading to more functional bone grafts and (ii) better understand how the different parameters (e.g. nutrient distribution, flow shear stress and fluid flow) inside dynamic bioreactors influence *in vitro* tissue morphogenesis [65]. A constant collaboration between modeling and experimental work is also required as experimental procedures inform models, which, in turn, suggest improvements to the *in vitro* protocols.

Numerical models on bioreactor systems and the tissues cultured in them can also be categorized into three main groups including continuum, discrete, and continuum macro-scale models. Initially, continuum models provide a description of the averaged behavior of the system and naturally represent large populations of cells and accommodate the fluid dynamics theory with the Navier–Stokes (NS) equations. They are based on the assumption that the cells and tissues are a number of continuous materials able to occupy the same space. Homogenization techniques can be applied to derive the constitutive relationships describing the properties of and interactions between the material phases. On the other hand, discrete models consider cells as a combination of separate entities. Computational algorithms, which define cell function using preset criteria, are applied to model the evolution of the cell lineage. Complex functions including intercellular interaction and signaling processes, movement of individual cells, as well as phenomena such as cellular division and self-organization into clusters can be evaluated using discrete models. However, the aforementioned modeling categories have both advantages and disadvantages. For instance, although discrete models preserve a high level of detail about the stochastic nature of cell-level processes, their main limitation is finding the required values for the unknown parameters, which are nearly impossible to measure experimentally at the level of individual cells. Continuum macro-scale models provide significant information about the mechanical and chemical environment even for complex geometries [66]. They allow for simplifications to analytical or asymptotic solutions, reducing the dependence on numerical computations and facilitating comparisons to experiments. The main disadvantage of these models is the loss of individual cell-scale data. It should be mentioned that as each category of models approaches the modeling of the key culture components differently, one modeling approach may be more appropriate than the others for a specific setup.

### 4.3.4 P4 Medicine

As the nature of the human body system is very complex, comprehensive computational models are required in order to achieve the needs of P4 medicine. To develop such models, new sensors, methods, platforms, and unique biomarkers for diagnosis and therapeutic outcome prediction have been presented [67–72].

In order to develop disease-specific models applicable to personalizing therapeutic interventions, it is essential to incorporate biomarkers from DNA sequencing and improve the quality of data [67]. Biomarkers play a crucial role in clinical medicine and personalized medicine. Modern omics technologies (e.g. genome sequencing) allow molecular profiling of individual patients. There are many definitions of biomarkers, but they all share the notion that they can objectively assess, quantify, and sometimes prognosticate the medical state of a patient [69]. This definition considers not only changes in patients' internal parameters but also interaction with external parameters (e.g. lifestyle, treatment, and environmental exposure). However, in some cases, biomarkers based on these technologies are unable to follow the progression of a disease or response to therapy. Also, for specific diseases, such as cancer, cell heterogeneity in a single tumor makes detection of low-level mutations difficult [68]. Computational models are considered a new conceptual approach to biomarker discovery and design. Being able to integrate a large amount of information, including dynamic information, computational models can simulate disease evolution and response to therapy with high sensitivity and specificity. By populating these models with personal data, they can be highly individualized and provide advanced new tools in personalized medicine [70].

Numerical modeling has been successfully applied to study the dynamic behavior of biological systems, especially the signal transduction networks (STNs) that control main cellular processes (e.g. cell cycle, cell growth, inflammation, processes disturbed by disease, and cell death). These models allow the systematic assessment of STN perturbations and adaptive responses through *in silico* simulations. For instance, in [70], a neuroblastoma study was presented which revealed that dynamic mathematical models can serve as highly useful biomarkers for personalized medicine. A significant advantage of this approach is that the models can identify STN control nodes. This is an important distinction versus biomarker signatures, which identify correlations rather than causalities.

In addition, computational modeling approaches have been used to assist the process of stratification biomarker discovery for cardiovascular diseases and cancer evolution [71–73]. Concerning cancer evolution, the first models in cancer essentially aimed at describing the growth kinetics of the tumors [73]. Beyond the simple description of growth, the models were further upgraded to account for more detailed information such as the shape of the tumor, the state of the tumor cells (proliferative, quiescent/dormant, or necrotic), the nature and state of the environment (vascularized, degraded, fibrous, hypoxic, acidic, etc.) and also to predict how all these elements evolve when therapy is also considered [74]. For the time being, few clinical studies have been presented including therapies which have been defined or optimized from a mathematical model, while it is most usual to establish models that retrospectively predict how the tumor would grow or explain why a therapy was successful or not.

## References


1 Moeendarbary, E. and Harris, A.R. (2014). Cell mechanics: principles, practices, and prospects. *WIREs Syst. Biol. Med.* 6: 371–388.

2 Brodland, G. (2015). How computational models can help unlock biological systems. *Semin. Cell Dev. Biol.* 47, 48: 62–73.

3 Dao, M., Lim, C.T., and Suresh, S. (2003). Mechanics of the human red blood cell deformed by optical tweezers. *J. Mech. Phys. Solids* 51: 2259–2280.

4 Cai, X., Xie, J., Yao, Y. et al. (2017). Angiogenesis in a 3D model containing adipose tissue stem cells and endothelial cells is mediated by canonical Wnt signaling. *Bone Res.* 5: 17048, https://doi.org/10.1038/boneres.2017.48.

**5** Jackson, R.C., Di Veroli, G.Y., Koh, S.-B. et al. (2017). Modelling of the cancer cell cycle as a tool for rational drug development: a systems pharmacology approach to cyclotherapy. *PLoS Comput. Biol.* 13 (5): e1005529, https://doi.org/10.1371/journal.pcbi.1005529.

**6** Ramis-Conde, I., Chaplain, M.A.J., and Anderson, A.R.A. (2008). Mathematical modelling of cancer cell invasion of tissue. *Math. Comput. Modell.* 47: 533–545.

**7** Simbawa, E. (2017, 3676295). Mechanistic model for cancer growth and response to chemotherapy. *Comput. Math. Methods Med.* 2017: 1–7. https://doi.org/10.1155/2017/3676295.

**8** Enderling, H. and Rejniak, K.A. (2013). Simulating cancer: computational models in oncology. *Front. Oncol.* 3, 233: 1–2.

**9** Metzcar, J., Wang, Y., Heiland, R., and Macklin, P. (2019). A review of cell-based computational modelling in cancer biology. *JCO Clin. Cancer Inform.* 3: 1–13, https://doi.org/10.1200/CCI.18.00069.

**10** Scheiner, S., Pivonka, P., Smith, D.W. et al. (2014). Mathematical modelling of postmenopausal osteoporosis and its treatment by the anti-catabolic drug denosumab. *Int. J. Numer. Methods Biomed. Eng.* 30 (1): 1–27.

**11** Martin, J.A., Anderson, D.D., Goetz, J.E. et al. (2017). Complementary models of post-traumatic osteoarthritis reveal cellular responses to contact stress that damage articular cartilage and present targets for intervention. *J. Orthop. Res.* 35 (3): 515–523.

**12** Guo, M., Cai, Y., Yao, X., and Li, Z. (2018). Mathematical modelling of atherosclerotic plaque destabilization: role of neovascularization and intraplaque hemorrhage. *J. Theor. Biol.* 450: 53–65. https://doi.org/10.1016/j.jtbi.2018.04.031.

**13** Rajagopal, V., Holmes, W.R., and Lee, P.V.S. (2017). Computational modelling of single-cell mechanics and cytoskeletal mech anobiology. *Wiley Interdiscip. Rev. Syst. Biol. Med.* 10 (2): e1407.

**14** Rodriguez, M.L. and Sniadecki, N. (2014). Computational modelling of cell mechanics. In: *Computational Modelling of Biomechanics and Biotribology in the Musculoskeletal System* (ed. Z. Jin), 93–140. Woodhead Publishing. https://www.sciencedirect.com/science/article/pii/B9780857096616500048.

**15** Charras, G.T., Yarrow, J.C., Horton, M.A. et al. (2005). Non equilibration of hydrostatic pressure in blebbing cells. *Nature* 435: 365–369.

**16** Fung, Y. (1993). *Biomechanics: Mechanical Properties of Living Tissues. New York: Springer.*

**17** Wakatsuki, T., Schwa, B., Thompson, N.C., and Elson, E.L. (2001). Effects of cytochalasin D and latrunculin B on mechanical properties of cells. *J. Cell Sci.* 114: 1025–1036.

**18** Cuerrier, C.M., Benoit, M., Guillemette, G., and Grandbois, M. (2009). Real-time monitoring of angiotensin II-induced contractile response and cytoskeleton remodelling in individual cells by atomic force microscopy. *Pflügers Arch.* 457: 1361–1372.

**19** Balland, M., Desprat, N., Icard, D. et al. (2006). Power laws in microrheology experiments on living cells: comparative analysis and modelling. *Phys. Rev. E Stat. Nonlin. Soft Matter Phys.* 74: 021911.

**20** Steward, R.L., Cheng, C.-M., Jonathan, D.Y. et al. (2011). Mechanical stretch and shear flow induced reorganization and recruitment of fibronectin in fibroblasts. *Sci. Rep.* 1: 1–12.

**21** Harris, A.R., Bellis, J., Khalilgharibi, N. et al. (2013). Generating suspended cell monolayers for mechanobiological studies. *Nat. Protoc.* 8: 2516–2530.

**22** Lu, H., Koo, L.Y., Wang, W.M. et al. (2004). Microfluidic shear devices for quantitative analysis of cell adhesion. *Anal. Chem.* 76: 5257–5264.

**23** Gossett, D.R., Tse, H.T.K., Lee, S.A. et al. (2012). Hydrodynamic stretching of single cells for large population mechanical phenotyping. *Proc. Natl. Acad. Sci. U.S.A.* 109: 7630–7635.

**24** Fu, J., Wang, Y., Yang, M.T. et al. (2010). Mechanical regulation of cell function with geometrically modulated elastomeric substrates. *Nat. Methods* 7: 733–736.

**25** Saier, M.H. and Stiles, C.D. (1975). Introduction: cell structure and function. In: *Molecular Dynamics in Biological Membranes*. New York, NY: *Heidelberg Science Library. Springer*. https://link.springer.com/chapter/10.1007/978-1-4613-9399-3_1.

**26** Brown, G., Butler, P., Chang, D. et al. (2012). Cellular and molecular bioengineering: a tipping point. *Cell. Mol. Bioeng.* 5 (3): 239–253.

**27** Lim, C.T., Zhou, E.H., and Quek, S.T. (2006). Mechanical models for living cells – a review. *J. Biomech.* 39 (2): 195–216.

**28** Winslow, R.L., Trayanova, N., Geman, D., and Miller, M.I. (2012). Computational medicine: translating models to clinical care. *Sci. Transl. Med.* 4 (158): 1–21.

**29** Hatzikirou, H., Chauviere, A., Bauer, A.L. et al. (2012). Integrative physical oncology. *Wiley Interdiscip. Rev. Syst. Biol. Med.* 4: 1–14.

**30** Deisboeck, T.S., Wang, Z., Macklin, P., and Cristini, V. (2011). Multiscale cancer modelling. *Annu. Rev. Biomed. Eng.* 13: 127–155.

**31** Macklin, P., Edgerton, M.E., Thompson, A.M., and Cristini, V. (2012). Patient-calibrated agent-based modelling of ductal carcinoma in situ (DCIS): from microscopic measurements to macroscopic predictions of clinical progression. *J. Theor. Biol.* 301: 122–140.

**32** Aldridge, B.B., Burke, J.M., Lauffenburger, D.A., and Sorger, P.K. (2006). Physicochemical modelling of cell signalling pathways. *Nat. Cell Biol.* 8: 1195–1203.

**33** Bhatia, S.N. and Ingber, D.E. (2014). Microfluidic organson-chips. *Nat. Biotechnol.* 32: 760–772.

**34** Novak, R., Didier, M., Calamari, E. et al. (2018). Scalable fabrication of stretchable, dual channel, microfluidic organ chips. *J. Vis. Exp.* 140: e58151, https://doi.org/10.3791/58151.

**35** A. Sontheimer-Phelps, B. A. Hassell, and D. E. Ingber, "Modelling cancer in microfluidic human organs-on-chips", *Nat. Rev. Cancer*, https://doi.org/10.1038/s41568-018-0104-6, 2019.

**36** Zhang, B. and Radisic, M. (2017). Organ-on-a-chip devices advance to market. *Lab Chip* 17: 2395–2420.

**37** Kimura, H., Sakai, Y., and Fujii, T. (2018). Organ/body-on-a-chip based on microfluidic technology for drug discovery. *Drug Metab. Pharmacokinet.* 33 (1): 43–48. *doi: 10.1016/j.dmpk.2017.11.003. Epub 2017 Nov 13. PMID: 29175062.*

**38** Volpe, A., Paiè, P., Ancona, A. et al. (2017). A computational approach to the characterization of a microfluidic device for continuous size-based inertial sorting. *J. Phys. D: Appl. Phys.* 50 (25): 255601. https://doi.org/10.1088/1361-6463/aa7086.

**39** Boy, D., Gibou, F., and Pennathur, S. (2008). Simulation tools for lab on a chip research: advantages, challenges, and thoughts for the future. *Lab Chip* 8 (9): 1424.https://doi.org/10.1039/b812596c.

**40** Osher, S. and Sethian, J. (1988). Fronts propagating with curvature-dependent speed: algorithms based on Hamilton-Jacobi formulations. *J. Comput. Phys.* 79: 12–49.

**41** Peyret, R. (2002). *Spectral Methods for Incompressible Viscous Flow*. New York, https://doi.org/10.1007/978-1-4757-6557-1: Springer-Verlag.

**42** Lin, H., Storey, B.D., Oddy, M.H. et al. (2004). Instability of electrokinetic microchannel flows with conductivity gradients. *Phys. Fluids A* 16.

**43** Cheng, B., Lin, M., Huang, G. et al. (2017). Cellular mechanosensing of the biophysical microenvironment: a review of mathematical models of biophysical regulation of cell responses. *Phys. Life Rev.* 22, 23: 88–119. https://doi.org/10.1016/j.plrev.2017.06.016.

**44** Tse, J.M., Cheng, G., Tyrrell, J.A. et al. (2011). Mechanical compression drives cancer cells toward invasive phenotype. *Proc. Natl. Acad. Sci. U.S.A.* 109 (3): 911–916.

**45** Case, L.B., Baird, M.A., Shtengel, G. et al. (2015). Molecular mechanism of vinculin activation and nanoscale spatial organization in focal adhesions. *Nat. Cell Biol.* 17: 547–555.

**46** Liu, J., Wang, Y., Goh, W.I. et al. (2015). Talin determines the nanoscale architecture of focal adhesions. *Proc. Nat. Acad. Sci. U.S.A.* 112: 4864–4873.

**47** Qian, J., Liu, H., Lin, Y. et al. (2013). A mechanochemical model of cell reorientation on substrates under cyclic stretch. *PLoS One* 8 (6): e65864. https://doi.org/10.1371/journal.pone.0065864.

**48** Kaunas, R. and Deguchi, S. (2011). Multiple roles for Myosin II in tensional homeostasis under mechanical loading. *Cell. Mol. Bioeng.* 4 (2): 182–191.

**49** Besser, A. and Schwarz, U.S. (2007). Coupling biochemistry and mechanics in cell adhesion: a model for inhomogeneous stress fiber contraction. *New J. Phys.* 9 (11): 425–425.

**50** Kong, D., Ji, B., and Dai, L. (2008). Stability of adhesion clusters and cell reorientation under lateral cyclic tension. *Biophys. J.* 95 (8): 4034–4044.

**51** Yang, L., Erdogan, B., Ao, M. et al. (2017). Biomechanics of cell reorientation in a three-dimensional matrix under compression. *Exp. Cell. Res.* 350 (1): 253–266.

**52** Tao, J. and Sun, S.X. (2015). Active biochemical regulation of cell volume and a simple model of cell tension response. *Biophys. J.* 109 (8): 1541–1550.

**53** Davies, P.F. (1995). Flow-mediated endothelial mechanotransduction. *Physiol. Rev.* 75 (3): 519–560.

**54** Civelekoglu-Scholey, G., Orr, A.W., Novak, I. et al. (2005). Model of coupled transient changes of Rac, Rho, adhesions and stress fibers alignment in endothelial cells responding to shear stress. *J. Theor. Biol.* 232 (4): 569–585.

**55** Groh, C.M., Hubbard, M.E., Jones, P.F. et al. (2014). Mathematical and computational models of drug transport in tumours. *J. R. Soc. Interface* 11, https://doi.org/10.1098/rsif.2013.1173.

**56** Hubbard, M.E., Jove, M., Loadman, P.M. et al. (2017). Drug delivery in a tumor cord model: a computational simulation. *R. Soc. Open Sci.* 4 (5): 170014. https://doi.org/10.1098/rsos.170014.

**57** Steuperaert, M., D'Urso Labate, G.F., Debbaut, C. et al. (2017). Mathematical modelling of intraperitoneal drug delivery: simulation of drug distribution in a single tumor nodule. *Drug Delivery* 24 (1): 491–501.

**58** Díaz-Zuccarini, V. and Lawford, P.V. (2010). An in-silico future for the engineering of functional tissues and organs. *Organogenesis* 6 (4): 245–251.

**59** Song, M.J., Dean, D., and Knothe Tate, M.L. (2012). Computational modelling of tissue engineering scaffolds as delivery devices for mechanical and mechanically modulated signals. In: *Computational Modeling in Tissue Engineering. Studies in Mechanobiology, Tissue Engineering and Biomaterials*, vol. 10 (ed. L. Geris), 127–143. Berlin, Heidelberg: Springer. https://doi.org/10.1007/8415_2012_138.

**60** Song, M., Brady-Kalnay, S., Dean, D., and Knothe Tate, M. L. (2012). Relating stem cell shape to fate commitment by mapping cell surface strains *in situ*. *ORS Annual Meeting*. Moscone West, San Francisco, California: ORS: Orthopaedic Research Society.

**61** Chang, H. and Knothe Tate, M.L. (2011). Structure-function relationships in the stem cell's mechanical world B: emergent anisotropy of the cytoskeleton correlates to volume and shape changing stress exposure. *Mol. Cell Biomech.* 8: 297–318.

**62** Inoue, Y., Deji, T., Shimada, Y. et al. (2010). Simulations of dynamics of actin filaments by remodelling them in shear flows. *Comput. Biol. Med.* 40 (11, 12): 876–882.

**63** Song, M.J., Dean, D., and Knothe Tate, M.L. (2010). *In situ* spatiotemporal mapping of flow fields around seeded stem cells at the subcellular length scale. *PLoS One* 5 (9) https://doi.org/10.1371/journal.pone.0012796.

**64** Burova, I., Wall, I., and Shipley, R.J. (2019). Mathematical and computational models for bone tissue engineering in bioreactor systems. *J. Tissue Eng.* 10, https://doi.org/10.1177/2041731419827922.

**65** Sladkova, M. and de Peppo, G. (2014). Bioreactor systems for human bone tissue engineering. *Processes* 2 (2): 494–525.

**66** O'Dea, R.D., Byrne, H.M., and Waters, S.L. (2012). Continuum modelling of *in vitro* tissue engineering: a review. In: *Computational Modeling in Tissue Engineering, Studies in Mechanobiology, Tissue Engineering and Biomaterials*, vol. 10 (ed. L. Geris). Berlin/Heidelberg: Springer.

**67** Reza Soroushmehr, S.M. and Najarian, K. (2016). Transforming big data into computational models for personalized medicine and health care. *Dialogues Clin. Neurosci.* 18 (3): 339–343.

**68** Kruglyak, K.M., Lin, E., and Ong, F.S. (2016). Next-generation sequencing and applications to the diagnosis and treatment of lung cancer. *Adv. Exp. Med. Biol.* 123–136.

**69** Strimbu, K. and Tavel, J.A. (2010). What are biomarkers? *Curr. Opin. HIV AIDS* 5 (6): 463–466.

**70** Kolch, W. and Fey, D. (2017). Personalized computational models as biomarkers. *J. Pers. Med.* 7 (3) https://doi.org/10.3390/jpm7030009.

**71** Syed, Z., Stultz, C.M., Scirica, B.M., and Guttag, J.V. (2011). Computationally generated cardiac biomarkers for risk stratification after acute coronary syndrome. *Sci. Transl. Med.* 3 (102) https://doi.org/10.1126/scitranslmed.3002557.

**72** Younesi, E. and Hofmann-Apitius, M. (2013). From integrative disease modelling to predictive, preventive, personalized and participatory (P4) medicine. *EPMA J.* 4 (1): 23. https://doi.org/10.1186/1878-5085-4-23.

**73** Lesart, A.-C., van der Sanden, B., Hamard, L. et al. (2012). On the importance of the submicrovascular network in a computational model of tumor growth. *Microvasc. Res.* 84 (2): 188–204.

**74** Jarrett, A.M., Lima, E.A.B.F., Hormuth, D.A. et al. (2018). Mathematical models of tumor cell proliferation: a review of the literature. *Expert Rev. Anticancer Ther.* 1–16, https://doi.org/10.1080/14737140.2018.1527689.

# 5

# Computational Modeling at Tissue Level

## 5.1 Introduction

Biological tissues are complex structures subjected to a number of external stimuli such as electrical signals, mechanical forces, and heating. The structure of biological tissues plays an important role in their response to those stimuli. The cells composing these tissues are able to sense the stimuli and modify or adapt the tissue matrix structure. Understanding the physical behavior of tissues and their adaptation are important from a scientific perspective and in many clinical areas such as cardiovascular diseases, musculoskeletal disorders, and skin cancer.

The term tissue refers to a group of cells with a similar structure working together for a specific function. The cells within a tissue have common morphological features and are arranged in such a manner that achieves the tissue's functions. More specifically, a tissue is composed of: (i) a certain group of cells (osteoblasts, fibroblasts, epithelial cells, etc.) from the same origin, forming the "building blocks" of the tissue, (ii) an extracellular matrix (ECM), forming the "cement" of the tissue, offering a 3D structural and biochemical support, and (iii) in most cases, a vascular network that allows transport of oxygen and nutrients [1].

Although there are several types of cells in the human body, they are organized into four main categories of tissues: epithelial, connective, muscle, and nervous tissues [2]. Figure 5.1 presents the four categories of human tissues. Each one of these categories is characterized by specific functions which contribute to the overall health of the body. The cell type, cell density, as well as the type of matrix may vary significantly depending on the tissue type. A disruption of the tissue structure might be caused due to injury or disease and can be identified by a histological examination.

*Multiscale Modelling in Biomedical Engineering*, First Edition. Dimitrios I. Fotiadis, Antonis I. Sakellarios, and Vassiliki T. Potsika.

Published 2023 by John Wiley & Sons, Inc.

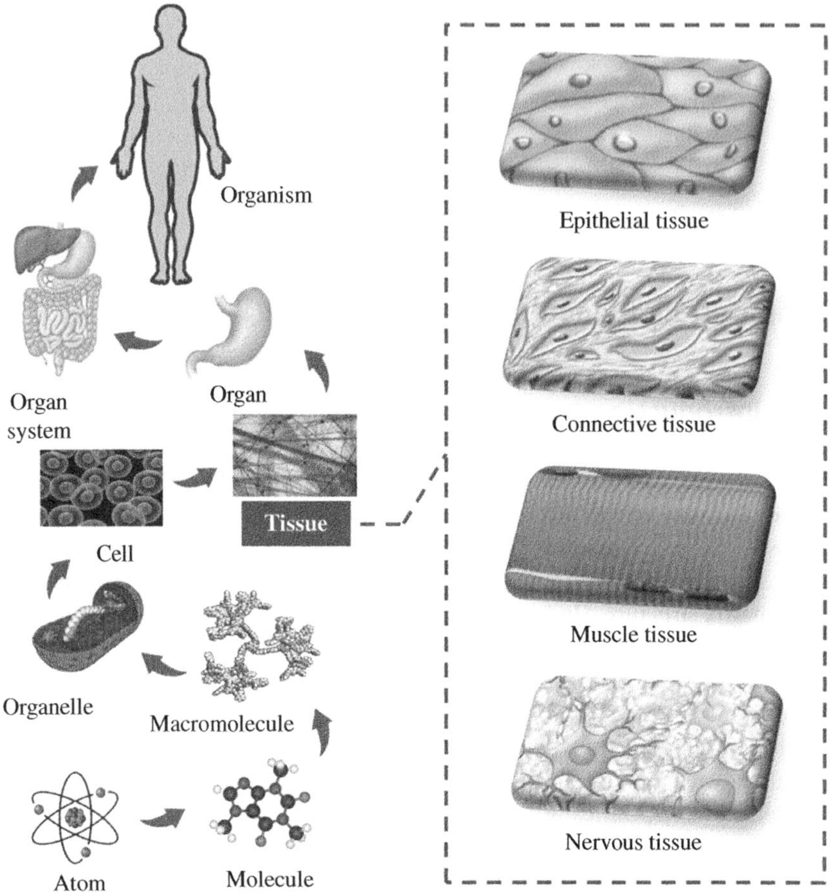


**Figure 5.1** The four types of human tissues: epithelial, connective, muscle, and nervous tissue. *Source:* SciePro/Adobe Stock.

The epithelial tissue or epithelium consists of the group of cells which covers the exterior surfaces of the body, lines internal cavities, and passageways, and forms certain glands. An example is the epidermis, the outermost layer of the skin. The epithelial tissue has a number of functions including protection against abrasion, invasion by pathogens, chemical stress, and radiation damage. There are four main categories of epithelial tissues: (i) squamous, (ii) stratified squamous, (iii) columnar, and (iv) cuboidal, which are further discussed in Section 5.2.1.

The connective tissue binds the cells and organs together aiming to protect, support, and integrate the different parts of the body. The cells of connective tissue include fibroblasts, adipocytes, macrophages, mast cells, and leucocytes. Compared to epithelial tissue, which is composed of cells closely packed with little or no extracellular space in between, connective tissue cells are dispersed in a matrix. This matrix usually includes a large amount of extracellular material produced by the connective tissue cells that are embedded within it. The properties of the connective tissue depend on variations in the composition of the ECM. For instance, if the matrix is calcified, it can form bone or teeth.

The muscle tissue is excitable, responds to stimulation, and functions to produce force and motion. Muscle cells contain myosin and protein filaments of actin which slide past one another, producing a contraction which modifies the shape and length of the cell. There are three different types of muscle tissue including smooth, cardiac, and skeletal tissue. The smooth muscle fibers can be found in walls of hollow visceral organs (except for the heart) and appear spindle-shaped. The cardiac muscle cells are located in the walls of the heart and appear striated. Skeletal muscle fibers are located in the muscles attached to the skeleton and can be divided into two subtypes, namely fast and slow twitch fibers. Cardiac and smooth muscles contract without conscious thought (involuntary), whereas the skeletal muscles contract upon command.

Nervous or neural tissue is the main tissue component of the nervous system which controls and regulates body activity and functions. The nervous tissue is also excitable, allowing the propagation of electrochemical signals in the form of nerve impulses that communicate between different regions of the body. It is made up of different types of nerve cells having an axon which is the long stem-like part of the cell that sends action potential (AP) signals to the next cell. The nervous tissue can be grouped into two subcategories known as neurons and neuroglia. Neurons or nerves transmit electrical impulses. On the other hand, neuroglia have many functions such as supporting and protecting neurons, but they are not able to transmit electrical pulses.

Due to the complexity of the different types of tissue and the need to quantify the tissue properties and response in healthy and pathological conditions, several researchers have used computer modeling techniques to simulate tissue behavior. Nevertheless, biological tissues differ in many ways from typical engineering materials. First, biological tissues are extremely heterogeneous within a single body and among different media. Second, the information on tissue structure is mainly derived from histology or imaging modalities. Third, tissues are hierarchical structures with many different scales, which are more complex compared to the typical engineering media. Finally, tissues are able to modify their structure in response to external stimuli.

Sections 5.2–5.5 describe the composition and properties of the different tissue types and cutting-edge research directions of the current literature on computational modeling.

## 5.2 Epithelial Tissue

### 5.2.1 Composition and Properties of Epithelial Tissue

The epithelial tissues are essentially large sheets of cells covering all the surfaces of the body exposed to the outside world and lining the outside of organs [2]. They form extensive and thin sheets varying from 1 to 10 cells in thickness and they are usually supported by connective tissue. Epithelial and connective tissues are separated by a membrane with a thickness of ~40 nm. The epithelial layer's thickness is up to ~500 μm and may be folded or flat. Stem cells usually reside in "niches" which are at the bottom of intestinal crypts and, in skin, at the base of rete ridges [3]. Figure 5.2 depicts the four different types of epithelial cells classified according to their shape in the normal vertical section as squamous, stratified squamous, columnar, and cuboidal.

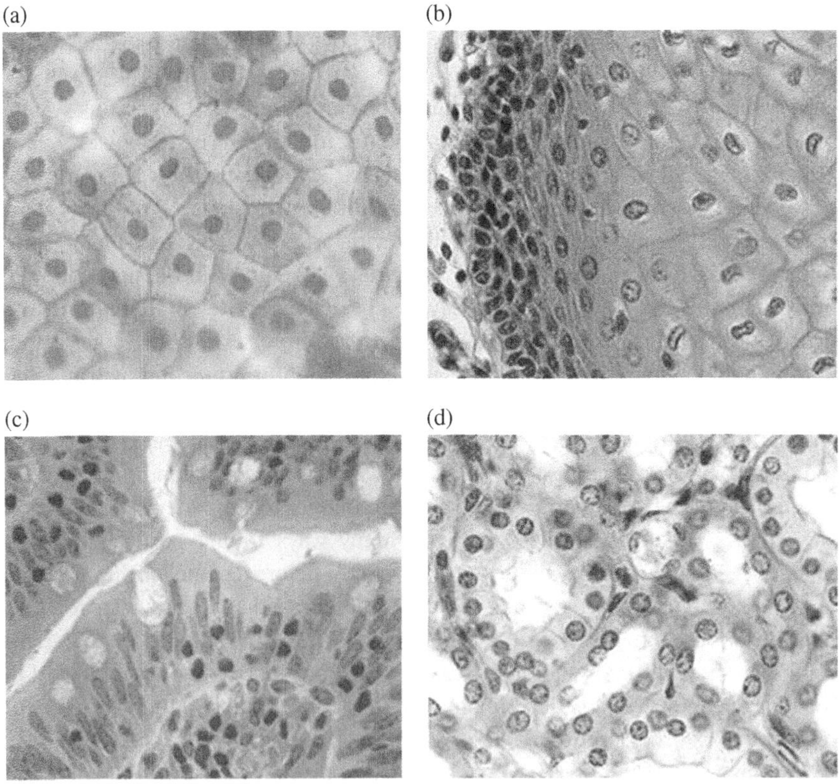


**Figure 5.2** The four categories of epithelial tissues: (a) squamous, (b) stratified squamous, (c) columnar, and (d) cuboidal. *Source:* DrWD40/Adobe Stock.

Epithelial cells are connected both mechanically and biochemically. There are three basic categories of connections which allow varying degrees of interaction between the cells: tight junctions, anchoring junctions, and gap junctions [2]. Tight junctions between the cell membranes stop the passage of molecules through the epithelial layer. This enables the epithelial cells to act as selective barriers. An anchoring junction includes several types of cell junctions which help stabilize epithelial tissues. Anchoring junctions are common on the lateral and basal surfaces of cells providing flexible and strong connections. Gap junctions allow small molecules and ions to pass between the cytoplasm of neighboring cells. These junctions allow electrical and metabolic coupling of adjacent cells, which coordinates function in large groups of cells [2, 3].

The cell shape and the nature of adhesion to its neighbors depend on both the location of the tissue (e.g. bladder and skin) and the position of the cell within the tissue. The mechanical environment of the cell, which is partly determined by interactions with neighboring cells and the substrate, and partly by externally applied forces (e.g. muscle layers in the wall of the intestine), can alter gene expression as a result of mechanotransduction [3]. A key issue is to understand and model the role of tissue mechanics and the impact of forces, displacements, and their time evolution on the coordinated changes of cell shape and position which determine morphogenetic flows at the tissue level [4].

### 5.2.2 Computational Modeling of Epithelial Tissue

During the last decade, remarkable progress has been made in experimental studies of epithelial dynamics due to the increased molecular understanding of epithelial cell biology and advances in imaging modalities. Computational modeling offers a complementary tool to experimental research, which could potentially contribute to the understanding of the role of cellular signaling in morphogenetic processes and mechanics [5]. Although there is extensive literature on computational modeling of epithelial tissues at different hierarchical levels, there are essentially two main approaches: (i) to start at the level of the tissue (i.e. a length scale of the order of 1 mm) and develop generalized equations for tissue behavior (a continuum approach), (ii) to start at the level of the cell (i.e. a length scale of the order of 10 μm) and study the tissue behavior as an emergent property of cellular behavior (an individual-based approach) [3].

Continuum mechanics describes tissue mechanical state and shape changes in terms of strain and stress. It accounts for individual cell properties using only a few spatially averaged material parameters. Several models describe epithelial tissues using continuum mechanics [4, 6]. One precondition is the existence of a mesoscopic scale defining a domain over which cell properties are well defined. This description further relies on the assumption that the tissue mechanical state can be quantified, at the same mesoscopic scale, by two variables: (i) the stress characterizing in which directions, and to what extent, the domain is under tension or under compression, and (ii) the strain characterizing how far the domain geometry is from the geometry of a relaxed state. Such description has the advantage of accounting for individual cell properties using only a few spatially averaged parameters, which determine, for instance, how fast and through which succession of states the domain reacts to external solicitations [4].

At the cellular level, tissue deformation is determined by changes in cell position, shape, and number, which are triggered by biochemical forces and signaling generated by the cells. While the tissue stress can affect cell morphogenetic processes through the changes in molecular activity and localization, cell morphogenetic processes generate stress. Nevertheless, the mechanisms by which the shape of a tissue evolves following these multiscale feedback processes remain unclear. To this end, a two-dimensional continuum model of epithelial tissue mechanics was formulated in [7] based on cellular-level mechanical ingredients and cell morphogenetic processes, including cellular shape changes and cellular rearrangements. This model incorporated stress and deformation tensors, which can be compared with experimental data. Focusing on the interplay between cell shape changes and cell rearrangements, the dynamical behavior underlying passive relaxation, active contraction-elongation, and tissue shear flow was elucidated, including a mechanism for contraction-elongation, whereby tissue flows perpendicularly to the axis of cell elongation. This study provided an integrated scheme for the understanding of the orchestration of morphogenetic processes in individual cells to achieve epithelial tissue morphogenesis.

Moreover, a popular approach used to describe tightly packed epithelial cell sheets is the vertex model framework [8–12]. Vertex models are a class of models that consider cells as individual objects, approximated by two-dimensional polygons representing cellular interfaces, in which each vertex moves in response to forces due to growth, pressure, and interfacial tension, within each cell. Vertex models are used to study cellular processes within epithelia, including cell motility (the capacity of cells to translocate onto a solid substratum), adhesion (the process by which cells interact and attach to neighboring cells through specialized molecules of the cell surface), mitosis (a process of cell duplication, or reproduction, during which one cell gives rise to two genetically identical daughter cells), and delamination (the process by which individual epithelial cells detach from an epithelial layer). The most common method of simulating vertex models is to solve equations of motion for each vertex. According to [8], it is standard to make the following two simplifying approximations: first, that motion may be described by considering each vertex to be embedded in a viscous medium that applies a drag force on it with mobility coefficient $\eta$; and second, that inertia is vanishing.

This leads to first-order dynamics, with the evolution of the position $\mathbf{x}_i$ of vertex $i$ determined by:

$$\eta \frac{d\mathbf{x}_i}{dt} = \mathbf{F}_i, \tag{5.1}$$

where $\mathbf{F}_i(t)$ denotes the total force acting on vertex $i$ at time $t$. The mobility coefficient $\eta$ determines the timescale over which mechanical relaxation occurs. One difference among vertex models in the literature lies in the definition of the force $\mathbf{F}_i$ in Eq. (5.1). The choice of form for the function $\mathbf{F}_i$ reflects which forces are thought to dominate epithelial mechanics for the system being studied. Some commonly modeled forces include tension or elastic forces due to the combined action of a cell's actomyosin cortex and adherens junctions, or pressure, due to hydrostatic pressure in some cases or some type of membrane conservation in others. Vertex models will most probably play an important role in the future to gain a deeper understanding of how forces generated inside cells affect the shape and mechanics of epithelial sheets [11]. As three-dimensional studies open new perspectives for a more realistic representation of human tissues, it will also be interesting to explore in more detail the mechanical interplay between epithelia and the ECM which requires close collaborations between experimentalists and theoretical biophysicists [11, 12].

Finally, another recent study investigated how epithelial tissues develop indicating the need for a mechanical model coupling the inside to the outside of the cell [13]. An epithelial tissue model was developed based on insect embryos, the epithelial blastoderm, which forms as a single layer on top of the yolk. In the epithelial blastoderm, cellular boundaries exactly coincide with the Voronoi tessellation of their nuclei (a Voronoi tessellation or diagram is a partition of a plane into regions close to each of a given set of objects). The same behavior was shown in a simple mechanical model of the tissue, based on cell growth and mechanical interaction.

## 5.3 Connective Tissue

### 5.3.1 Composition and Properties of Connective Tissue

Connective tissue is the most abundant and diverse type of human tissue [14]. It can be broadly subdivided into connective tissue proper and specialized connective tissue. Connective tissue proper consists of (Figure 5.3a,b): (i) loose connective tissue which holds organs in place and attaches epithelial tissue to other underlying tissues, and (ii) dense connective tissue which helps in muscles' attachment to

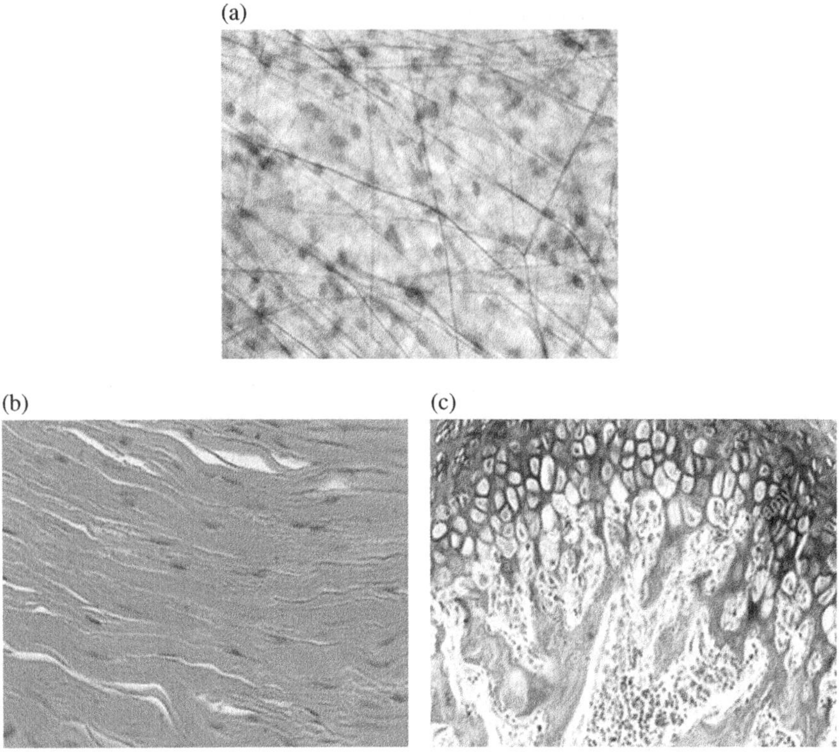


**Figure 5.3** The three main categories of connective tissues: (a) loose, (b) dense, and (c) specialized connective tissue.

bones and link bones together at the joints. Special connective tissue consists of reticular connective tissue, adipose tissue, cartilage, bone, and blood (Figure 5.3c).

Although connective tissue is diverse, the main components of all connective tissues are ground substance, fibers, and cells. The ground substance is a mixing complex between glycosaminoglycans, proteoglycans, and glycoproteins (mainly laminin and fibronectin). The purpose of the ground substance is to allow for the exchange of cellular nutrients between cells and capillaries [15].

Fibers are another main component of connective tissue found in the ECM. There are three basic fiber types in all connective tissues and the amount of the presence of each type reflects the classification and function of the specific tissue. The three fiber types are: (i) collagenous fibers which are large, strong fibers, composed of Type I collagen that provides high tensile strength to the ECM. They are found in dense and loose connective tissue and (ii) reticular fibers which are delicate, thin fibers composed of Type III collagen which cross-link to form a supporting meshwork in the reticular lamina of the basement membrane. They are found in soft tissues such as the liver, bone marrow, spleen, and lymph nodes, (iii) elastic fibers which are thin, branching fibers made of elastin that provide stretch and recoil to the ECM. They are found in tissues such as aorta, lungs, skin, and vocal cords [14].

Common cells of the connective tissue are fibroblasts, macrophages, adipocytes, leukocytes, and mast cells [14]. In most connective tissues, the matrix constituents are secreted by fibroblasts, but in several certain specialized types of connective tissues, like cartilage and bone, these components are secreted by chondroblasts and osteoblasts [16].

The mechanical properties of connective tissue have been investigated in several studies [17–19]. The mechanical strength of connective tissue varies significantly, from the stiffness and hardness of bone to the typical soft tissue behavior that tendons exhibit when stretched. Tendons resist tension and do not stretch making them ideal for linking muscle to bone. Cartilage resists compression. Large blood vessels can withstand stretch and recoil in response to changes in blood pressure. Connective tissue varies in mechanical strength as far as the organization and density of the collagen fibers are concerned. Tendon consists of a high ratio of collagen to number of cells, while the collagen fibers are organized in parallel arrays along the lines of tension. This provides maximal resistance to external forces. The dermis of skin contains a large amount of collagen fibers that are less organized and oriented in multiple directions. This allows skin to resist tension in different directions but limits its overall mechanical strength. Connective tissue in organs contains much less collagen and is more cellular. Organs, such as the small intestine, are structurally weaker than tendon. This is due to the fact that they require connective tissue to provide metabolic and immune support implying that the connective tissue must contain blood vessels, macrophages, lymphocytes. These cells are absent in the tendon allowing it to pack in more fibers and their presence in the connective tissue of most organs means fewer fibers [17]. The in-depth understanding of the connective tissue's mechanical properties is a major research direction which might provide answers to significant clinical problems such as the effect of ageing and osteoarthritis on the mechanical properties of cartilage and bone in the human knee joint [20].

As the hierarchical bone structure and material properties are further described in Chapter 8, this subsection focuses mainly on ligaments, tendons, and cartilage mechanical features. Ligaments and tendons are fibrous connective tissues with a highly hierarchical structure, from collagen fibers, to fibrils and fascicules. Tendons help execute joint motion by transmitting mechanical forces (tensions) from muscles to bones. Ligaments create joints for bones and provide stability to the joints. Unlike muscles, which are active tissues and can produce mechanical forces, tendons, and ligaments are passive tissues and cannot actively contract to generate forces. Around many joints of the human body, there is insufficient space to attach more than one or a few muscles. This requires that to accomplish a certain task, one or a few muscles must share the burden of generating and withstanding large loads with intensities (stress) even larger at regions closer to the bone attachments where the cross-sectional areas of the muscles are small. As compared to muscles, tendons are stiffer, have higher tensile strengths, and can endure larger stresses. More specifically, around the joints, the space is limited, and the attachment of muscles to bones is accomplished by tendons. Tendons can sustain large loads with small deformations. In this way, the muscles are able to transmit forces to bones without consuming energy to stretch tendons. The mechanical properties of ligaments and tendons depend on their composition which can differ significantly. The most common means of evaluating the mechanical response of tendons and ligaments is the uniaxial tension test. Tendons are considered to function in the body at strains of up to around 0.04 (yield strain). Tendons rupture at strains of about 0.1 (ultimate strain), or stresses of about 60 MPa (ultimate stress) [21].

The composition and structure of ligaments depend on their position and function in the body. They are composite materials, such as tendons, containing crimped collagen fibers surrounded by ground substance. As compared to tendons, they often contain a greater proportion of elastic fibers that accounts for their higher extensibility but lower strength and stiffness. The mechanical properties of ligaments are qualitatively similar to those of tendons. Like tendons, they are viscoelastic and exhibit hysteresis, but deform elastically up to strains of about 0.25 (about five times as much as the yield strain of tendons) and stresses of about 5 MPa. They rupture at a stress of about 20 MPa [21].

Cartilage is a two-phase material consisting of about 75% water and 25% organic solid which covers the articulating surfaces of bones at the diarthrodial (synovial) joints. The cartilage primary function is to facilitate the relative movement of articulating bones. Cartilage reduces: (i) stresses applied to bones by increasing the area of contact between the articulating surfaces and (ii) bone wear by reducing the effects of friction. Cartilage is composed of cells called chondrocytes, surrounded by an ECM in which the chondrocytes are located in spaces called gaps. The chondrocytes synthesize and secrete the organic components of the ECM, which are essentially collagen, hyaluronic acid, proteoglycans, and glycoproteins. The characteristics of the matrix distinguish hyaline and fibrous cartilage. There is also elastic cartilage, in which the elastin is a part of the ECM [22]. Collagen fibers are relatively strong and stiff in tension, while proteoglycans are strong in compression. The solid–fluid composition of cartilage makes it a viscoelastic material. The mechanical properties of cartilage under different loading conditions have been examined using various testing techniques. For instance, the response of the human patella to compressive loads has been investigated by using an indentation test in which a small cylindrical or hemispherical indenter is pressed into the articulating surface, and the resulting deformation is recorded [21].

### 5.3.2 Computational Modeling of Connective Tissue

Several studies have been presented in the current literature aiming to evaluate connective tissue using computational approaches [23–34]. In [23], the possibility to identify damage mechanisms in the tissue structure was analyzed using *in silico* models for both the anterior cruciate ligament, with fascicules forming a helix with its longitudinal axis, and the patellar tendon, with fascicules co-aligned with its longitudinal axis. Tissues of interest were modeled as cylinders submitted to uniaxial tension. The damage of tissue was modeled as either a reduction of collagen volume fraction with increased strain, assuming the number of collagen fibers sustaining load decreases as fibers fail, or a reduction of the modulus of the fibers, assuming prefailure damage of the fibers. Each damage mechanism was associated with a damage variable with different fiber stretch thresholds for damage initiation and assuming a linear variation of damage until an arbitrary failure point. The apparent behavior of the modeled tissues was significantly different as damage thresholds, damage mechanisms, and types of fascicules varied. This preliminary work showed that using a structural constitutive model to describe occurrence and propagation of structural damage in an *in silico* model of hierarchical connective tissues is a framework that can clearly differentiate between different values of damage threshold and different damage mechanisms for tissue with co-aligned or helical fascicules [23].

Wound healing was examined in [24] and numerical models were presented to investigate the biology of connective tissue repair and healing and identify key cellular processes and matrix characteristics critical to the design of soft tissue grafts. Polymer meshes of various fiber diameters at the nano (390 and 740 nm) and micro scales (1420 nm) were produced to model the various stages of repair post-tendon injury. Alignment was also introduced in the group with 740 nm diameter to model matrix undergoing biological healing rather than scar formation. The response of human tendon fibroblasts on these model substrates was evaluated over time as a function of fiber diameter and alignment. It was found that the repair models of unaligned nanoscale fibers enhanced collagen synthesis and cell growth, while these outcomes were significantly reduced in the mature repair model consisting of unaligned micron-sized fibers. Also, it was shown that the early-stage repair model of unaligned nanoscale fibers elicits a response characteristic of the proliferative phase of wound repair, while the more mature model consisting of unaligned micron-sized fibers is more representative of the remodeling phase by supporting cell organization while suppressing growth and biosynthesis. Interestingly, introduction of fiber alignment in the nanofiber model alters fibroblast response from repair to healing, implicating matrix alignment as a critical design factor for circumventing scar formation and promoting biological healing of soft tissue injuries [24]. In this direction, the objective of [25] was to conduct a brain fissure-splitting simulation considering the damage to connective tissues. Assuming that connective tissues are homogeneously distributed in a brain fissure, a damage model of connective tissue was proposed. In order to verify the splitting brain fissure simulation, longitudinal fissure splitting experiments were conducted by using porcine brains with different splitting velocities. By comparing the experimental data and the simulation results, it was concluded that numerical simulations can reproduce the reaction force of the brain fissure splitting with connective tissues [25].

Concerning the study of cartilage function and mechanical behavior, computational simulations may provide a significant means to better understand cartilage function in healthy and pathologic cases and enhance the development of the tissue engineering domain [26–31]. Three main trends have been reported it the literature for cartilage modeling. The first deals with models according to the order of intent, starting with causal confirmation, then predictive accuracy, and at the highest order of intent, determination of effect (models predicting propagation of effects across radically different scales) [29]. The second trend focuses on models of tendon mechanics across scales, with obvious application to cartilage, given reasonably analogous networks of collagen support a mechanical load in both types of tissue [30]. The last trend highlights multiscale models based on the Mixture Theory, the multiphase properties of which are ideal for cartilage [31]. Multiscale approaches based on Mixture Theory present a significant opportunity to bridge spatial and temporal scales in modeling soft tissues and may enable more predictive models.

More specifically, in [27], a mathematical model of tissue-engineered cartilage was developed under cyclic compressive loading. More specifically, a fully coupled mathematical formulation was presented accounting for simultaneous simulation of solute transport and uptake, cell proliferation, ECM production, and remodeling mechanical properties accounting for the effect of mechanical stimulation on the rates of synthesis and proliferation was developed and validated for several amplitudes and frequencies of unconfined cyclic compression. The model showed a good agreement with the experimental data and allowed to capture the main phenomena associated with spatial and temporal growth of tissue-engineered cartilage under short-term loading and estimate the distributions of variables not experimentally measured. This formulation can be used as well to predict the effect of other culture conditions and to optimize experimental designs. Nevertheless, the complex properties of the cartilage render the development of computational models of the tissue more complicated. Cartilage is not only mechanically nonlinear but also multiphasic, anisotropic, viscoelastic, and spatially heterogeneous. To capture these effects, image-driven constitutive modeling has been presented to exploit image-derived data and improve predictions of the intratissue responses of cartilage [26]. In [28], a methodology was established to segment, quantify, and generate mechanical models of developing cartilage to explore how variations in geometry and material properties affect strain distributions. Confocal microscopy was used to derive imaging data of multicellular geometries from embryonic day E16.5 and postnatal day P3 and finite element simulations were conducted. The geometries based on confocal images and isolated, single-cell models were compressed by 5% and the equivalent von Mises strain of cells and ECM were compared. It was found that cells had similar strains at both time points, suggesting that the stiffness and organization of cartilage changes during development to maintain a constant strain profile within cells. In contrast, the ECM at P3 took on more strain than at E16.5. The isolated, single-cell geometries underestimated both cell and ECM strain and were not able to capture the similarity in cell strain at both time points.

In the research area of tendon mechanics, several mathematical models have been presented at the tissue-, fiber-, or fibril-level and can be further categorized into two general types according to [32]: (i) the phenomenological approach, which uses a mathematical formulation that can closely match the mechanical response of tendon, but often with parameters that lack a clear physical interpretation, and (ii) the microstructural approach, which describes gross tendon mechanical behavior by combining or generalizing the mechanical behavior of different components in tendon (e.g. collagen fiber distribution). Many models, of both types, are developed based on a continuum theory, wherein the mathematical formulation used to represent tendon behavior is based on a continuum assumption on the tissue scale. In addition, several different modeling schemes have been developed to evaluate mechanical properties of individual tendon constituents, or their interactions, on the molecular level [32]. On the tissue scale, phenomenological approaches have been widely used to evaluate tendon mechanics.

Tendon exhibits the following stages of the stress–strain curve under quasi-static tensile loading: (i) an initial low-strain toe region, (ii) a higher-strain linear region, followed by failure. Because waviness (or crimp) of collagen fibrils and a helical superstructure of tendon are thought to contribute to the toe region, a few models have sought to address the influence of these structural features on tendon mechanics. As an example, the model described in [30] proposes that the straightening stretch ratio $\lambda_c$ (i.e. the tissue stretch at which the fiber straightens and begins to bear load) of fibers in tendon is distributed based on a Weibull three-dimensional fibril orientation probability distribution function:

$$P_{\mathrm{w}}\left(\lambda_{\mathrm{s}}\right)=\frac{\beta}{\delta}\left(\frac{\lambda_{\mathrm{s}}-\gamma}{\delta}\right)^{\beta-1}\exp\left\{-\left(\frac{\lambda_{\mathrm{s}}-\gamma}{\delta}\right)^{\beta}\right\},\quad \lambda_{\mathrm{s}}>\gamma$$

and

$$P_{\mathrm{w}}\left(\lambda_{\mathrm{s}}\right)=0,\quad \lambda_{\mathrm{s}}\leq\gamma, \tag{5.2}$$

where $\beta(\beta > 0)$, $\delta(\delta > 0)$, and $\gamma(\gamma > 0)$ are shape, scale, and location parameters, respectively. In this approach, the longitudinal tissue-level stress can be computed by integrating the contribution of all fibers over $\lambda_{\mathrm{S}}$:

$$\sigma_{\mathrm{t}}\left(\lambda_{\mathrm{t}}\right)=\int_{\gamma}^{\lambda_{\mathrm{t}}}P_{\mathrm{w}}\left(\lambda_{\mathrm{s}}\right)\sigma_{33}\left(\frac{\lambda_{\mathrm{t}}}{\lambda_{\mathrm{s}}}\right)\mathrm{d}\lambda_{\mathrm{s}},\ \lambda_{\mathrm{t}}>\gamma$$

and

$$\sigma_{\mathrm{t}}\left(\lambda_{\mathrm{t}}\right)=0,\ \lambda_{\mathrm{t}}\leq\gamma, \tag{5.3}$$

where $\sigma_{33}(\lambda_{\mathrm{t}}/\lambda_{\mathrm{S}})$ is the normal stress of the fiber and $\lambda_{\mathrm{t}}$ is an integration variable. After fitting parameters to experimental data via least-squares minimization, this model could predict the response of tendon in the nonlinear toe region.

Finally, the application palette of computational modeling to derive new knowledge for connective tissue is rather wide. For instance, several computational models study the phenomenon of the pelvic floor dysfunctions focusing mainly on the anatomy of the pelvic viscera, dense fibromuscular ligaments, and the muscles [32, 33]. Therefore, it can be concluded that different clinical sectors can benefit from the numerical study of connective tissues varying from orthopedics to obstetrics and gynecology applications.

## 5.4 Muscle Tissue

### 5.4.1 Composition and Properties of Muscle Tissue

Muscle tissue differs according to the various functions and locations in the body. In mammals, there are the following three types of muscle tissue with different morphological and functional characteristics: (i) the skeletal or striated muscle containing bundles of very long, multinucleated cells with cross-striations. Their contraction is quick, forceful, and usually under voluntary control, (ii) the smooth or non-striated muscle which consists of collections of fusiform cells which lack striations and have slow, involuntary contractions, (iii) the cardiac muscle, also known as semi-striated, has cross-striations and is composed of elongated, often branched cells bound to one another at structures called intercalated discs which are unique to cardiac muscle. Contraction is involuntary, vigorous, and rhythmic (Figure 5.4). Smooth and cardiac muscle tissues contract involuntarily and may be activated both through interaction of the central nervous system (CNS), as well as by receiving innervation from peripheral plexus or endocrine (hormonal) activation. Skeletal muscle contracts only voluntarily, upon influence of the CNS.

Skeletal muscle is attached to bones and its contraction makes possible locomotion, facial expressions, posture, and other voluntary movements of the body. More specifically, 40% of the body mass consists of skeletal muscle. Skeletal muscles generate heat as a product of their contraction and thus participate in thermal homeostasis. Shivering is an involuntary contraction of skeletal muscles in response to perceived lower-than-normal body temperature. Skeletal muscle tissue is organized in bundles surrounded by connective tissue. Skeletal

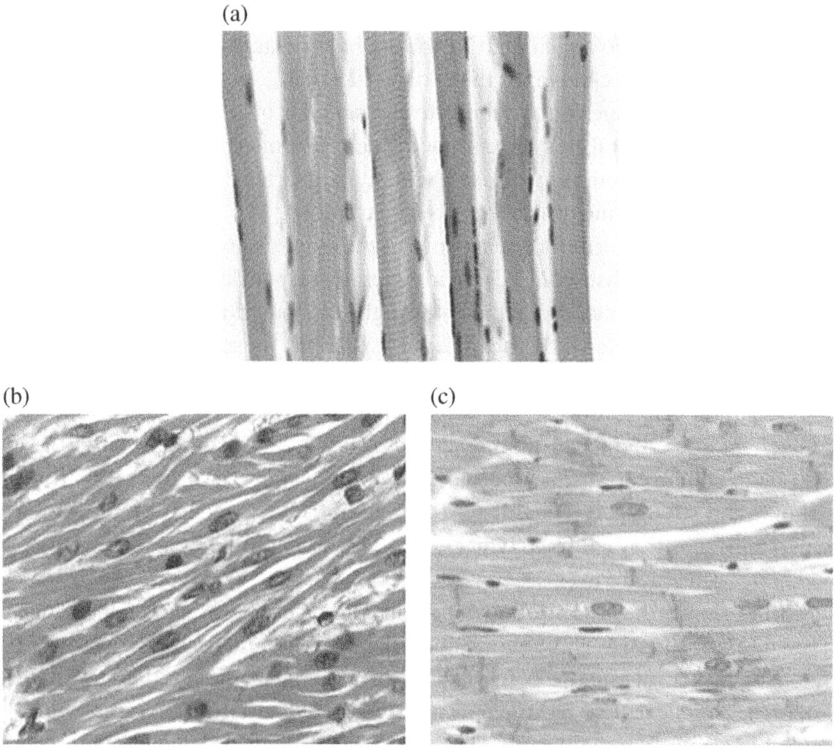


**Figure 5.4** The three main categories of muscle tissues: (a) skeletal, (b) smooth, and (c) cardiac tissue.

muscle cells show a striated appearance with many nuclei squeezed along the membranes. The striation is due to the regular alternation of the contractile proteins actin and myosin, along with the structural proteins that couple the contractile proteins to connective tissues. The muscle cell, or myocyte, is built from myoblasts derived from the mesoderm. Myocytes remain relatively constant throughout life. The cells are multinucleated as a result of the fusion of many myoblasts that fuse to form each long muscle fiber [34].

Cardiac muscle forms the contractile walls of the heart. The cells of cardiac muscle, known as cardiomyocytes, also appear striated. Unlike skeletal muscle fibers, cardiomyocytes are single cells typically with a single centrally located nucleus. A principal characteristic of cardiomyocytes is that they contract on their own intrinsic rhythms without any external stimulation. Cardiomyocytes attach to one another with specialized cell junctions known as intercalated discs. Intercalated discs include both gap junctions and anchoring junctions. Attached cells form long, branching cardiac muscle fibers which are an electrochemical and mechanical syncytium which enables the cells to synchronize their actions. The cardiac muscle is under involuntary control and is responsible to pump blood through the body. The attachment junctions hold adjacent cells together across the dynamic pressure changes of the cardiac cycle [34].

Smooth muscle tissue contraction is responsible for involuntary movements in the internal organs. It forms the contractile component of the digestive, urinary, and reproductive systems as well as the airways and arteries. Each cell is spindle-shaped with a single nucleus and no visible striations [34].

Muscle tissues perform four main important functions of the body: (i) to produce movement; for instance, skeletal muscles are responsible for locomotion and manipulation, while blood circulates through the body due to the rhythmically beating cardiac muscle of the heart and the smooth muscle in the walls of the blood vessels helping to maintain blood pressure, (ii) to maintain posture and body position. These muscles function almost continuously, making one tiny adjustment after another to counteract the never-ending downward pull of gravity, (iii) to stabilize joints. As muscles pull on bones to trigger movement, they stabilize and strengthen the joints of the skeleton, (iv) to generate heat. Muscles generate heat as they contract, which plays a significant role in maintaining normal body temperature [35].

Muscle tissues have the following main characteristics: (i) excitability, also termed responsiveness, is the ability of a cell to receive and respond to a stimulus by changing its membrane potential. In the case of a muscle, the stimulus is usually a chemical – for example, a neurotransmitter released by a nerve cell, (ii) contractility is the ability to shorten forcibly when adequately stimulated, (iii) extensibility is the ability to extend or stretch. Muscle cells shorten when contracting, but they can stretch, even beyond their resting length, when relaxed and (iv) elasticity is the ability of a muscle cell to recoil its resting length after stretching [35].

When an isolated muscle is stretched in the laboratory, it is possible to measure the change in muscle length and, indirectly, the tension that develops in the muscle. The results can be plotted on a passive length-tension curve. A typical passive length-tension curve is shown in Figure 5.5a. The passive length-tension curve shows the length at which the relaxed muscle first develops internal tension, denoted with $L_1$ in Figure 5.5a. This is often called the muscle's resting length, or sometimes simply its length. Another important mechanical property of muscle is its stiffness. Stiffness is the ratio of the tension developed in the muscle when it is stretched to the amount the muscle lengthens. Thus, the slope of the passive length-tension curve represents the muscle's stiffness. The stiffness of the muscle determines how much the muscle will lengthen given a certain change in tension and how much tension will develop in the muscle as it is lengthened. In some studies, the measurements of tension, change in length, and stiffness are normalized so that the resulting values reflect only the properties of the muscle tissue and not the size of the muscle. In this case, the tension is divided by the cross-sectional area of the muscle (called stress) and the change in length is divided by the initial length of the muscle (called strain). It should be noted that muscle

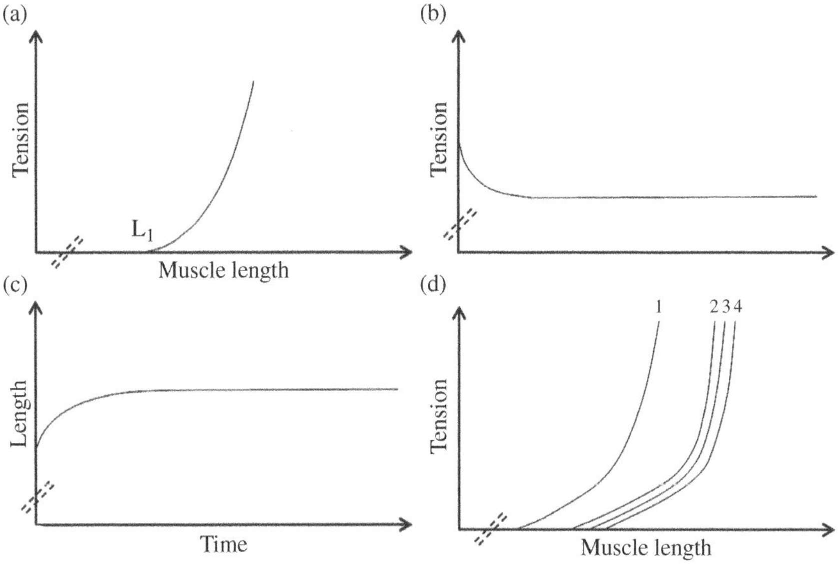


**Figure 5.5** (a) A passive length-tension curve for muscle ($L_1$ is the length at which the muscle first develops tension), (b) stress relaxation, (c) creep, (d) the effect of cyclic loading (numbers 1–4 denote the order of cycles).

length is also dependent on the mechanical properties of tendon with which the muscle is connected in series [36] (see Section 5.4.2.4). The mechanical properties of relaxed muscle may be provided by several of the anatomical structures that make up the muscle. The components of muscle that are most likely to determine its length and stiffness are the intramuscular connective tissue and intracellular structures.

The muscle is said to behave as a "viscoelastic" material. If the muscle is stretched to a given length and that length is maintained, then the tension will be seen to decrease over time until a new steady state force is reached (called stress relaxation, see Figure 5.5b). Alternatively if the muscle is stretched to a certain tension and the tension is maintained, then the length of the muscle will increase over time until a new steady state length is reached (called creep, Figure 5.5c). If the muscle is repeatedly stretched each subsequent stretch will demonstrate different length-tension curves (Figure 5.5d). This viscous deformation is probably the major source of the increase in muscle length seen immediately following muscle stretching.

### 5.4.2 Computational Modeling of Muscle Tissue

Muscle models are used in several fields of science such as medical science, biology, biomechanics, and physiology. These models deal with two different objectives. First, muscle tissue computational modeling is used to gain an insight into the structure and functioning of muscles and to improve the general understanding of contraction dynamics. Second, muscle models are used as "tools" in the field of applied research to answer questions on the four main important functions which muscle tissues perform for the body as described in Section 5.4.1.

#### 5.4.2.1 Computational Modeling of Skeletal Muscle Tissue

Finite-element modeling of skeletal muscle is not simple for several reasons. First, each muscle has a unique, complicated shape that makes the generation of accurate geometrical representations challenging. Further, muscle tissue's constitutive properties are highly nonlinear, nearly incompressible, active, and anisotropic. Muscles have fibers with complicated fiber trajectories that must be defined in order to incorporate the anisotropic properties. Muscle fibers insert into tendons that inter-digitate within muscles. There are complex boundary conditions and contact conditions between muscles, bones, and other structures. All of these complexities require careful treatment of a robust modeling pipeline [37].

One of the most important aspects of the 3D muscle model, which often gets limited attention, is the attachment between muscle tissue and the tendons/aponeuroses. In order for the 3D muscle simulation to predict realistic deformations, stresses, and forces, the tendinous attachments must be carefully modeled. This is particularly challenging for two reasons: (i) the interface between the muscle and tendon materials needs to be perfectly aligned and (ii) often the aponeuroses do not appear clearly in MR images. So, they must be estimated based on other information or experimental measurements. Because the tendon and aponeurosis attachments also define the origin and insertion areas of the muscle fascicles, the definition of the fiber trajectories is another stage in which these areas can be evaluated. If the areas of attachment are not defined accurately, this will lead to errors in the orientation and length of the fascicles. Once the geometrical representation is established, the geometry must be meshed. Three-dimensional models require volumetric meshes, and in particular, muscle simulations that involve significant deformation and highly nonlinear material properties which are highly sensitive to mesh quality. The next step in the process is to define the arrangements of fibers within the area of the model corresponding to muscle tissue. This step is critical because it defines the fiber orientation for each element in the muscle material, which is an important input to the constitutive model. This is considered a unique problem compared to other finite-element models because each muscle has a different arrangement of fibers, many of which can be complicated with multiple areas of attachment, and curvatures that vary throughout the muscle. There are multiple approaches that can be used to define fibers, and the decision depends on the complexity of the fiber map (which depends on modeling decisions and chosen muscle), and what type of data are used as input and/or validation for the process [37].

More specifically, a novel multiscale modeling framework for skeletal muscles based on analytical and numerical homogenization methods was presented in [38] to study the mechanical muscle response at finite strains under three-dimensional loading conditions. First, an analytical microstructure-based constitutive model was developed and numerically implemented using a finite element software. The analytical model took into account explicitly the volume fractions, the material properties, and the spatial distribution of muscle's constituents by using homogenization techniques to bridge the different length scales of the muscle structure. Next, a numerical homogenization model was developed based on the periodic eroded Voronoi tessellation (a Voronoi diagram is a partition of a plane into regions close to each of a given set of objects) to virtually represent skeletal muscle microstructures. The eroded Voronoi unit cells were then resolved by finite element simulations and used to assess the analytical homogenization model. The application of periodic boundary conditions was necessary in order to be able to extract a homogenized response out of the numerical unit cell. These conditions directly imply that the proposed unit cell can be infinitely repeated in the three directions to provide a representative muscle structure, whereby requiring only the solution of a reduced unit cell to estimate the homogenized response. The periodic boundary conditions are expressed as:

$$\mathbf{u} = \left(\hat{\mathbf{F}} - \delta\right) \cdot \mathbf{X} + \mathbf{u}^*, \tag{5.4}$$

where the second-order tensor $\hat{\mathbf{F}}$ denotes the average deformation gradient of the unit cell which is equal to the applied deformation gradient, $\mathbf{X}$ denotes the spatial coordinates in the reference configuration, $\delta$ is the second-order identity tensor, and $\mathbf{u}^*$ is a periodic displacement field. The volume average true stress tensor in the unit cell $\hat{\boldsymbol{\sigma}}$ is calculated as:

$$\hat{\boldsymbol{\sigma}} = \sum_{\mathrm{i}} \boldsymbol{\sigma}_{\mathrm{i}} V_{\mathrm{i}} \Big/ \sum_{\mathrm{i}} V_{\mathrm{i}}, \tag{5.5}$$

where i and $V_{\mathrm{i}}$ are the true stress tensor evaluated at the centroid and the volume of the ith finite element, respectively. The volume average nominal (1st Piola-Kirchhoff) stress tensor $\hat{\mathbf{t}}$ is then computed by the relation:

$$\hat{\mathbf{t}} = \hat{\mathrm{J}}\ \hat{\mathbf{F}}^{-1} \cdot \hat{\boldsymbol{\sigma}}, \tag{5.6}$$

where $\hat{\mathrm{J}} = \det \hat{\mathbf{F}}$ and $\hat{\mathbf{F}}$ has been defined via the periodic boundary conditions in Eq. (5.4). All loadings are characterized by the macroscopic applied average deformation gradient $\hat{\mathbf{F}}$. For the problem of uniaxial tension in direction i, the deformation gradient is of the form:

$$\hat{\mathbf{F}} = \hat{\lambda} \mathbf{e}_i \mathbf{e}_i + \hat{\lambda}_{\mathrm{t}} \left(\mathbf{e}_j \mathbf{e}_j + \mathbf{e}_k \mathbf{e}_k\right), \quad \forall i,j,k = 1,2,3 \ \text{ and } \ i \neq j \neq k, \tag{5.7}$$

where $\left(\hat{\lambda}, \hat{\lambda}_{\mathrm{t}}\right)$ are the average axial and transverse stretch ratios and ($\mathbf{e}_i$, $\mathbf{e}_j$, $\mathbf{e}_k$) are the base vectors along the coordinate axes. For the problem of simple shear, the deformation gradient is of the form:

$$\hat{\mathbf{F}} = \delta + \hat{\gamma} \mathbf{e}_i \mathbf{e}_j, \quad \forall i,j = 1,2,3 \ \text{ and } \ i \neq j, \tag{5.8}$$

where $\hat{\gamma}$ is the average amount of shear on the $i$–$j$ plane.

The material parameters of the analytical model were identified successfully by the use of available experimental data [38]. The analytical model was found to be in very good agreement with the numerical model for the full range of loadings, and a wide range of different volume fractions and heterogeneity contrasts between muscle's constituents.

Tissue engineering approaches also investigate numerically the role of muscle regeneration and adaptation to advance muscle tissue regeneration strategies. In [39], a new agent-based computational model of skeletal muscle inflammation and regeneration was presented following acute muscle injury. This model simulated the recruitment and cellular behaviors of key inflammatory cells (e.g. neutrophils and M1 and M2 macrophages) and their interactions with native muscle cells (muscle fibers, satellite stem cells, and fibroblasts) that result in the clearance of necrotic tissue and muscle fiber regeneration. The ability of the model to track key regeneration metrics was demonstrated during both unencumbered regeneration and in the case of impaired macrophage function. Also, the model was applied to simulate regeneration enhancement when muscle is primed with inflammatory cells prior to injury. This was a putative therapeutic intervention which had not been investigated experimentally in the past. Computational modeling of muscle regeneration, pursued in combination with experimental analyses, was found to provide a quantitative framework for evaluating and predicting muscle regeneration, enabling the rational design of therapeutic strategies for muscle recovery. These models reflect the next generation of computational models in the field of skeletal muscle injury and disease.

#### 5.4.2.2 Computational Modeling of Smooth Muscle Tissue

Smooth muscle tissue is a complex organization of multiple cell types regulated by numerous signaling molecules such as neurotransmitters, hormones, and cytokines. Smooth muscle contractile function can be regulated via distribution and expression of the contractile and cytoskeletal proteins, and activation of any of the second messenger pathways regulating them [40].

For instance, certain smooth muscles, such as the detrusor of the urinary bladder, exhibit a variety of spikes that differ as far as their amplitudes and time courses are concerned. The origin of this diversity is poorly understood but is often attributed to the syncytial nature of smooth muscle and its distributed innervation [41, 42]. An electrophysiological model of the detrusor smooth muscle (DSM) spike based on the Hodgkin–Huxley formalism was first reported in [43]. However, there were key conflicts between the components used for this model and experimentally elucidated DSM electrophysiology. For instance, the properties of the ion channels were not validated based on experimental data, such as ionic currents recorded under voltage clamp conditions and current–voltage curves derived from these. More recently, a DSM spike model has been reported which addresses the aforementioned issues [44]. Specifically, DSM ionic currents were modeled using parameters directly drawn from previously published data. Furthermore, these currents were tuned in respect of their amplitudes and their dynamic profiles under voltage clamp conditions to achieve an optimal match to experimentally recorded signals. The known complement of ion channels for DSM cells was integrated to generate the spike-type AP of the DSM. In physiology, an AP occurs when the membrane potential of a specific cell location rapidly rises and falls. Due to this depolarization, adjacent locations similarly depolarize. APs were elicited both by external current injection and physiologically realistic inputs represented by synaptic potentials. It was found that the emergent computational spikes matched quite closely the experimentally recorded ones, while the model was able to reconstruct various spontaneous AP shapes recorded in DSM [41].

In [45], it was demonstrated that a homogeneous syncytium consisting of smooth muscle cells with identical membrane properties can produce APs of varying shapes. A close examination of the differences between the experimental and simulated AP features and their correlations provided insight into the biophysical environment of the detrusor syncytium, such as the density of innervation, the size of the bundles, and the strength of gap junction coupling between cells. It could also be inferred that the fraction of the propagated AP is very low in the detrusor. This work presented a different approach aiming to investigate the properties of a smooth muscle syncytium. More specifically, an understanding of different aspects of the syncytium could be obtained by classifying the cells based on the variations displayed in their electrical activities. Further, this mapping between the AP shape and the syncytial properties could also be used to delineate the syncytial changes during disease and provide a better understanding of pathological conditions.

Finally, a structural constitutive model that characterizes the active and passive responses of biological tissues with smooth muscle cells was proposed in [46]. The model was established considering that the connected collagen fibers and the contractile units in smooth muscle cells are the active tissue component, while the collagen fibers not connected to the smooth muscle cells are the passive tissue component. An evolution law describing the deformation of the active tissue component over time was developed based on the sliding filament theory. According to this law, the contraction force was considered to be the sum of a motor force that initiates contraction, a viscous force that describes the actin–myosin filament sliding, and an elastic force that accounts for the deformation of the cross-bridges. The mechanical response of the collagen fibers was considered to be governed by the fiber recruitment process implying that collagen fibers support load and behave as a linear elastic material only after becoming taut. This structural constitutive model was assessed with published passive and active, uniaxial and biaxial experimental data of pig arteries.

#### 5.4.2.3 Computational Modeling of Cardiac Muscle Tissue

Mathematical modeling is a powerful tool to investigate the orchestrated and complex biological process of cardiac electrical activity. Starting from the pioneering work of Hodgkin and Huxley (1952) on the nerve axon model, several increasingly sophisticated models have been developed for the propagation of electrical signals in the cardiac tissue separating between models for cardiac cell electrophysiology, and macroscopic tissue-level models based on continuum mechanics [47–50].

Through the proper exploitation of experimental data from key aspects of cardiac electrophysiology, systems biology simulations might complement empirical findings, provide quantitative insight into physiological and pathophysiological mechanisms of action, and guide new hypotheses to better comprehend this complex biological system and establish novel cardiotherapeutic approaches. Force balance equations for an elastic continuum medium are employed to describe large deformations of the myocardium under fluid pressure, the surrounding organs and its own contraction. Such a framework has to be coupled with the macroscopic bidomain or monodomain equations accounting for the propagation of the electric potential and ionic currents [48].

Simulations of cardiac tissue remodeling and growth are still in their early stages because they are driven by rule-based phenomenological models and require further development of theory and implementation [49]. At the same time, the physiology and drivers of remodeling and growth are often characterized by pathway topology and not quantitative dynamics, which limits their utility in quantitative predictive cardiac modeling. The ability to predict how the heart will remodel in response to disease progression or to treatment is an essential element in making long-term predictions related to clinical outcome and intervention success and will substantially increase the applications of simulations in cardiology.

An appropriate mathematical framework for understanding and modeling cardiac tissue growth must include large deformation elasticity theory. Since cardiac tissue, like all biological tissues, is highly anisotropic, the description of material properties by constitutive relations (both stress–strain relations and growth "laws") must be based on material axes which are defined by the underlying tissue microstructure [50]. With appropriate microstructural constitutive models, inverse finite element analysis can be used to analyze the factors that lead to impaired mechanical behavior in heart failure. Computational modeling of cardiac mechanics can provide valuable insights into the mechanisms enhancing physiological and pathophysiological cardiac processes, as well as serve as a noninvasive method to assess myocardial tissue mechanical properties which provide specific biomarkers for cardiac performance.

#### 5.4.2.4 Musculotendon Models

It has already been mentioned in Section 5.4.1 that muscle length depends also on the mechanical properties of tendon with which the muscle is connected in series. There are two broad classes of musculotendon models: (i) cross-bridge models and (ii) Hill-type models. Despite the advantage of the cross-bridge models to be derived from the fundamental structure of muscle, they include many parameters which are rarely used in muscle-driven simulations including several muscles and are difficult to be measured. Hill-type models are widely used in muscle-driven simulations. Musculotendon models which are both computationally fast and biologically accurate are required to simulate human movement.

In [51], the computational speed and biological accuracy of three musculotendon models were examined. Initially, the equilibrium musculotendon model was evaluated that is commonly used in muscle-driven simulations of movement. Then, two alternative models were assessed: (i) a damped equilibrium model, and (ii) a rigid-tendon model. The computational speed of each musculotendon model was assessed by simulating musculotendon dynamics over the operational range of the muscle using constant activation and sinusoidal displacement tests. The biological accuracy of the equilibrium, damped equilibrium, and rigid tendon models was evaluated by comparing the simulated musculotendon forces to those measured experimentally from isolated rat soleus and cat soleus muscles. It was shown that the equilibrium and damped equilibrium models produce similar force profiles but have different computational speeds. The damped equilibrium model produced forces that compare favorably with those observed in maximally activated biological muscle and simulated faster than the equilibrium model regardless of the use of an explicit or implicit integrator. The rigid-tendon model was fast and accurate when simulating a maximally activated muscle with a short tendon.

## 5.5 Nervous Tissue

Nervous tissue, which is also called neural tissue or nerve tissue, is the main tissue component of the nervous system. The nervous system regulates and controls body functions and activity and consists of two parts: (i) the CNS comprising the brain and spinal cord, and (ii) the peripheral nervous system (PNS) comprising the branching peripheral nerves. The nervous system is considered a complex interconnected network associated with axons – specialized neural fibers. These fibers compose the basis for communication between functionally distinct regions of the CNS and relay sensorimotor signals in the PNS. Remarkably, there are more than 160 000 km of axons in the adult nervous system [52].

Nervous tissue is grouped into two main categories: (i) neurons and (ii) neuroglia. Neurons, or nerves, transmit electrical impulses, while neuroglia do not have this capacity [2]. More specifically, neurons are the cells considered to be the basis of nervous tissue. They are responsible for the electrical signals that communicate information about sensations, and which produce movements in response to those stimuli, along with inducing thought processes within the brain. The behavior of neurons depends on their structure or shape. The three-dimensional shape of these cells makes the immense numbers of connections within the nervous system possible. There are many neurons in the nervous system and many different types of neurons. Neuroglia or glial cells are cells that support neurons, supply them with nutrients, and get rid of dead cells and pathogens such as bacteria. They also form insulation between neurons so that electrical signals do not get crossed, and can also support the formation of synaptic connections between neurons.

The microstructure of neural tissues is extremely complex and this is also reflected in their mechanical properties. The tissues of the CNS, encompassing the brain and spinal cord, are anatomically and mechanically distinct from those of the PNS (nerve roots, ganglia, and peripheral nerves). The CNS tissues are bathed in cerebrospinal fluid and are protected by a series of collagenous membranes, the meninges. The meninges are made up of the arachnoid mater, pia mater, and dura mater. Both brain and spinal cord tissues are heterogeneous, with white and grey matter regions having various structural arrangements and constituents. This gives rise to the complex, nonlinearly viscoelastic mechanical behavior of these tissues [53]. The peripheral neural tissues are also protected by an external collagenous membrane, known as a nerve sheath.

The following subsections present computational models of brain, spinal cord, and peripheral nerve tissues. Although the mechanical behavior of the brain tissue has been extensively studied in the literature, the heterogeneous, nonlinear viscoelastic behavior and substantial changes to tissue behavior postmortem have made such data difficult to be interpreted. The spinal cord and peripheral nerves have been studied to a more limited extent, and only recently their mechanical behavior has begun to be elucidated [54].

### 5.5.1 Computational Modeling of Brain Tissue

Brain tissue is one of the most important tissues in the human body, as well as the most compliant and complex tissue. While long underestimated, increasing evidence confirms that mechanics plays a critical role in modulating brain function and dysfunction. Mechanical models that capture the complex behavior of nervous tissues and are accurately calibrated with reliable and comprehensive experimental data are key to carrying out reliable predictive simulations. Mathematical modeling and computational simulations of the brain tissue are useful for both biomedical and clinical communities and cover a wide range of applications ranging from predicting disease progression and estimating injury risk to planning surgical procedures. Computational simulations, based on the field equations of nonlinear continuum mechanics, could potentially provide important insights into the underlying mechanisms of brain injury and disease that go beyond the possibilities of traditional diagnostic tools [54].

With the exponential growth of computational power, there has been an increasing interest in the use of the finite element method for structural modeling of the human head. The constitutive material models of brain tissue and the finite element analysis of blast-induced traumatic brain injury (TBI) have been discussed in the literature. So far, many numerical models of the head have been developed and this topic is still being expanded. However, only a few of these models have been verified in comparison with experimental studies carried out on postmortem human subjects. According to the literature, several studies have investigated the mechanical response of brain tissues using shared finite-element nodes between the brain and skull models. However, in this modeling approach, the skull-brain connection does not allow a complete understanding of the biomechanics of the head. In general, there is no clear view in the literature on how to model brain tissue [55–57]. For instance, the choice of the material of the model can affect the brain kinematics in relation to the skull, while only a few numerical models have considered bridging veins [55, 58, 59]. The authors of these studies stated that the behavior of the brain-tissue system as a whole can be significantly affected by the inclusion of these elements.

In a more recent study [55], a numerical model of the brain was developed that mimics the actual behavior of brain tissue during mechanical loading based on medical imaging methods. Computational simulations were performed using the finite element method and the brain's material properties were modeled using the hyperelastic and viscoelastic constitutive law. It was found that the Mooney–Rivlin model (a hyperelastic material model where the strain energy density function is a linear combination of two invariants of the left Cauchy–Green deformation tensor) was the most suitable for studying the biomechanical response of the brain. This approach allowed the authors to assess the changes in the mechanical and geometrical parameters of the brain tissue caused by the impact of mechanical loads. The developed model was verified through comparison with experimental studies on postmortem human subjects described in the literature, as well as through numerical tests. Based on the current research, the authors identified important aspects of the modeling of brain tissue which influence the assessment of the actual biomechanical response of the brain for dynamic analyses.

In another study [60], a high-fidelity three-dimensional computational model of brain injury biomechanics was developed and the contours of strain and strain rate at the grey matter-white matter boundary were mapped. Diffusion tensor imaging abnormalities in a cohort of 97 TBI patients were also mapped at the grey matter-white matter boundary, while 51 healthy subjects served as controls. Different causes of brain tissue injury were considered such as road traffic accidents and sports injuries. Experimental work on the mechanics of brain tissue has shown that brain tissue deformation has a nonlinear relationship with force and that brain tissue stiffens as the rate of deformation increases. The authors of [60] implemented these two important properties in their computational model by using a hyper-viscoelastic material model combining hyperelasticity and viscoelasticity. The nonlinear isochoric (volume preserving) response of brain tissue was modeled with the Ogden hyperelastic model with a strain energy function of the form:

$$\Psi^{\infty} = \sum_{p=1}^{n} \frac{\mu_p}{\alpha_p}\left(\lambda_1^{\alpha_p} + \lambda_2^{\alpha_p} + \lambda_3^{\alpha_p} - 3\right), \tag{5.9}$$

where $\lambda_1$, $\lambda_2$, and $\lambda_3$ are principal stretches of the brain tissue, $\mu_p$ and $\alpha_p$ are material constants, which have been determined for brain tissue through curve fitting to experimental data obtained from stretching brain tissue. The rate-dependent behavior of brain tissue was modeled with a convolution integral as follows:

$$S(t) = S^{\infty} + \int_0^t G(t-T)\frac{\partial E(T)}{\partial T}\mathrm{d}T, \tag{5.10}$$

where $S^{\infty}$ denotes the long-term (or the equilibrium state) second Piola-Kirchhoff stress tensor, obtained from the Ogden strain energy function given in Eq. (5.9) and $G(t)$ is the relaxation function represented with a Prony series of the form:

$$G(t) = \sum_{i=1}^{n} G_i e^{\frac{-t}{\tau_i}}, \tag{5.11}$$

where $\tau_i$ and $G_i$ are material constants, which have been determined through fitting the material model to the results of transient experiments on brain tissue. The subarachnoid space, between the arachnoid and pia membranes, was occupied by delicate connective tissue filaments and intercommunicating channels containing the cerebrospinal fluid. The nonlinear mechanical behavior of the subarachnoid space and ventricles was modeled with the hyperelastic model ($n=1$). The major extensions of the dura mater (e.g. falx and tentorium) and the pia mater, which envelopes the brain, were modeled with the hyper-viscoelastic material model. The computational models in [60] revealed that high strain values are most prominent at the depths of sulci. In addition, the volume fraction of sulcal regions exceeding brain injury thresholds was found to be significantly higher compared to that of gyral regions. Strain and strain rates were highest for a road traffic accident and sports injury. Strain was greater in the sulci for all injury types, but strain rate was larger only in the road traffic and sporting injuries. Based on diffusion tensor imaging, converging imaging abnormalities were observed within sulcal regions with a significant decrease in fractional anisotropy in the patient group compared to controls within the sulci. It was found that brain tissue deformation induced by head impact loading is larger in sulcal locations, where pathology is identified in the case of chronic traumatic encephalopathy. Also, the nature of initial head loading might have a significant effect on the pattern and magnitude of the injury. Clarifying this relationship is a key factor to enhance our understanding of the long-term effects of head impacts and contribute to the improvement of protective strategies (e.g. helmet design).

More recently [61], transcranial near-infrared stimulation (tNIRS) has been proposed as a tool to modulate cortical excitability. Nevertheless, the underlying mechanisms are not well understood, where the heating effects on the brain tissue need further study due to the increased near-infrared (NIR) absorption by fat and water. To this end, in [61], investigation of the interaction of light with the chromophores that are responsive to photons in the red-NIR spectral region has been performed. Due to the increased absorption of longer wavelengths by water in the tissue, it was considered that NIR light interaction with neural tissue may have effects of photobiomodulation and photothermal neurostimulation which requires attention for rational dosing of tNIRS as a result of the biphasic dose response. Although tNIRS is considered a modulator of cortical excitability in a healthy human brain, the exact pathways of neuromodulation have not been defined. The role of computational modeling was crucial to isolate photothermal effects from photobiomodulation during tNIRS with 810 nm while comparing that with the red spectrum (630 and 700 nm) used for low-level light therapy (LLLT) and giving insight into the mechanisms underlying neuromodulation. The main objective was to better comprehend the extent of optically induced tissue heating (primarily due to water and fat absorption) during tNIRS. A temperature increase was found in the scalp below 0.25 °C and a minimal temperature increase in the gray matter less than 0.04 °C at 810 nm. Similar heating was found for 630 and 700 nm used for LLLT. So, photothermal effects were considered not to evolve in the brain tissue.

### 5.5.2 Computational Modeling of the Spinal Cord Tissue

TBI and spinal cord injury (SCI) are major causes of death and disability in the developed world. Despite the ongoing experimental and modeling research approaches aiming to provide new knowledge on the mechanics of tissue and cell damage typically observed in such events, the evolution of stress, strain, and their corresponding loading rates on the damage level are elusive. More specifically, the direct relations between brain and spinal cord tissue or cell damage, and electrophysiological functions have not been unraveled yet. Mechanical modeling approaches focus mainly on mechanistic-based damage criteria and stress distribution implying that simulated function-based damage criteria have not been determined. In [62], a new multiscale model was proposed of myelinated axon associating electrophysiological impairment to structural damage as a function of strain and strain rate. This multiscale approach lays new foundations on damage assessment directly relating electrophysiological properties and neuron mechanics and provides a link between subsequent functional deficits and mechanical trauma. A three-dimensional finite element model of the spinal cord was developed with two components: (i) the model of the white matter and (ii) the model of the grey matter. Due to insufficient resolution of the MRI images of the spinal cord, exact measurements of the grey matter could not be performed. It was shown that numerical modeling using the finite element method might be applicable in the analysis of SCI in humans.

Although essential for understanding SCI and TBI at the cellular level (or directly at the cognitive scale in the latter reference), these approaches suffer from a lack of applicability to tissue-scale electrophysiology scenarios. To this end, in [62], a tissue-level framework was presented to describe the coupled mechanical and electrophysiological behavior of axonal-based tissues. This framework combined a large deformation mechanical constitutive model with an electrophysiological tissue model. The mechanical part accounts for strain rate dependency, viscous effects, and continuum mechanical damage and was developed within a thermodynamically consistent framework. The electrophysiological behavior was described by the FitzHugh–Nagumo-based model [63, 64] which included anisotropic conduction following fiber orientation. The mechanical model was first calibrated against experimental measurements of the stress–strain response of spinal cord tissue under a wide range of strain rate conditions, and the electrophysiological parameters were obtained from the literature. Then, the complete framework was calibrated to investigate the mechanoelectrophysiological coupling in spinal cord samples subjected to tension-free relaxation tests by considering electrophysiological damage as a function of mechanical energetic terms. The predictive capability of the framework was revealed based on the complex scenario of mechanoelectrophysiological coupling in a blast-loaded spinal cord. Finally, the mechanical parameters were identified for the white matter and an analysis of the influence of axonal dispersion on anisotropic AP conduction was presented. It was shown that the mechanoelectrophysiological framework could allow for the modeling of the mechanics of CNS tissues and their translation into electrophysiological dysfunction through energetic terms.

The constitutive equation used to characterize and model spinal tissues can significantly influence the conclusions from experimental and computational studies [65]. Therefore, researchers must make critical decisions regarding the balance of computational efficiency and predictive accuracy necessary for their purposes. The objective of this study is to quantitatively compare the fitting and prediction accuracy of linear viscoelastic (LV), quasi-linear viscoelastic (QLV), and (fully) nonlinear viscoelastic (NLV) modeling of spinal-cord-pia-arachnoid-construct (SCPC), isolated cord parenchyma, and isolated pia-arachnoid complex (PAC) mechanics in order to enhance decision making. Experimental data were collected during dynamic cyclic testing of each tissue condition in order to fit each viscoelastic formulation. These fitted models were then applied to predict independent experimental data from stress-relaxation testing. It was shown that relative fitting accuracy does not directly reflect relative predictive accuracy, which reveals the need for material model validation based on predictions of independent data. Concerning the isolated cord and SCPC, although the NLV formulation was found to best predict the mechanical response to arbitrary loading conditions, it required significantly larger computational time. Also, the QLV formulation was found to be the best predictor of the mechanical response of the PAC under arbitrary loading conditions.

### 5.5.3 Computational Modeling of Peripheral Nerves

With growing interest in peripheral nerve stimulation to treat diseases, it is important to leverage the capacity of computational modeling to explore broad ranges of geometrical and electrical parameter values for the nerve and electrode, while being careful about the accuracy and underlying assumptions of the chosen values [66–69]. Peripheral nerve models are considered computational models of one or more axons. Several models have been established in the literature and variation in models typically pertains to the specific ionic currents and their conductance through the axon membrane. Another common variation refers to the method used to model myelination. Models of peripheral nerves are nonlinear, require solving systems based on differential equations and are fast approximation techniques which can provide new knowledge on axonal response to varied stimulation. In the neurosciences, peripheral nerve models are more widely applied to study single neurons, while in neural and biomedical engineering, they are mainly used in large population models to evaluate the global response to electrical nerve stimulation or for nerve recording, primarily for use in functional electrical stimulation of neuroprosthetic or neuromodulation systems [69].

Bioelectronic medicine modulates the activity patterns of peripheral nerves and is a new way to treat diverse medical conditions from epilepsy to rheumatism. To enable a faster, more detailed analysis of peripheral nerve stimulation and recording and limiting experimentation load, computational models incorporating experimental insights are of great importance. In [66], a peripheral nerve stimulator was presented that combines biophysical axon models and numerically solved and idealized extracellular space models in one environment. More specifically, the extracellular space was modeled as a three-dimensional resistive continuum governed by the electro-quasistatic approximation of the Maxwell equations. Finite element models for different media were established for various precomputed distributions (homogeneous, nerve in saline, nerve in cuff) and imported into the simulator. Axons, on the other hand, were modeled as one-dimensional chains of compartments. Unmyelinated fibers were based on the Hodgkin–Huxley model, while for myelinated fibers, the model proposed by [67] was adapted to smaller diameters. An iterative algorithm was developed to position fibers along the nerve with a variable tortuosity fit to imaged trajectories, obtaining thus realistic axon shapes. The model was validated with data from the stimulated rat vagus nerve. Simulation results indicated that tortuosity alters recorded signal shapes and increases stimulation thresholds. It was also reported that this model could be adapted to different nerves [66].

Finally, in [69], a systematic investigation of the effects of values and methods for modeling the perineurium and endoneurium on activation and block thresholds was presented, which provides critical guidance on the electrical parameter values for future modeling studies and may help address discrepancies between modeling and experimental data. Three-dimensional finite element models of cuff electrodes and compound peripheral nerves were developed to block thresholds of model axons and quantify activation. In addition, a two-dimensional finite element model of a bundle of axons was developed to investigate the bulk transverse endoneurial resistivity. It was concluded that the temperature and frequency should be considered for the perineurium resistivity, and can be modeled as a sheet resistance boundary condition calculated using an appropriate perineurial thickness. Also, the longitudinal endoneurial resistivity was found to have a greater impact on thresholds than the transverse resistivity.

## 5.6 Conclusion

This chapter indicated the significant role that computational modeling can play to provide new knowledge to existing clinical and experimental information at the tissue level. The key computational studies of the current literature were discussed, while the biological and mechanical behavior of the different tissue types was presented. It can be concluded that computational modeling approaches can open new horizons and reveal supplementary quantitative and qualitative information on the various body functions at the tissue level.

## References


1 Cools, P., Ghobeira, R., Van Vrekhem, S. et al. (2016). Non-thermal plasma technology for the improvement of Scaffolds for tissue engineering and regenerative medicine – a review, Chapter 7. In: *Plasma Science and Technology: Progress in Physical States and Chemical Reactions*. London, United Kingdom: IntechOpen. https://www.intechopen.com/chapters/49662 doi: 10.5772/62007.

2 OpenStax, Anatomy & Physiology. (2016) OpenStax CNX. http://cnx.org/contents/14fb4ad7-39a1-4eee-ab6e-3ef2482e3e22@8.24

3 Smallwood, R. (2009). Computational modelling of epithelial tissues. *Wiley Interdiscip. Rev. Syst. Biol. Med.* 1 (2): 191–201. https://doi.org/10.1002/wsbm.18.

4 Bonnet, I., Marcq, P., Bosveld, F. et al. (2012). Mechanical state, material properties and continuous description of an epithelial tissue. *J. R. Soc. Interface* 7 (75): 2614–2623.

5 Osborn, J.M., Fletcher, A.G., Pitt-Francis, J.M. et al. (2017). Comparing individual-based approaches to modelling the self-organization of multicellular tissues. *PLoS Comput. Biol.* 13 (2): e1005387. https://doi.org/10.1371/journal.pcbi.1005387.

6 Kida, N. and Morishita, Y. (2018). Continuum mechanical modelling of developing epithelial tissues with anisotropic surface growth. *Finite Elem. Anal. Des.* 144: 49–60.

7 Ishihara, S., Marcq, P., and Sugimura, K. (2017). From cells to tissue: a continuum model of epithelial mechanics. *Phys. Rev. E* 96 (2) https://doi.org/10.1103/physreve.96.022418.

8 Fletcher, A.G., Osterfield, M., Baker, R.E., and Shvartsman, S.Y. (2014). Vertex models of epithelial morphogenesis. *Biophys. J.* 106 (11): 2291–2304. https://doi.org/10.1016/j.bpj.2013.11.4498.

9 Kursawe, J., Baker, R.E., and Fletcher, A.G. (2017). Impact of implementation choices on quantitative predictions of cell-based computational models. *J. Comput. Phys.* 345: 752–767.

10 Odell, G.M., Oster, G., and Burnside, B. (1981). The mechanical basis of morphogenesis. I. Epithelial folding and invagination. *Dev. Biol.* 85: 446–462.

11 Alt, S., Ganguly, P., and Salbreux, G. (2017). Vertex models: from cell mechanics to tissue morphogenesis. *Philos. Trans. R. Soc. London, Ser. B* 372 (1720): 20150520. https://doi.org/10.1098/rstb.2015.0520.

12 Barton, D.L., Henkes, S., Weijer, C.J., and Sknepnek, R. (2017). Active Vertex Model for cell-resolution description of epithelial tissue mechanics. *PLoS Comput. Biol.* 13 (6): e1005569. https://doi.org/10.1371/journal.pcbi.1005569.

13 van Drongelena, R., Vazquez-Faci, T., Huijbena, T.A.P.M. et al. (2018). Mechanics of epithelial tissue formation. *J. Theor. Biol.* 454: 182–189.

14 Nezwek, T.A. and Varacallo, M. (2019). *Physiology, Connective Tissue*. StatPearls Publishing LLC.

15 Young, R.A. (1984). The ground substance of connective tissue. *J. Physiol. (Lond.)* 16 (5, 6): 325–350.

16 Kusindarta, D.L. and Wihadmadyatami, H. (2018). The role of extracellular matrix in tissue regeneration. In: *Tissue Regeneration*. London, United Kingdom: IntechOpen. https://www.intechopen.com/chapters/60312 doi: 10.5772/intechopen.75728.

17 Goh, K.L., Listrat, A., and Béchet, D. (2014). Hierarchical mechanics of connective tissues: integrating insights from nano to macroscopic studies. *J. Biomed. Nanotechnol.* 10 (10): 2464–2507.

18 Hubbard, R.P. (1984). Mechanical behavior of connective tissue. In: *Concepts and Mechanisms of Neuromuscular Functions* (ed. P.E. Greenman). Berlin/Heidelberg: Springer. https://doi.org/10.1007/978-3-642-69779-1_7.

19 Petrovich Omelyanenko, N., Ilyich Slutsky, L., and Pavlovich Mironov, S. (2016). *Connective Tissue: Histophysiology, Biochemistry, Molecular Biology*. CRC Press.

**20** Peters, A.E., Akhtar, R., Comerford, E.J., and Bates, K.T. (2018). The effect of ageing and osteoarthritis on the mechanical properties of cartilage and bone in the human knee joint. *Sci. Rep.* 8: 5931, https://doi.org/10.1038/s41598-018-24258-6.

**21** Özkaya, N., Leger, D., Goldsheyder, D., and Nordin, M. (2016). Mechanical properties of biological tissues. In: *Fundamentals of Biomechanics*, 361–387. Cham. https://doi.org/10.1007/978-3-319-44738-4_15: Springer.

**22** Landínez-Parra, N.S., Garzón-Alvarado, D.A., and Carlos Vanegas-Acosta, J. (2012). Mechanical behavior of articular cartilage. In: *Injury and Skeletal Biomechanics*. London, United Kingdom: IntechOpen. https://www.intechopen.com/chapters/38122 doi: 10.5772/48323.

**23** Mengoni, M. (2019). Computational modelling of tendons and ligaments: a pathway to understanding damage mechanisms. *Orthop. Proc.* 101-B (SUPP_2).

**24** Lee, N.M., Erisken, C., Iskratsch, T. et al. (2017). Polymer fiber-based models of connective tissue repair and healing. *Biomaterials* 112: 303–312, https://doi.org/10.1016/j.biomaterials.2016.10.013.

**25** Chen, X., Sase, K., Tsujita, T., and Konno, A. (2019). Numerical model of connective tissue for splitting brain fissure simulation. In: *Proceedings of the 2019 IEEE/SICE International Symposium on System Integration*, Paris, France, 118–123. doi: 10.1109/SII.2019.8700442.

**26** Wang, X., Neu, C.P., and Pierce, D.M. (2019). Advances toward multiscale computational models of cartilage mechanics and mechanobiology. *Curr. Opin. Biomed. Eng.* 11: 51–57, https://doi.org/10.1016/j.cobme.2019.09.013.

**27** Bandeiras, C. and Completo, A. (2017). A mathematical model of tissue-engineered cartilage development under cyclic compressive loading. *Biomech. Model. Mechanobiol.* 16: 651–666.

**28** Lycke, R., Walls, M., and Calve, S. (2019). Computational modelling of developing cartilage using experimentally derived geometries and compressive moduli. *J. Biomech. Eng.* 141 (8): 081002.

**29** Bhattacharya, P. and Viceconti, M. (2017). Multiscale modelling methods in biomechanics. *Wiley Interdiscip. Rev. Syst. Biol. Med.* 9, https://doi.org/10.1002/wsbm.1375.

**30** Lake, S.P. and Fang, F. (2016). Modelling approaches for evaluating multiscale tendon mechanics. *Interface Focus* 6: 20150044, https://doi.org/10.1098/rsfs.2015.0044.

**31** Feng, Y., Boukhris, S.J., Ranjan, R., and Valencia, R.A. (2015). *Multiscale Modelling in Biomechanics and Mechanobiology*. London: Springer-Verlag.

**32** Bhattarai, A. and Staat, M. (2018). Modelling of soft connective tissues to investigate female pelvic floor dysfunctions. *Comput. Math. Methods Med.* 1–16. https://doi.org/10.1155/2018/9518076, 2018.

**33** Baah-Dwomoh, A., McGuire, J., Tan, T., and De Vita, R. (2016). Mechanical properties of female reproductive organs and supporting connective tissues: a review of the current state of knowledge. *Appl. Mech. Rev.* 68 (6): 060801. https://doi.org/10.1115/1.4034442.

34 Gordon Betts, J., Young, K.A., Wise, J.A. et al. (2013). *Anatomy and Physiology*. Texas: OpenStax.

35 Marieb, E.N. and Hoehn, K.N. (2015). *Human Anatomy & Physiology*, Global Edition. UK: Pearson Education Limited.

36 Herbert, R. (1988). The passive mechanical properties of muscle and their adaptations to altered patterns of use. *Aust. J. Physiother.* 34 (3): 141–149.

37 Blemker, S.S. (2017). Chapter 17 – Three-dimensional modelling of active muscle tissue: the why, the how, and the future. In: *Translational Epigenetics, Biomechanics of Living Organs*, vol. 1 (ed. Y. Payan and J. Ohayon), 361–375. Academic Press. https://doi.org/10.1016/B978-0-12-804009-6.00017-1.

38 Spyrou, L.A., Brisard, S., and Danas, K. (2019). Multiscale modelling of skeletal muscle tissues based on analytical and numerical homogenization. *J. Mech. Behav. Biomed. Mater.* 92: 97–117.

39 Martin, K.S., Virgilio, K.M., Peirce, S.M., and Blemker, S.S. (2016). Computational modelling of muscle regeneration and adaptation to advance muscle tissue regeneration strategies. *Cells Tissues Organs* 202 (3, 4): 250–266.

40 Eddinger, T.J. (2014). Smooth muscle-protein translocation and tissue function. *Anat. Rec.* 297 (9): 1734–1746.

41 Manchanda, R., Appukuttan, S., and Padmakumar, M. (2019). Electrophysiology of syncytial smooth muscle. *J. Exp. Neurosci.* 13, 117906951882191. https://doi.org/10.1177/1179069518821917.

42 Appukuttan, S., Brain, K.L., and Manchanda, R. (2015). A computational model of urinary bladder smooth muscle syncytium: validation and investigation of electrical properties. *J. Comput. Neurosci.* 38 (1): 167–187.

43 Korogod, S.M., Osorio, N., Kulagina, I.B., and Delmas, P. (2014). Dynamic excitation states and firing patterns are controlled by sodium channel kinetics in myenteric neurons: a simulation study. *Channels* 8: 536–543.

44 Mahapatra, C., Brain, K.L., and Manchanda, R.A. (2018). A biophysically constrained computational model of the action potential of mouse urinary bladder smooth muscle. *PLoS One* 13: e0200712.

45 Appukuttan, S., Padmakumar, M., Young, J.S. et al. (2018). Investigation of the syncytial nature of detrusor smooth muscle as a determinant of action potential shape. *Front. Physiol.* 9: https://doi.org/10.3389/fphys.2018.01300.

46 Tan, T. and De Vita, R. (2015). A structural constitutive model for smooth muscle contraction in biological tissues. *Int. J. Non Linear Mech.* 75: 46–53. https://doi.org/10.1016/j.ijnonlinmec.2015.02.009.

47 Hodgkin, A.L. and Huxley, A. (1952). A quantitative description of membrane current and its application to conductance and excitation in nerve. *J. Physiol.* 117: 500–544.

**48** Ruiz-Baier, R., Ambrosi, D., Pezzuto, S. et al. (2013). Activation models for the numerical simulation of cardiac electromechanical interactions. In: *Computer Models in Biomechanics* (ed. G. Holzapfel and E. Kuhl). Dordrecht. https://doi.org/10.1007/978-94-007-5464-5_14: Springer.

**49** Mayourian, J., Sobie, E.A., and Costa, K.D. (2018). An introduction to computational modelling of cardiac electrophysiology and arrhythmogenicity. *Exp. Model. Cardiovasc. Dis.* 17–35. https://doi.org/10.1007/978-1-4939-8597-5_2.

**50** Wang, V.Y., Hussan, J.R., Yousefi, H. et al. (2017). Modelling cardiac tissue growth and remodelling. *J. Elast.* 129 (1, 2): 283–305.

**51** Millard, M., Uchida, T., Seth, A., and Delp, S. (2013). Flexing computational muscle: modelling and simulation of musculotendon dynamics. *J. Biomech. Eng.* 135 (2): 021004/1-11.

**52** Struzyna, L.A., Harris, J.P., Katiyar, K.S. et al. (2015). Restoring nervous system structure and function using tissue engineered living scaffolds. *Neural Regen. Res.* 10 (5): 679–685.

**53** Bilston, L.E. (2016). The influence of microstructure on neural tissue mechanics. In: *Structure-Based Mechanics of Tissues and Organs* (ed. G. Kassab and M. Sacks). Boston, MA. https://doi.org/10.1007/978-1-4899-7630-7_1: Springer.

**54** Budday, S., Ovaert, T.C., Holzapfel, G.A. et al. (2019). Fifty shades of brain: a review on the mechanical testing and modelling of brain tissue. *Arch. Comput. Meth. Eng.*, https://doi.org/10.1007/s11831-019-09352-w.

**55** Ratajczak, M., Ptak, M., Chybowski, L. et al. (2019). Material and structural modelling aspects of brain tissue deformation under dynamic loads. *Materials* 12 (2): 271. https://doi.org/10.3390/ma12020271.

**56** De Rooij, R. and Kuhl, E. (2016). Constitutive modelling of brain tissue: current perspectives. *Appl. Mech. Rev.* 68 (1): 010801. https://doi.org/10.1115/1.4032436.

**57** Zhao, W., Choate, B., and Ji, S. (2018). Material properties of the brain in injury-relevant conditions – experiments and computational modelling. *J. Mech. Behav. Biomed. Mater.* 80: 222–234. https://doi.org/10.1016/j.jmbbm.2018.02.005.

**58** Raul, J.-S., Deck, C., Willinger, R., and Ludes, B. (2008). Finite-element models of the human head and their applications in forensic practice. *Int. J. Legal Med* 122: 359–366.

**59** Roth, S., Raul, J.-S., and Willinger, R. (2010). Finite element modelling of paediatric head impact: global validation against experimental data. *Comput. Methods Programs Biomed* 99: 25–33.

**60** Ghajari, M., Hellyer, P.J., and Sharp, D.J. (2017). Computational modelling of traumatic brain injury predicts the location of chronic traumatic encephalopathy pathology. *Brain* 140 (2): 333–343. https://doi.org/10.1093/brain/aww317.

**61** Bhattacharya, M. and Dutta, A. (2019). Computational modelling of the photon transport, tissue heating, and cytochrome C oxidase absorption during transcranial near-infrared stimulation. *Brain Sci.* 9 (8): 179. https://doi.org/10.3390/brainsci9080179.

**62** Garcia-Gonzalez, D. and Jerusalem, A. (2018). Energy based mechano-electrophysiological model of CNS damage at the tissue scale. *J. Mech. Phys. Solids*, https://doi.org/10.1016/j.jmps.2018.12.009.

**63** FitzHugh, R. (1961). Impulses and physiological states in theoretical models of nerve membrane. *Biophys. J* 1: 445–466.

**64** Nagumo, J., Arimoto, S., and Yoshizawa, S. (1962). An active pulse transmission line simulating nerve axon. *Proc. IRE* 50: 2061–2070.

**65** Ramo, N.L., Troyer, K., and Puttlitz, C. (2019). Comparing predictive accuracy and computational costs for viscoelastic modelling of spinal cord tissues. *J. Biomech. Eng.*, https://doi.org/10.1115/1.4043033.

**66** Lubba, C.H., Le Guen, Y., Jarvis, S. et al. (2018). PyPNS: multiscale simulation of a peripheral nerve in Python. *Neuroinformatics*, https://doi.org/10.1007/s12021-018-9383-z.

**67** McIntyre, C.C., Richardson, A.G., and Grill, W.M. (2002). Modelling the excitability of mammalian nerve fibers: influence of afterpotentials on the recovery cycle. *J. Neurophysiol.* 87 (2): 995–1006.

**68** N. A. Pelot, C. E. Behrend, W. M. Grill, "On the parameters used in finite element modelling of compound peripheral nerves" *J. Neural Eng.* https://doi.org/10.1088/1741-2552/aaeb0c, 2018.

**69** Schiefer, M. (2014). Peripheral nerve models. In: *Encyclopedia of Computational Neuroscience* (ed. D. Jaeger and R. Jung). New York, NY. https://doi.org/10.1007/978-1-4614-7320-6_213-3: Springer.

# 6

# Macroscale Modeling at the Organ Level

## 6.1 Introduction

The basic unit of all living beings is the cell. A tissue is composed of cells with similar functions and structures. An organ is a structure composed of two or more tissue types and performs a specific function for the body. Organs like the lungs, the heart, the stomach, the skin, the kidneys, and the liver consist of two or more tissue types organized to perform a specific function. The lungs bring in oxygen and limit carbon dioxide, the heart pumps blood, and the skin serves as a barrier which protects internal structures from the external environment. An organ system is a group of organs that interact to conduct a specific body function. The organs of the digestive system work together to disintegrate food and absorb the end products into the bloodstream providing energy and nutrients for the body's cells. In human beings, there are 11 distinct organ systems forming the basis of anatomy. These are the respiratory system, the digestive system, the cardiovascular system, the urinary system, the integumentary system, the skeletal system, the muscular system, the endocrine system, the lymphatic system, the nervous system, and the reproductive system. The lymphatic system also encompasses a functional system called the immune system, which is composed of a number of mobile cells that act to protect the body from foreign substances. Each organ system is characterized by specific regulatory mechanisms and complex structural organization leading to complex transient, nonlinear, and intermittent behavior. Medical specialists traditionally focus on a single physiological system. For instance: (i) cardiologists examine the heart and evaluate electrocardiogram (ECG) signals and other parameters, (ii) brain neurologists examine brain wave electroencephalography (EEG) signals, magnetic resonance imaging (MRI) and other factors, (iii) pulmonologists check lung structure and function, as well as

*Multiscale Modelling in Biomedical Engineering*, First Edition. Dimitrios I. Fotiadis, Antonis I. Sakellarios, and Vassiliki T. Potsika.

Published 2023 by John Wiley & Sons, Inc.

probe respiratory patterns. Nevertheless, the human organism is considered an integrated network, where multicomponent organ systems continuously interact across different spatiotemporal scales and through various feedback mechanisms to control and improve their function [1].

The analysis of multiple anatomical structures requires the development of new approaches specifically tailored to the context of medical imaging. The availability of large imaging datasets, together with the continuous increase in computing power, has been instrumental to the progress of medical image analysis in general, and to the development of efficient and sophisticated approaches to model the different organ systems and interorgan interactions. These challenges include the simultaneous characterization of the interorgan relations together with the particular locality of each organ, the use of complex anatomical and pose priors, or the need for geometrical constraints that prevent overlapping between organs, among others [2].

The next sections (sections 6.2–6.11) present an analysis of the function of the different organ systems and the current trend of the use of computational modeling approaches at the macroscale level.

## 6.2 The Respiratory System

The main function of the lungs is to keep air in contact with blood so that oxygen can diffuse into and carbon dioxide out of the blood. In addition, the lungs have important nonrespiratory functions such as clearing particles from the airways and converting or inactivating prostaglandins, circulating polypeptides, etc. [3].

From an anatomical point of view, the trachea branches into two main bronchi and these then subdivide into lobar bronchi, segmental bronchi, and so on for 27 generations. The first 19 create the conducting airways that do not have any gas-exchange function. Generations 1–7 correspond to the cartilaginous airways or bronchi and 8–19 to the membranous airways or nonrespiratory bronchioles that end as terminal bronchioles. Generations 20–23 are composed of respiratory bronchioles and 24–27 are alveolar ducts and both generations participate in gas exchange. The branches arising from a single terminal bronchiole are collectively known as the terminal respiratory unit. The bronchial arteries provide the airways with their blood supply and arise from the aorta and its branches. Together the cartilaginous bronchi and the nonrespiratory bronchioles make up the bulk of the anatomical dead space (air within them does not take part in gas exchange). The cartilaginous airways vary from the trachea and larger bronchi to bronchi approximately 2 mm in diameter and are supported by incomplete rings of smooth muscle and cartilage. The nonrespiratory bronchioles lack cartilage and secretory glands but do have smooth muscles in their walls. Concerning the larger bronchi, smooth-muscle contraction brings together tips of U-shaped cartilages. In the medium- and smaller-sized bronchi and bronchioles, contraction reduces both the caliber and the length of the airways due to the distribution of the muscle fibers in their walls. Therefore, airway smooth-muscle contraction produces increased rigidity, as well as narrowing of the airways [3].

### 6.2.1 Computational Modeling of the Respiratory System

From a mathematical perspective, the lung consists of a complex network of branching compliant tubes spanning a wide range of scales and flow regimes from almost laminar diffusion-dominated processes to highly turbulent and pulsatile flows. Further, at the terminal branches of the airway tree, flow enters into a sponge-like microstructure of lung tissue whose material behavior is highly non-linear and whose topology is optimized for efficient gas exchange with small embedded blood vessels. Recently, various efforts have been made in mechanical and mathematical modeling of the lung to: (i) study isolated effects which are hard to assess in a subject *in vivo*, (ii) to advance medical imaging and functional diagnostics by incorporating computational models which are based on the underlying physical characteristics, and (iii) to finally assist in patient-specific treatment planning and optimization.

In this direction, advanced computational models of the respiratory system are presented in [4] with a special focus on approaches that are able to tackle the interaction between flow and tissue components, which is necessary to accurately represent the underlying physics of the lung. These models are known as coupled models and present strategies for sensible and target-oriented dimensional reduction. Four suitable coupled approaches are presented introducing their underlying modeling idea and assumptions, their novelty against previous methods, possible scenarios of application, and limitations in clinical practice. In brief, the first model in [4] was the fully resolved three-dimensional lung model (3D/3D). This is a fluid–structure interaction (FSI) problem extended by the idea that all fluid flow that enters or leaves an outlet of the deformable fluid domain has to change the volume of the associated parenchymal region inducing deformation. The volumetrically coupled 3D/3D model induces a consistent deformation of the fluid and structural domains, which is governed by the current flow characteristics and not prescribed from additional information such as imaging data recorded under different or ambiguous flow conditions. Further, correct boundary conditions are applied at the last fully resolved generation of airways of the deformable fluid domain. The second model was the reduced-dimensional structure and fully resolved three-dimensional fluid (0D/3D). This is a classical fluid or FSI problem extended by a nonlinear boundary condition for the outlet pressure dependent on current flow and tissue volume. For efficiency, the boundary condition is linearized within the system of equations of the fluid/FSI problem and can be solved within the same Newton step, i.e. no further iteration loop is needed for computing the boundary pressure. We put the coupling of a fully resolved deformable fluid domain with resistive airway trees also in this model category as it shows basically the same effect on the deformable fluid domain. Whether such models are able to model expiration depends on the availability of storage capacity at the terminal ends of the tree [5, 6]. The novelty against previous approaches seems to be limited; however, it has a great impact in the field of respiratory biomechanics. The third model is the fully resolved three-dimensional structure and reduced-dimensional fluid (3D/0D). This is essentially a combination of pressure and flow computed from well-known reduced-dimensional fluid models with volume changes and tractions of a fully resolved structure or poroelastic medium. This concept is valuable when mainly structural aspects of the lung are to be investigated. In comparison with the 3D/3D model, the 3D/0D approach is approximately 10 times faster for the same patient-specific geometry and boundary conditions [7]. Finally, the fourth model is the reduced-dimensional lung model (0D/0D). The last presented 0D/0D model is essentially a dimensionally reduced relationship between pressure and flow in the compliant airways and in the viscoelastic acini equipped with all properties that are necessary to accurately describe the underlying physics of the lung. Extensions include interdependency, a literature-based thorax boundary condition, flexible airways able to collapse and reopen [8], and an acinar model that is valid over the entire physiological pressure range [9]. The quality of presented lung models is extended via regional validation against clinical measurements. This validation is performed using temporal highly resolved electrical impedance tomography monitoring. This detailed and, for the first time, dynamic regional validation generates further trust in the presented mathematically derived approaches [4].

In another study [10], the aim of the authors was to understand respiratory airflow in different age groups in the age-specific treatment of respiratory disorders. This study examined numerically the age-related impact on inspiratory and expiratory airflow dynamics. In this direction, four-generation lung airway models were established under normal breathing conditions. Tracheobronchial airway models of infants (6 months old), children (5 years old), and adults (25 years old) from the sixth to ninth generations were developed. Computational fluid dynamics (CFD) was applied to solve the equations characterizing the airflow. The air within the human respiratory tract was considered to be a homogeneous, Newtonian, and incompressible fluid. Airway generations were labeled as G0–G6, where G0 is the trachea and G6 is the sixth generation. The Womersley numbers at the inlet of G6 were about 0.25, 0.34, and 0.44 for infants, children, and adults, respectively, during inspiration under normal conditions. These indicate that the unsteady effects of the flow fields are relatively minor. Thus, a steady airflow was assumed in the airway model G6–G9. In the regions of the airway models, the Reynolds number is sufficiently low assuming laminar flow. The governing equations for the airflow are the continuity and Navier–Stokes equations for a viscous incompressible Newtonian fluid. In vector notation, these are:

$$\nabla \cdot \mathbf{u} = 0, \tag{6.1}$$

$$\rho\left(\mathbf{u} \cdot \nabla \mathbf{u}\right) = -\nabla p + \mu \nabla^2 \mathbf{u}, \tag{6.2}$$

where $\mathbf{u}$ is the velocity vector, $p$ is the static pressure, $\rho$ is the density, and $\mu$ is the dynamic viscosity of the air. Taking US Standard Atmosphere Air Properties Data at the sea level, the density and dynamic viscosity of the air are taken to be $\rho = 1.225\,\text{kg/m}^3$ and $\mu = 1.7894 \times 10^{-5}$ kg/(ms). Concerning the boundary conditions, the velocity inlet and pressure outlet conditions were used at the inlets and outlets of the airway models, respectively, while the no-slip condition was imposed on the walls for inspiratory and expiratory airflow modeling. In this study, the breathing parameters, respiratory rate and tidal volume were derived from [11] and the inspiration-to-expiration time ratio was adopted from [12]. The Weibel model for flows in lung airways was applied to derive the corresponding flow rates and the inlet velocities. It was found that as age increases, airflow velocity, pressure, and wall shear stress decrease for both inspiration and expiration in this particular subregion of the respiratory tract [10]. During inspiration, the splitting of velocity streamlines at bifurcations increases with age. The opposite situation evolves during expiration, and it also increases with age. The level of splitting and merging of streamlines reflects the influence of respiratory mechanics in the age groups. Finally, it was concluded that the computational models give insight into new information on patterns and characteristics of age-dependent respiratory airflow in the sixth to ninth generations of tracheobronchial airways and can be used for other generations.

## 6.3 The Digestive System

The digestive system starts from the mouth to the anus. It is able to work with peristaltic movements-automatic compression and relaxation food pipe which pushes the food particles through the digestive system. The important parts of the digestive system are the mouth, teeth, tongue, esophagus, liver, gallbladder, pancreas, stomach, small intestine, large intestine (colon), rectum, and anus. The mouth is the entry point of the digestive system where the breakdown of the food substance takes place mechanically, e.g. chewing and crushing. There are four different types of teeth with different functions. The incisors which are placed at the front of the mouth have a sharp biting surface serving for cutting or shearing food into small chewable pieces. The canines are situated at the "corners" of the dental arches. They have a sharp, pointed biting surface. Their function is to grip and tear food. The premolars differ from the incisors and canines as they have a flat biting surface. Their function is to tear and crush food. The molars are used to chew, crush, and grind food.

The tongue secretes the salivary glands which produce the bolus, mix with the food, and make it easy to digest. The esophagus is the main part of the food track. It allows the down-way movement of food substances and this movement is called alimentary canal or peristalsis movement. The liver creates the bile from the bile duct, which cleanses and purifies the blood which is coming from the small intestine. The gallbladder stores and absorbs the chemicals, nutrients, and the remaining liquid is sent back to the small intestine. The pancreas secretes the enzyme called the insulin which breaks the food substance down into basic glucose. The sphincter is a connecting tube between stomach and esophagus. The stomach is capable of holding food substances of about 4 L and mixes and digests the food with hydrochloric acid. The stomach and its mechanics play a key role not only in the digestion of food, one of the most essential processes in living organisms, but also in drug administration. Around 70% of all drugs are administered orally, and their processing and effectiveness thus depend mainly on gastric mechanics [13]. The small intestine (6–7 m in length but tiny-sized pipe-like structure) is the most important digestive organ which digests almost 90% of the food substances. There are three parts – duodenum, jejunum, and ileum. In the large intestine (colon) (1.5 m in length), the remaining water and useful substances are absorbed; the waste substance is excreted out to the rectum and to the anus. The large intestine has ascending colon, descending colon, and transverse colon. The feces are stored in the rectum (8-in. base part of the large intestine). The anus waste products are liquid and solid. The liquid waste excretes in the urinary bladder and solid wastes go out via the anus.

The digestive system works cooperatively with the other body systems. An indicative case is an interrelationship between the digestive and cardiovascular systems. Arteries provide the digestive organs with processed nutrients and oxygen, while the role of veins is to drain the digestive tract. These intestinal veins are unique and constitute the hepatic portal system. They do not return blood directly to the heart as this blood is diverted to the liver where its nutrients are off-loaded for processing before blood completes its circuit back to the heart. Also, the digestive system provides nutrients to the vascular tissue and heart muscle to support their function. The interrelationship of the digestive and endocrine systems is also critical. Hormones secreted by several endocrine glands, as well as endocrine cells of the pancreas, the stomach, and the small intestine, contribute to the control of digestion and nutrient metabolism. In turn, the digestive system provides the nutrients to fuel endocrine function. Diseases of the digestive system are disorders of the digestive tract, which is also known as the gastrointestinal tract. In digestion, food and drink are broken down into small parts (called nutrients) that the body can use as energy and building blocks for cells. The digestive tract is made up of the esophagus, stomach, liver, large and small intestines, pancreas, and gallbladder.

### 6.3.1 Computational Modeling of the Digestive System

This subsection focuses on mathematical and computational modeling of the stomach which is an emerging field of biomechanics where several complex phenomena, such as gastric electrophysiology, fluid mechanics of the digesta, and solid mechanics of the gastric wall, need to be addressed [13]. During gastric digestion, food is disintegrated according to a complex process of mechanical and chemical effects. Although the process of chemical digestion is usually assessed by using *in vitro* analysis, the difficulty in reproducing the stomach motility and geometry has obstructed a good understanding of the local fluid dynamics of gastric contents [14].

In general, the fluid flow in the gastric lumen is assumed to show laminar and incompressible behavior being expressed by a balance of mass and a balance of momentum in Eulerian form:

$$\nabla \mathbf{u} = 0, \tag{6.3}$$

$$\rho\left(\frac{\partial \mathbf{u}}{\partial t} + \mathbf{u}\cdot\nabla\mathbf{u}\right) = -\nabla p + \nabla\cdot\tau + \rho\mathbf{g}, \tag{6.4}$$

with fluid density $\rho$, velocity field $\mathbf{u}$, pressure $p$, gravity $\mathbf{g}$, and viscous stress tensor $\boldsymbol{\tau}$. The viscous stress tensor $\boldsymbol{\tau}$ is typically expressed in terms of the strain rate tensor:

$$\mathbf{E} = \frac{1}{2}\left(\nabla\mathbf{u} + \left(\nabla\mathbf{u}\right)^{\mathrm{T}}\right). \tag{6.5}$$

As a first approximation, dilute digesta can be modeled as a Newtonian fluid, which is characterized by a linear relationship of the strain rate tensor $\mathbf{E}$ and the viscous stress tensor $\boldsymbol{\tau}$ via a dynamic viscosity $\mu$:

$$\tau = 2\mu\mathbf{E}. \tag{6.6}$$

The fluid viscosity of gastric juice in the stomach typically varies from 0.01 to 2 Pa·s. Experiments *in vitro* showed that the viscosity of digesta decreases within 40 minutes from a maximum of 17 Pa·s to 2.2 Pa·s caused by intragastric dilution with gastric juice. While the assumption of a Newtonian fluid may be appropriate as a reasonable approximation for dilute digesta, there are obviously many cases where it appears not applicable. Then a non-Newtonian fluid with a nonlinear relation between strain rate and shear stress has to be assumed, typically with shear thinning behavior.

Examination of gastric fluid mechanics *in vivo* and *ex vivo* is difficult and often nearly impossible due to technical hurdles and ethical concerns, especially in humans. *In vitro* models have greatly helped advance our understanding of gastric fluid mechanics but can hardly mimic real gastric geometry and motility. The first CFD models of the stomach were developed around 15 years ago. After the first attempts in two dimensions, in 2007, 3D computational models have also been proposed [14, 15]. The goal of [14] was to use CFD to develop a 3D model of the shape and motility pattern of the stomach wall during digestion and use it to characterize the fluid dynamics of gastric contents of different viscosities. To this end, a geometrical model of an average-sized human stomach was presented. The motility of the model was characterized by a series of antral-contraction waves of up to 80% relative occlusion. The flow field that develops within the stomach was modeled as a laminar and incompressible fluid flow of a continuous liquid phase. Under these flow conditions, the conservation of mass and momentum within the system were given by:

$$\frac{\partial u_i}{\partial x_i} = 0, \tag{6.7}$$

$$\frac{\partial u_i}{\partial t} + u_j \frac{\partial u_i}{\partial x_j} = -\frac{1}{\rho}\frac{\partial P}{\partial x_i} + \frac{\mu_\alpha}{\rho}\frac{\partial^2 u_i}{\partial x_j \partial x_j}, \tag{6.8}$$

in which $u_i$ is the velocity component in the $x_i$ direction, $t$ is the time, $\rho$ is the density, and $\mu_\alpha$ is the dynamic viscosity. The flow field within the model strongly depends on the viscosity of gastric contents. By increasing the viscosity, the formation of the two flow patterns considered the main mechanisms driving digestion (the retropulsive jet-like motion and eddy structures) was significantly reduced, while a significant pressure field increase was predicted. These results were in agreement with experimental data in the literature revealing that contrary to the traditional idea of a rapid and complete homogenization of the meal, gastric contents associated with highly viscous meals are poorly mixed. This study shows the capability of CFD to provide a better comprehension of the fluid dynamics of the gastric contents and indicates its potential to develop a fundamental understanding of the mechanisms involved in the digestion process [14].

Recently, a promising smooth particle hydrodynamics computational model developed originally for the intestine was also applied to a stomach-like geometry examining gastric motility and emptying [16, 17]. These computational models revealed important patterns in gastric flow. For low viscosity dominant retrograde jets in the antrum, circulatory flow between the antral contraction waves (ACWs), and a so-called "stomach road," along which gastric emptying occurs predominantly, were reported. In contrast, for higher viscosity, more ordered patterns prevail. The importance of buoyancy and antral recirculation for gastric mixing was disclosed by different techniques. Food disintegration has been studied so far only using highly simplified models (e.g. assuming a fixed amount of erosion each time a particle is passing through an ACW). Despite their great merits, all current computational models of the human stomach suffer from specific limitations as follows: (i) the gastric wall is modeled as rigid and its deformation during ACWs is kinematically prescribed. It is experimentally well confirmed that altered properties of the digesta or also altered gastric geometry can significantly change gastric flow, motility, and emptying. Current models with a prescribed wall motion can by design not account for these dependencies and do thus not allow predictive simulations of the impact of substantial parameter changes in the stomach. Such predictive simulations would require coupled multiphysics models accounting for interactions between flow of the digesta and viscoelasticity and electrophysiology of the wall, (ii) current models have never been applied with person-specific *in vivo* geometries and associated data about gastric motility (i.e. muscular contractions), (iii) mechanical and chemical food disintegration is currently not modeled explicitly. The effect of gastric juice is generally neglected despite its significant impact on food disintegration and, thus, also on the time-dependent viscosity of the digesta and gastric flow [13].

## 6.4 The Cardiovascular System

The main role of the cardiovascular system is to support mass transport, which refers to the transport of oxygen, nutrients, carbon dioxide, hormones, waste products, etc., within the body. The main components of the cardiovascular system are three: (i) the heart, which serves as the pump, (ii) the blood, which is the conducting medium, and (iii) the vasculature functioning as the conduit through which the blood flows. Blood is pumped by the heart toward the arteries, and after tissue irrigation, the veins ensure the filling of the cardiac chambers. This circulation system ensures human body cell survival by carrying all the necessary nutrients and removing the waste products of tissue metabolism. The cardiopulmonary and renovascular systems are considered closely related systems. Several numerous "special" circulations can be observed within the vasculature such as the coronary, cerebral, pulmonary, and fetal [5, 6].

### 6.4.1 Computational Modeling of the Cardiovascular System

Computational models of the heart cover most of the biophysical complexity of the individual patient's cardiac pathology. As a result, models linking cellular electrophysiology, myocardial tissue mechanics, and system hemodynamics have become promising platforms for virtual-patient simulations and for *in silico* evaluation of novel diagnostic and therapeutic strategies [18–22].

Patient-specific computational fluid dynamic models are being used to address aortic aneurysms, coronary stenosis, cardiac valves, and congenital heart disease. Biventricular patient-specific models of electromechanics have been applied to heart failure, left ventricular assist devices, and cardiac resynchronization therapy. Patient-specific models of electrophysiology have shown promise regarding genetic mutations, ablation therapy, and clinical classification criteria [23].

The construction of a patient-specific model typically involves a workflow in which patient data are merged with equations and other external data. External data can be any data that is not personalized and can be obtained from a variety of sources such as experimental results, clinical studies, and the literature These pipelines involve obtaining information from the patient such as age, sex, survey results, and even physician diagnoses. In addition, measurements are taken from the patient using various instruments including sophisticated imaging modalities. The development of patient-specific electrophysiological heart models became possible through the availability of cardiac images from patients, usually MRI or CT, and was substantially propelled by the movement toward personalized medicine. Clinical MRI data are used to visualize the structural remodeling in atria and ventricles and, therefore, have been used in computational modeling. Atlases of ventricular geometry and shape have been assembled by averaging 3D cardiac image datasets from individuals and generating a mean 3D cardiac image or shape representative of cardiac anatomy [19, 20]. Together, these streams represent the patient's raw data which together with external data and equations governing the physical process being modeled are used to develop the patient-specific model [23].

Two medical devices have recently been marketed in the USA, which include patient-specific cardiovascular models. First, the Heartflow® fractional flow reserve derived from computed tomography (FFRCT) is a postprocessing software for the clinical quantitative and qualitative analysis of image data for clinically stable symptomatic patients with coronary artery disease [24]. The workflow involves generating a personalized mathematical geometrical representation of the coronary arteries and performing CFD simulations using lumped parameter models of the heart, systemic circulation, and smaller downstream coronary arteries as boundary conditions. Specifically, it provides FFRCT, a mathematically estimated quantity, computed from simulated pressure, velocity, and blood flow information derived from a 3D computational model created from the patient's static coronary CT images. Second, the Medtronic CardioInsight® Cardiac Mapping System is a noninvasive mapping system for beat-by-beat, multichamber, 3D mapping of the heart [25]. This device includes solving the classic electrocardiographic "inverse problem," i.e. computing the dipole sources on the heart surface from multiple potential measurements from the body surface [26]. This is accomplished by CardioInsight® via a workflow that involves computing the personalized torso and epicardial heart surfaces from CT images, and then computing the virtual electrograms on the heart surface using body surface potential signals recorded from > 200 electrodes from a vest worn by the patient [27]. Both Heartflow® FFRCT and CardioInsight® Cardiac Mapping System use CT images to generate a representation of the patient's anatomy and solve the governing equations using models with many nonpersonalized parameters.

In Chapter 9, computational multiscale modeling approaches of the cardiovascular system are discussed in detail at different hierarchical levels. Therefore, in this subsection, only fundamental information was provided. Nevertheless, it should be noted that although simulations have made inroads in the clinic, a number of barriers to their widespread adoption remain. Computational studies in cardiology are currently targeted at a narrow range of heart diseases for which the effect of treatment is well-characterized, imaging and invasive measurements are routinely made and treatment can be evaluated with a short-term measure of success [18]. The advancement of algorithms and approaches for high-speed simulations to enable cardiac modeling has become a routine clinical tool. Finally, communicating model results to cardiologists and caregivers in a simple format and under the right conditions will be critical for the adoption of model predictions in clinical decision-making.

## 6.5 The Urinary System

The urinary system is composed of the kidneys, bladder, ureters, and urethra. It is also known as the urinary tract or renal system. Its main function is to eliminate waste from the body, regulate blood volume and blood pressure, and control blood pH and levels of electrolytes and metabolites. More specifically, the kidneys filter the blood to remove metabolic wastes and then modify the resulting fluid, which allows these organs to maintain fluid, electrolyte, acid–base, and blood pressure homeostasis. This process creates urine, which is a fluid consisting of electrolytes, water, and metabolic wastes. The kidneys receive approximately one-fourth of the total cardiac output from the right and left renal arteries (about 1200 ml/min), which branch from the abdominal aorta. Then, the remaining organs of the urinary system transport, store, and eliminate urine from the body [28].

### 6.5.1 Computational Modeling of the Urinary System

Models of the lower urinary tract are used to understand better the physiological and pathological functions of the tract and to gain insight into the relative importance of different components. Different types of models have been presented in the literature such as: (i) computational models that describe mathematically the whole urinary tract or components, (ii) physical models useful especially in testing medical devices, and (iii) tissue-engineering models. The purpose of modeling is first described in terms of the ability of models to predict the properties of the system of interest, using components that have a physiological interpretation, and to gain insight into the relative importance of different components [29].

Urine flow through a complex structure such as the urinary tract depends upon the interacting physical properties of its component parts. The purpose of mathematical models is to reconstruct the function of the whole tract by simulating in a linear or nonlinear fashion the properties of the component parts so that the relative importance of different components may be assessed. Initial approaches were to treat urine flow through the urinary tract as a fluid dynamics problem. This was especially amenable with respect to the ureter and urethra that could be treated as collapsible (distensible) tubes surmounted by a variable pressure head, supported or not by peristaltic movements [30, 31]. From these studies, several conclusions have been drawn, e.g. in a closed tube, the active and passive mechanical properties of the tube determine flow, whereas in open tubes, outlet resistance dominates, and with peristalsis also present, the magnitude and frequency of individual waves were also crucial. Similar models described the passive properties of the urinary bladder during filling, without including a detailed representation of structure and function. These models accurately reproduced the pressure and volume response of the bladder to both slow and stepwise filling. Subsequently, active contractile properties of the bladder wall were added in a similar, global manner, and it was possible to estimate those factors that were the principal determinants of pressure. In this case, the size and shape of the bladder, as well as its intrinsic contractile strength is significant and embodied in such concepts as the Watts factor. These studies have permitted nomograms and algorithms to be constructed that help, for example, to define the extent of functional bladder outflow obstruction and estimate detrusor contractile function. A benefit of such an approach is that certain parameters may be defined to aid the clinician in describing the presence and extent of the lower urinary tract dysfunction and propose strategies of management.

Attention may then focus on a more detailed description of component parts, such as the interaction of contractile bladder function with the resistance of the urethra during normal voiding [32]. An example of the approach is illustrated here where the lower urinary tract is modeled as a contractile organ described by the Hill equation contracting against an outflow resistance. The relationship between the velocity $u$, of muscle contraction versus a load $F$, is:

$$\frac{F}{F_0}+\frac{a}{F_0}(u+b)=\left(1+\frac{a}{F_0}\right)b, \tag{6.9}$$

where $F_0$ is the load at zero velocity and $a$, $b$ are constants. For a number of muscles, the ratio $a/F_0$ is about 0.25 and replacing $F$ by pressure, $P$ and $u$ by flow $Q$ yields the bladder output relation:

$$\left(P+\frac{P_0}{4}\right)\left(Q+\frac{Q_0}{4}\right)=\frac{5}{16}P_0Q_0. \tag{6.10}$$

The simplest urethral pressure-flow function is:

$$P = P_{\text{op}} + RQ, \tag{6.11}$$

where $P_{\text{op}}$ is the urethral opening pressure and $R$ is the outlet resistance, assumed to be constant.

More recent studies [33] deal with numerical modeling of kidney hemodynamics. The kidney is responsible for maintaining the stability of the body's extracellular environment. This is executed by regulating the solute concentrations in the blood, the volume of extracellular fluid, the acid–base balance, and blood pressure in the body, as well as excreting urea and other metabolic waste products. The kidney also plays a role in the body's endocrine system, secreting hormones such as renin as part of the renin–angiotensin system, erythropoietin to regulate red blood cell production and thrombopoietin to regulate platelet production. The human kidney is made up of approximately $1\times10^6$ nephrons, convoluted tubules that adjust the solute levels of the body's blood plasma. Nephrons are supplied with blood by a network of blood vessels, and filter the blood, removing metabolic wastes, and reabsorbing the solutes and hormones needed by the body [34]. Through regulation of the extracellular fluid volume, the kidneys provide important long-term regulation of blood pressure. At the level of the individual functional unit (the nephron), flow and pressure control involves two different processes that both generate oscillations. The nephrons are arranged in a complex branching structure that delivers blood to each nephron and, at the same time, provides a basis for interaction between adjacent nephrons. The functional consequences of this interaction have not been well understood. For this reason, experimental data and a new modeling approach were presented in [33] to clarify this problem. To resolve details of microvascular structure, 3D data from more than 150 afferent arterioles were collected in an optically cleared rat kidney. Using these results together with published μCT data, an algorithm for generating the renal arterial network was developed. Then, a mathematical model was introduced describing blood flow dynamics and nephron-to-nephron interaction in the network. The model included an implementation of electrical signal propagation along a vascular wall. Simulation results revealed that the renal arterial architecture plays an important role in maintaining adequate pressure levels and the self-sustained dynamics of nephrons.

In [34], systems of nephrons, the operational units of the kidney, were modeled and the dynamics of such systems were explored. Nephron behavior can fluctuate widely and, under certain conditions, become chaotic. However, the behavior of the whole kidney remains remarkably stable and blood solute levels are maintained under a wide range of conditions even when many nephrons are damaged or lost. A network model was used to investigate the stability of systems of nephrons and interactions between nephrons. More sophisticated dynamics were explored including the observed oscillations in single nephron filtration rates and the development of stable ionic and osmotic gradients in the inner medulla which contribute to the countercurrent exchange mechanism. This model was used to explore the effect of changes in input parameters including hydrostatic and osmotic pressures and concentrations of ions, such as sodium and chloride. The intrinsic nephron control, tubuloglomerular feedback, was included and the impact of coupling between nephrons is assessed in 2-, 8-, and 72-nephron models. This is considered a novel approach to modeling multinephron systems, which combines graph automata and complex network approaches into a single model, allowing the connectivity and interactions in the model to be altered interactively.

## 6.6 The Integumentary System

The integumentary system is the organ system that protects the body from damage and includes the skin and its appendages. The skin combines various tissues which perform important functions for human survival. The skin helps maintain body temperature, receives stimuli from the environment, and stores chemical compounds. The skin also acts as a protective covering keeping underlying tissues from bacterial invasion and harmful light rays and from drying out [35]. At the meso-macroscopic level, skin is generally considered a multilayer assembly made up of three main distinct structures: the epidermis, dermis, and hypodermis [36].

The outermost level, the epidermis, consists of a specific constellation of cells known as keratinocytes, which function to synthesize keratin, a long, threadlike protein with a protective role. The epidermis – which is avascular – is a terminally differentiated stratified squamous epithelium about 200 μm thick. Ninety percent of the cells contained in the epidermis are keratinocytes which undergo mitosis at the 0.5–1 μm-thick epidermal basement membrane (also known as basal lamina) [37].

The dermis, which is the middle layer, is fundamentally composed of the fibrillar structural protein known as collagen. The dermis lies on the subcutaneous tissue, or panniculus, with small lobes of fat cells called lipocytes. The thickness of these layers varies considerably, depending on the geographic location of the anatomy of the body [35]. The living epidermis is connected to the underlying dermis through a 3D interlocking wavy interface, called the dermal–epidermal junction (DEJ) which is the basal lamina. Papillae are the protrusions of the papillary dermis into the epidermis. These finger-like structures increase the contact surface area between the reticular dermis and the living epidermis and are thus believed to favor biochemical mass exchanges between these layers, e.g. transport of oxygen or nutrients. The DEJ controls the transit of biomolecules between the dermis and epidermis according to their dimension and charge. It allows the passage of migrating and invading cells under normal (i.e. melanocytes and Langerhans cells) or pathological (i.e. lymphocytes and tumor cells) conditions [36]. Collagen fibers account for approximately 70% of the weight of dry dermis. Collagen fibers in the papillary and subpapillary dermis are thin and sparsely distributed, while reticular fibers are thick, densely distributed, and organized in bundles. Fibrils are typically very long, 100–500 nm in diameter featuring a cross-striation pattern with a 60–70 nm spatial periodicity. The diameter of thick collagen bundles can span from 2 to 15 μm [38].

The subcutaneous tissue, also called the hypodermis, is the lowermost layer of the integumentary system in vertebrates. The types of cells found in the hypodermis are fibroblasts, macrophages, and adipose cells. In several animals (e.g. hibernating mammals), the hypodermis composes an important insulating layer and/or food store. In several plants, the hypodermis forms a layer of cells under the epidermis of leaves which is often mechanically strengthened (e.g. in pine leaves) creating an extra-protective layer or a water storage tissue.

### 6.6.1 Computational Modeling of the Integumentary System

From the mechanical and material science points of view, the skin is mainly a multiphasic and multiscale structure which, as a result, encompasses a rich set of mechanical properties and constitutive behaviors. The biological nature of this structure implies that these properties are very dynamic, particularly over a lifetime, and there is a strong variability according to factors such as the body site, individuals, age, sex, exposure to specific environmental conditions, and ethnicity [36].

A variety of methods have been used to measure the mechanical properties of skin: e.g. uniaxial and biaxial tensile tests [39], multiaxial tests [40], application of torsion loads [41], indentation [42], suction [43], and bulge testing [44]. More recent techniques have focused on the experimental characterization of the mechanical properties of the epidermis [45, 46] which are particularly relevant for cosmetic and pharmaceutical applications. Besides intra- and interindividual biological variability as well as sensitivity to environmental conditions, the nature of these experimental techniques combined with their operating spatial scale (e.g. macroscopic or cellular levels) are the main reasons for such a wide variability in mechanical properties reported in the literature. Differences in the mechanical properties of the skin or those of its individual layers can span several orders of magnitude. Considering its complex hierarchical structure, the skin consolidates a wide range of viscoelastic phenomena including creep, hysteresis, relaxation, and strain-rate dependency [36].

According to [36], mathematical and computational models of the skin can be classified into three main categories: phenomenological, structural, and structurally based phenomenological models. Phenomenological models are based on the assumption that the skin is a homogeneous material where the microstructure and multiple phases and their associated mechanical properties are ignored. These phenomenological models aim to capture the overall – generally macroscopic – behavior of the tissue without accounting for the individual behavior of its elemental constituents and their mutual interactions [47]. Typically, if one considers mechanical behavior only, a phenomenological model is a set of mathematical relations that describe the evolution of stress as a function of strain [48]. Provided the formulation is appropriate, it is generally always possible to fit such a constitutive law to a set of experimental data. However, the main drawback of this approach is that the resulting constitutive parameters often do not have a direct physical interpretation, and the model effectively acts as a "black box" without the flexibility of integrating and studying mechanistic structural effects. Structural models of the skin simulate the tissue as a composite material consisting of an explicitly defined assembly of key microstructural elements such as collagen fibers arranged in bundles with a certain degree of crimp and dispersion, within a matrix mainly including proteoglycans. The way these structural elements interact can also be specified by developing appropriate equations (e.g. mutual shear interactions of collagen fibrils and fibers or small-range electromagnetic interactions between proteoglycans and collagen fibers [49]). In this approach, not only the mechanical properties of the individual basic structures need to be determined or known but also the way they are geometrically arranged to form the macroscopic tissue, and how they interact mechanically, thermally, or through any other type of physics. The overall mechanical properties of the tissue are the result of this – generally nonlinear – coupling between geometry and mechanics. Structural models can be viewed as geometrical assemblies of phenomenological models. For this reason, one could argue that, strictly speaking, structural models are not fully structural, and that this classification as a structural or phenomenological model is a matter of spatial scale. Only if models are built from the beginning from the first principles of quantum chemistry could one talk about structural models. In that respect, the constitutive parameters of phenomenological models describing the behavior of elemental microstructural components may also not have a direct physical interpretation but, typically, they do. For example, if collagen fibers are part of the formulation, their elastic modulus, waviness, and degree of dispersion around the main orientation are indeed constitutive parameters and have a direct physical interpretation. One of the main drawbacks of structural models lays in the necessity to have accurate information about the geometrical and material characteristics of each elemental building block as well as their mutual spatial and interfacial arrangement. This presents obvious experimental characterization challenges which, however, are progressively overcome as technologies in this field improve and new techniques emerge. Moreover, in a computational finite-element (FE) environment, structural models spanning several orders of magnitude in terms of length scale require significantly high mesh density to explicitly capture the geometry of the structural constituents. Inevitably, despite tremendous advances in the computing power of modern processors, this can lead to prohibitively computationally expensive analyses and lengthy run times. A judicious compromise between strictly phenomenological and structural models is a third class of models, formulated by combining certain characteristics of phenomenological models with those of structural ones. These models could be denoted under the general appellation of structurally based phenomenological models or structurally based continuum models. The continuum composite approach of Spencer is a good example of that idea [50]. In this type of constitutive formulation for fiber-reinforced composite materials, fibers are not explicitly modeled at the geometric level but their mechanical contribution to the overall (i.e. continuum) behavior is implicitly accounted for by strain energy density terms directly related to microstructural metrics (e.g. stretch along the fiber direction).

From a clinical point of view, within the last decade, computational studies have focused on different applications such as stretch-induced skin growth during tissue expansion. Tissue expansion is a common surgical procedure to grow extra skin for reconstructing birth defects, burn injuries, or cancerous breasts [51, 52]. For instance [52], deals with local skin flaps which have revolutionized reconstructive surgery. Mechanical loading is critical for flap survival as excessive tissue tension reduces blood supply and induces tissue necrosis. However, skin flaps had never been analyzed mechanically. To this end, in [52], the stress profiles of two common flap designs are examined, direct advancement flaps and double back-cut flaps. FE simulations were conducted following the common two-step clinical protocol of tissue expansion. In the first step, the skin was gradually grown to create skin flaps for defect repair. In the second step, the defect was removed; the flap was cut, and it was advanced to cover the defect region. The simulations predicted a direct correlation between regions of maximum stress and tissue necrosis. This indicates that elevated stress could be used as predictor for flap failure. This model was considered a promising step toward computer-guided reconstructive surgery with the goal to minimize stress, accelerate healing, minimize scarring, and optimize tissue use.

## 6.7 The Musculoskeletal System

The human musculoskeletal system, which is also called the locomotor system, is the organ system which enables movement using the skeletal and muscular systems. The musculoskeletal system supports the body's form, stability, and movement. Subsections 6.7.1–6.7.3 provide an introduction to the main functions and components of the skeletal system and muscular systems, as individual components. Also, an overall approach to computational modeling of the muscular-skeletal system is provided which will be analyzed in detail in Chapter 8 (Multiscale modeling of musculoskeletal system) at different hierarchical levels.

### 6.7.1 Introduction to the Skeletal System

The skeletal system performs vital functions: support and movement, protection, blood cell production, mineral homeostasis, and triglyceride storage [53]. Bones offer a region of attachment for tendons and ligaments and constitute a skeletal framework which can produce movement through the coordinated use of muscles, levers, tendons, and ligaments. The role of bones is to act as levers, while the role of muscles is to generate the forces required to move the bones. Bones are considered the boundaries of soft organs (e.g. the vertebral column surrounding the spinal cord, the cranium around the brain, and the ribcage containing the heart and lungs). Bones are the main reservoirs for minerals in the body, containing approximately 99% of the body's calcium, 85% of its phosphate, and 50% of its magnesium and contribute significantly to maintaining homeostasis of minerals in the blood. Blood cells consist of hemopoietic stem cells which are present in the red bone marrow. Babies are born with only red bone marrow; over time, this is replaced by yellow marrow due to a decrease in erythropoietin, the hormone responsible for stimulating the production of erythrocytes in the bone marrow. Yellow bone marrow acts as a potential energy reserve for the body; it consists largely of adipose cells, which store triglycerides (a type of lipid that occurs naturally in the blood) [53].

The skeletal system is formed of bones and cartilage, which are connected by ligaments to form a framework for the remainder of the body tissues. The skeleton provides a structure on which muscles can work. The size and shape of the bones, mechanics of joint articulations, and location of muscular attachments form an efficient system of levers and struts. Joints allow bones to articulate with each other, and the shape of the joint contributes to efficiency of movement. Cartilage acts as a shock absorber and protects joint surfaces from wear and tear.

From a macrostructural point of view, bone consists of the periosteum, bone tissue, bone marrow, blood vessels, and nerves. There are two different types of bone tissue: cancellous or spongy bone and compact or cortical bone, which differ in density, or how tightly the tissue is packed together. Three types of cells contribute to bone homeostasis: (i) osteoblasts which are bone-forming cells, (ii) osteoclasts which have the function of resorbing or breaking down bone, and (iii) osteocytes which are mature bone cells. The equilibrium between osteoblasts and osteoclasts maintains bone tissue [54].

Trabecular bone is found in the inner parts of bones and has a significantly porous structure. The porosity in trabecular bone ranges from 50 to 95%, usually found in cuboidal bones, flat bones, and at the ends of long bones. At the microstructure level, trabecular bone consists of three-dimensional cylindrical structures, called trabeculae, with a thickness of about 100 μm and a variable arrangement form. This porous network of trabecular bone includes pores filled with marrow which produces the basic blood cells and consists of blood vessels, nerves, and various types of cells.

Cortical bone composes the external surface of all bones and has a porosity of about 5–10%. From a microscopic point of view, it consists of cylindrical structures called osteons or Haversian systems with a diameter of about 10–500 μm formed by cylindrical lamellae surrounding the Haversian canal (size of 3–7 μm). The boundary between the osteon and the surrounding bone is called the cement line. The most significant type of porosity, known as vascular porosity, is formed by the Haversian canals (aligned with the long axis of the bone) and the Volkmann's canals (transverse canals connecting Haversian canals) with capillaries and nerves. Experimental studies have shown that the shear modulus of wet single femoral osteons is higher than that of the whole bone which suggests that microstructural effects are significant.

### 6.7.2 Introduction to the Muscular System

The muscles of the body compose the muscular system. The muscular system mainly consists of skeletal muscles, which are attached to bones and enable voluntary body movement. Specifically, when these muscles contract, they move the body, while they also maintain posture and help keep balance. The skeletal and muscular systems act together to enable body movement.

Smooth muscles in the walls of blood vessels contract to cause vasoconstriction, which may help conserve body heat. Relaxation of these muscles is the cause of vasodilation, which can assist the body to lose heat. Concerning the digestive system, smooth muscles contract in sequence to form a wave of muscle contractions known as peristalsis and squeeze food through the gastrointestinal tract. Peristalsis of smooth muscles enables the movement of urine through the urinary tract.

Cardiac muscle tissue can be located only in the walls of the heart. When the cardiac muscle contracts, the heartbeat is generated. The pumping action of the beating heart maintains blood flow through the cardiovascular system [55].

### 6.7.3 Computational Modeling of the Muscular-Skeletal System

Multiscale modeling of the muscular-skeletal system is a rapidly emerging research domain that has provided new insights into the function of this system, especially from a multiscale point of view. Bone and muscle have a hierarchical organization from nano- to macroscale, and in order to better understand their macroscopic behavior, in both physiological and pathological status, it is essential to study their hierarchical organization and properties [56].

The continuum approach is the most common method to describe systems at the macroscale and this is a very successful way to describe many structural problems of homogeneous materials [56]. In this approach, a system is modeled as a continuous material rather than being composed of particles, without explicitly accounting for a material's internal structure. One of the basic models in continuum mechanics is based on Hook's relation, in which the elongation of elements is proportional to the applied external force. When a force $F$ is applied to a beam of length $L$ and cross-sectional area $A$, the extension $\Delta u$ can be derived from the following equation:

$$\sigma = \frac{F}{A} = \frac{Lk}{A}\frac{\Delta u}{L} = E\varepsilon, \tag{6.12}$$

where $\sigma$, $E$, and $\varepsilon$ are the stress, Young's modulus, and strain, respectively. This linear relation between stress and strain governs the behavior of brittle materials in the elastic regime, where the deformation is reversible. However, the deformation of materials such as metals generally has a plastic deformation regime in which irreversible deformation occurs.

Griffith [57] and Irwin [58] developed linear elastic fracture mechanics. The Griffith criterion states that failure of materials with a flaw occurs when the stored strain energy is large enough to create two new surfaces under uniaxial loading. The energy release rate $G$, which represents the dissipation of stored strain energy per unit of newly created surface area during the crack propagation, is defined as:

$$G = \frac{\partial U}{\partial A}, \tag{6.13}$$

where $U$ is the stored strain energy available for crack propagation and $A$ is the surface area. A small crack can propagate when $G = 2\gamma$, where $\gamma$ is the surface energy (the energy necessary to create a new surface; two new surfaces are created as the crack advances). The failure strength of a linear system with a small crack is given as:

$$\sigma_c = \sqrt{(EG)/(\pi\alpha)}, \tag{6.14}$$

where $E$ and $a$ are Young's modulus and length of the crack, respectively. LEFM can describe the failure of brittle and homogeneous materials such as glass [57].

In multiscale modeling, Young's modulus and the energy release rate can be readily obtained from atomistic simulation by calculating the corresponding materials' elastic constants and surface energy. However, the energy release rate is not simply $2\gamma$ when the dissipated energy constitutes a large portion of the energy release rate, such as in heterogeneous materials. A better expression would be:

$$G = 2\gamma + G_{\text{diss}}. \tag{6.15}$$

The energy release rate of inhomogeneous materials can be obtained from atomistic simulations by measuring the total external work for failure of the systems [59].

The damages and deformations of the muscular-skeletal system are locally inhomogeneous, which is challenging to be described based on linear elastic fracture mechanics. The cohesive zone model has been widely adopted for describing damages on the basis of the critical energy release rate. The basic elastic properties and the energy release rate are key parameters to connecting particle-based simulations to continuum theory. However, a more systematic framework needs to be developed to bridge these disparate methods.

From a macroscopic point of view, bone can be considered a continuum without voids, with material properties assigned across elements based on empirical relationships between CT attenuation values, density, and Young's modulus [60]. Macroscale continuum FE models can run in a few minutes on a standard workstation but present a limited resolution and typically overlook anisotropic material properties. The macroscale continuum approach has been applied in several studies investigating modeling and remodeling of bone, using different stress and strain stimuli to guide the bone apposition and resorption algorithm. Predictive studies have successfully extended the material constitutive relationship to include orthotropy and anisotropy in 2D planar and 3D spacial models of the femur.

3D muscle modeling is based on the FE method which is used as a standard numerical tool for the simulation of solid materials to analyze different state variables (e.g. stress or strain distributions during muscle deformation). The solid to be investigated is discretized into a finite number of elements with specific properties. The majority of the FE muscle models combine one-dimensional modeling of the active forcc-producing muscle fibers (active muscle behavior) with a macroscopic characterization of the soft tissue (passive muscle behavior). One problem of actual FE muscle models is the usage of simplified, and therefore, in most cases, unrealistic muscle architectures [61].

However, the last two decades have witnessed a major shift, from treating bone and muscles only at the macroscale level to complex multiscale models [62].

Multiscale modeling of musculoskeletal system will be further discussed in Chapter 8, examining the mechanical properties of bone and muscles and numerical approaches at different hierarchical levels.

## 6.8 The Endocrine System

Hormones regulate several of the body's functions such as growth and development, electrolyte balances, metabolism, and reproduction. They are produced by several glands throughout the body [63]. The endocrine system serves as a chemical messenger comprising feedback loops of the hormones released by internal glands of an organism directly into the circulatory system, regulating distant target organs. Hormone behavior is essential and ubiquitous to maintain normal physiological body functions. Experimental approaches combined with mathematical modeling have revealed that this behavior results from regulatory processes which evolve at different organization levels and require continuous dynamic equilibration, particularly in response to stimuli [64]. In the body, the most significant endocrine glands are the adrenal glands and the thyroid gland.

The thyroid, also known as the thyroid gland, is an endocrine gland located at the front of the neck, below the Adam's apple, and is composed of two connected lobes. The thyroid isthmus is a thin band of tissue connecting the lower two-thirds of the lobes. The functional unit of the thyroid gland, from a microscopic point of view, is the spherical thyroid follicle, lined with follicular cells (thyrocytes), and occasional parafollicular cells surrounding a lumen containing colloid. Three hormones are released by the thyroid gland and more specifically triiodothyronine (T3), thyroxine (T4), and calcitonin, a peptide hormone. The thyroid hormones (T3 and T4) influence protein synthesis, the metabolic rate, and especially in children, they affect development and growth, while calcitonin plays a role in calcium homeostasis [65]. The thyroid-stimulating hormone (TSH) is secreted by the anterior pituitary gland. It is regulated by the thyrotropin-releasing hormone (TRH) and controls the release of T3 and T4.

The adrenal glands, which are found above the kidneys, are endocrine glands that produce a variety of hormones including the steroids aldosterone and cortisol, as well as adrenaline [66]. Each gland consists of an outer cortex producing an inner medulla and steroid hormones.

### 6.8.1 Computational Modeling of the Endocrine System

This section focuses on the major endocrine glands which are the thyroid gland and the adrenal glands and computational models which have been presented in the literature.

Computational models of the thyroid deal with iodide metabolism steps. Early research [67] used a three-compartment model to represent the pool dynamics for blood hormone, blood iodine, and intrathyroidal iodine. Nevertheless, this type of models could not accurately predict noneuthyroid behavior. In [68], the authors developed an 18-compartment model of human iodine metabolism, but only two compartments were devoted to describing intrathyroidal iodine dynamics. A multicompartmental model including three intrathyroidal pools was developed to quantify the effects of perchlorate on thyroid status and circulating thyroid hormones [69]. In [70, 71], researchers considered whole-body thyroid models and used feedback control concepts to prescribe hormone-replacement therapies and to understand the regulation of hormone pool sizes. A dynamical model of the hypothalamic–pituitary–thyroid axis (HPT-axis) was developed in [72] by fitting model parameters to match clinical data. However, the ability of this model to predict thyroid behavior for various physiological conditions was not examined. Recently, a more refined model for the pituitary–thyroid axis and thyroid-hormone metabolism was developed in [73] in order to study possible feedback structures that may play a role in the generation of TSH pulses in the human pituitary. Nevertheless, this analysis depends on a static intrathyroidal model. In a more recent study [74], recent molecular-level and clinical observations were used to create a computational thyroid model. Simulation and analysis revealed that this model captures known aspects of thyroid physiology, while specific features were identified which are responsible for hormonal regulation.

At the macroscale level, the FE analysis has been used to simulate pathological cases related to the thyroid based on imaging data. In [75], in order to provide a quantitative disclosure on the radiofrequency ablation (RFA)-induced thermal ablation effects within thyroid tissues, a 3D FE simulation strategy was developed based on an MRI-reconstructed model. The thermal lesion's growth was predicted and interpreted under the following treatment conditions: (i) single-cooled-electrode modality, and (ii) two-cooled electrode system. It was found that the growth of the thermal lesion is significantly affected by two factors including the position of the RF electrode and the thermal–physiological behavior of the breathing airflow. Other parametric research works revealed several valuable phenomena, e.g. due to the electrode's movement, thermal injury with varying severity would appear to the trachea wall. Alterations of the airflow mass revealed significant effects on the total heat flux of the thyroid surface, while modifications in breathing frequency led to minor effects that could be ignored. This study provided a better understanding of the thermal lesions of RFA within the thyroid domain, which will help guide future treatment of thyroid cancer. In [76], a computational model was developed in order to determine the necessary resolutions for thermal sensors that will be used to obtain thermal images of patients' necks using the FE method. In order to come up with a model that is simple enough to be implemented but not overly simplistic, several assumptions were made. The first assumption considered heat conduction through the skin to be the only method of heat transfer to the skin surface. Secondly, the heating and cooling effects of blood flow in the capillaries were neglected, as well as the interface between the thyroid and the inside surface of the skin. Since these assumptions would simplify the model to a hot spot under the skin surface, the model was modified to a 30 mm × 30 mm piece of skin, 25-mm thick with an embedded 10-mm diameter hot nodule. This was done to compensate for some of the loss in thermal capacity due to the assumptions.

Fewer studies have been presented for the evaluation of adrenal gland pathologies at the macroscopic level. A recent study [77] deals with microwave thermal ablation, a minimally invasive modality to treat 10–20 mm benign adrenal adenomas while preserving normally functioning adjacent adrenal tissue, and returning the gland to a normally functioning status that is under normal regulation. In contrast to applications for tumor ablation, where devices have been developed with the objective of maximizing the size of the ablation zone for treating large tumors, a challenge for adrenal ablation is to minimize thermal damage to non-targeted adrenal tissue and thereby preserve adrenal function. Methods were investigated for creating small spherical ablation zones of volumes in the range of 0.5–4 $cm^3$ for the treatment of benign adrenal adenomas using water-loaded microwave monopole antennas operating at 2.45 and 5.8 GHz. Coupled electromagnetic and bioheat transfer simulations and experiments in *ex vivo* tissue were employed to investigate the effect of frequency, applied power, ablation duration, and coolant temperature on the length and width of the ablation zone. Experimental results showed that small spherical ablation zones with diameters in the range of 7.4–17.6 mm can be obtained by adjusting the applied power and ablation duration. Multiway ANOVA analysis of the experimentally measured ablation zone dimensions demonstrated that frequency of operation and ablation duration are the primary parameters for controlling the ablation zone's length and width, respectively. Additionally, it was demonstrated that the coolant temperature provides another effective parameter for controlling the ablation zone's length without affecting the ablation zone width. This study demonstrated the feasibility of creating small spherical microwave ablation zones suitable for targeting benign adrenal adenomas.

## 6.9 The Lymphatic System

The lymphatic or lymphoid system is an organ system in vertebrates that belongs to both the immune and circulatory systems. It is created of a large network of lymphatic vessels, tissues, and organs. It is the second vascular system in higher vertebrates in addition to the blood vasculature and has several vital functions such as the regulation of tissue pressure, the absorption of dietary fat in the intestine, and immune surveillance. There is growing evidence that the lymphatic system also contributes to a number of diseases, such as lymphedema, cancer metastasis, and various inflammatory disorders [78–81]. The lymphatic system consists of lymphangions which are separated by valves and possessed of active, contractile walls to pump interstitial fluid from its collection in the terminal lymphatics back to the main circulation. At the level of the microcirculation, plasma filtrate drains into lymphatic capillaries, which converge to form collecting vessels. These possess contractile walls and irregularly spaced one-way valves; spontaneous contraction of the walls acts to pump lymph fluid along the network, and the contractions may be modulated by nerves or by autocrine factors [80]. The vessels converge and pump the lymph through lymph nodes to drain into the venous circulation. The importance of the lymphatic system can be seen by the effects of its dysfunction. The accumulation of fluid in the tissue, lymphedema, caused by the removal of vessels during surgery or occlusion by parasites, results in pain, disfigurement, incapacity, and reduced immunity. Despite its importance, there is limited information on the fluid dynamics of the lymphatic system. The lymphatic network is considered to be asymmetrical implying that the right-hand side of the head and thorax and the right arm drain into the right subclavian vein, while the lymphatic vessels of the rest of the body converge at the thoracic duct [82].

### 6.9.1 Computational Modeling of the Lymphatic System

Lymphatic valves facilitate the lymphatic system's role in maintaining fluid homeostasis. Malformed valves can be found in different forms of primary lymphedema, resulting in immune dysfunction and incurable tissue swelling. Due to their small size and operation in low-pressure and low-Reynolds-number environments, their experimental investigation is rather complicated. Mathematical models of these structures can give insight and complement experimentation [79].

The earliest computational model of the lymphatic system was published in [80, 81]. Their analysis assumes 1D flow along a succession of lymphangions, each of which is modeled as a single computational node. The Navier–Stokes equations (NSEs) describe the fluid mechanics of the laminar flow of lymph fluid as follows:

$$\nabla \cdot \mathbf{u} = 0, \tag{6.16}$$

$$\frac{\partial \mathbf{u}}{\partial t} + \mathbf{u} \cdot \nabla \mathbf{u} = -\frac{1}{\rho} \nabla p + \nu \nabla^2 \mathbf{u}, \tag{6.17}$$

where $u$ is the velocity, $t$ is the time, $\rho$ is the density, $p$ is the pressure, and $\nu$ is the viscosity. With the assumption that the flow is uniform across the cross section, and with the expression of the NSE in cylindrical coordinates, these equations simplify considerably. Next, the volumetric flow rate is denoted as $Q$, the pressure as $p$, and the radius as $\alpha$ of the vessel as dependent variables, in which case the NSE can be written as:

$$\frac{\partial Q}{\partial x} = -2\pi\alpha\frac{\partial \alpha}{\partial t} \tag{6.18}$$

$$\frac{\partial Q}{\partial t} = -\frac{\pi\alpha^2}{\rho}\frac{\partial}{\partial x}\left[p - \rho g z\right] + \frac{2\pi\alpha}{\rho}\tau\bigg|_{r=a} \tag{6.19}$$

The solution of these equations can be achieved by taking into account the influence of the elastic wall on the pressure inside, leading to another equation which connects the radius of the vessel, $\alpha$, and the internal pressure, $p_i$ as follows:

$$p_i = p_{\text{ext},i} + \frac{h_i}{\alpha_i}\left[\sigma_{\text{hoop},i} + \sigma_{\text{act},i}\right] \tag{6.20}$$

in which $h_i$ denotes the wall thickness at location $i$, $\sigma_{\text{hoop},i}$ the stresses in the wall due to the passive elasticity of the wall and $\sigma_{\text{act},i}$ due to the active contraction of the wall. A single computational cell is used to model each lymphangion in the chain, while the valve between neighboring lymphangion is modeled by a fixed resistance to flow (open status) and a constraint on $Q$ (closed status), with an associated opening resistance. Despite the fact that parts of the lymphatic anatomy had to be excluded due to a lack of available information or for simplicity, this was a pioneering work which remains the only model of the entire lymphatic circulation [82].

In [80], a detailed computational model of the behavior of a single lymphangion was presented based on and validated against experimental measurements of the elastic properties of lymph vessels measured in the laboratory. The model was established considering a vessel with a length of 2 cm with an unstretched radius of 1.25 mm with valves at each end. The finite difference method was used to develop an explicit time-marching algorithm. It was shown that the computational model can reproduce the pumping behavior of the real vessel based on a simple contraction function, suggesting that lymphatic pumping is governed by simple, fast contraction pulses traveling in the retrograde direction to the flow.

In [79], the first valve geometry was presented and reconstructed from confocal imagery and used in the construction of a subject-specific model in a closing mode. A framework is proposed to convert images into valve models. An FE analysis study was conducted to evaluate the consequences of smoothing the leaflet surface, the significance of the shear modulus, and the impact of wall motion on valve behavior. Smoothing is inherent to any analysis of imagery. The type of image segmentation and meshing may lead to the attenuation of features of high frequency. Smoothing causes the loss of surface area, as well as the loss of high-frequency geometric features which may limit stiffness. This study takes these effects into account based on a manual reconstruction and through manifold harmonic analysis, attenuating higher-frequency features to replicate lower-resolution images or lower degree-of-freedom reconstructions. In this direction, two specific metrics were considered: (i) transvalvular pressure required to close the valve, ($\Delta P_c$), and (ii) the retrograde volume displacement after closure. As the value of $\Delta P_c$ increases, the volume of lymph that will pass through the valve during closure also increases. Retrograde volume displacement after closure gives a metric of compliance of the valve and for the quality of the valve seal. In the case of the image-specific reconstructed valve, removing features with a wavelength longer than 4 μm caused changes in $\Delta P_c$. Varying the shear modulus from 10 to 60 kPa caused a 3.85-fold increase in the retrograde volume displaced. $\Delta P_c$ increased from 1.56 to 2.52 cm $H_2O$ due to the inclusion of a nonrigid wall.

Nevertheless, the numerical work carried out on the lymphatic system seems to be in its infancy and further research work is required in combination with experimental analysis in order to provide new insights into the lymphatic system functions. According to [82], although several computational models of the lymphatic system in the literature have been established based on continuum mechanics, this is not the only available solution for the modeling of biological systems. More specifically, a continuum-based approach is not the most suitable option to represent events of mechanobiological nature as alternatives to phenomenological continuum mechanics models can also be adapted, which are able to incorporate the activities of cells. Cellular automatic and agent-based models of cellular growth and cell–cell interaction have been presented in the literature. Additionally, the theory of active particles is an alternative option which assigns a state that defines the microscopic behavior of each particle extending, thus, the kinetic theory of particles. Therefore, macroscopic models of tissues can be established based on the underlying behavior of cells, cell proliferation, cell–cell interaction and cell death. These powerful mathematical means pave the way for the integration of events at the molecular, cellular, and tissue levels. This can be achieved by developing advanced computational models that can provide new knowledge on the domains of developmental biology, lymphangiogenesis, angiogenesis, tissue remodeling, and tumors growth as well as on the interaction of cancer cells with those of the immune system.

## 6.10 The Nervous System

The nervous system is a highly complex system which controls the actions and sensory information by transmitting signals among different parts of the human body. It is defined by the presence of a specific type of cell known as the neuron or nerve cell. The nervous system identifies environmental changes which affect the body, and then it collaborates with the endocrine system to respond to such events.

The nervous system of vertebrates consists of the peripheral nervous system (PNS) and the central nervous system (CNS). The CNS includes the spinal cord and the brain. The spinal cord is located in the spinal canal and the brain in the cranial cavity. The meninges protect the CNS is enclosed and compose a three-layered system of membranes, including a tough, leathery outer layer known as the dura mater. The skull protects the brain and the vertebra the spinal cord. Concerning the PNS, it is composed of nerves, which are enclosed bundles of long fibers or axons connecting the CNS with all the other parts of the human body. The nerves which transmit information from the body to the CNS are known as afferent or sensory. The nerves which transmit signals from the brain are known as efferent or motor nerves. The spinal nerves contribute to both functions and are known as mixed nerves. The PNS consists of three different systems: (i) the autonomic system, (ii) the somatic system, and (iii) the enteric nervous system. The autonomic nervous system consists of two subsystems known as the sympathetic and the parasympathetic nervous system. More specifically, the role of the sympathetic nervous system is to act in cases of emergencies to mobilize energy, while the parasympathetic nervous system acts when organisms remain in a relaxed state. The somatic nerves enable voluntary movement. The enteric nervous system controls the gastrointestinal system. The autonomic and enteric nervous systems function involuntarily [83, 84].

The vertebrate nervous system consists of two subareas known as gray matter and white matter [84]. The gray matter includes a high ratio of cell bodies of neurons, while the white matter is composed mainly of myelinated axons, and takes its color from the myelin. The gray matter is located in clusters of neurons in the spinal cord and brain, as well as in cortical layers that line their surfaces. The white matter includes all the nerves and much of the interior of the spinal cord and brain.

### 6.10.1 Computational Modeling of the Nervous System

Several numerical studies have been presented in the literature at the macroscopic level to provide insight into the complex nature and function of the CNS and PNS, especially in the cases of pathologies.

In [85], kinematic axonal and previously established FE models are coupled and applied to a pseudo-3D representative volume element of CNS white matter to study the multiscale mechanical behavior. Axonal injury is an important area for the prevention and treatment of traumatic spinal cord and brain injuries and further numerical evaluation is required. FE models of the head and/or brain have been widely applied to predict brain injury due to external mechanical loadings (e.g. direct impact and explosive waves). The proper determination of the material properties and the precise representation of the tissues' microarchitecture influence the accuracy of the numerical models. Additionally, the stress and strain fields are highly anisotropic and axon orientation dependent considering that the axonal microstructure for specific regions of the brain white matter is locally oriented. Additionally, mechanical strain has been identified as the proximal cause of axonal injury, which further demonstrates the importance of this multiscale relationship. An inverse FE procedure was established by combining the results of FE modeling with uniaxial testing to determine material parameters of spinal cord white matter. The required balance was achieved between experiment and simulation processes via optimization by minimizing the squared error between the simulated and experimental force-stretch curves. It was shown that the combination of experimental and FE analysis is a useful means for soft biological tissue evaluation, enabling the assessment of the axonal response to tissue-level loading and subsequent predictions of axonal damage.

An FE study of the dynamic response of brain based on two parasagittal slice models was presented in [86]. The use of numerical models enables the quantitative evaluation of biomechanical responses, such as stress and strain of brain tissues and intracranial pressure, as well as the further investigation of the mechanism of the head traumatic brain injury. The objective of this study was to investigate the influence of gyri and sulci on the response of human head under transient loading. In neuroanatomy, a sulcus is a groove or depression in the cerebral cortex which surrounds a gyrus and creates the folded appearance of the brain in humans. Two detailed parasagittal slice models with and without sulci and gyri were presented and the material properties were derived from the literature. It was found that there is no significant difference between the models with and without gyri and sulci as far as intracranial pressure is concerned. The equivalent stress below gyri and sulci in the model with gyri and sulci was found to be slightly higher than that in the counterpart model without gyri and sulci. Finally, the maximum principle strain in brain tissue was lower in the model with gyri and sulci, while the stress and strain distributions were modified due to the presence of gyri and sulci.

A 3D FE model of the cervical spine with spinal cord was presented in [87] to investigate differences in cord strain distributions during various column injury patterns. Although the spinal cord can be injured due to different spinal column injury patterns such as fracture dislocation or burst fracture, the relationship between column injury pattern and cord damage requires further investigation. The study of strain distributions is essential to derive mechanism-specific characteristics of spinal cord damage and provide guidelines for clinical treatment. Model predictions demonstrated focal strains in contusion and dislocation, while those in distraction were more uniformly distributed throughout the cord. It was found that dislocation mechanism is associated with compressive lateral strains and increased strains in the lateral columns, as compared to contusion. This was the first 3D model of a spinal column with a spinal cord that includes verification of cord displacements under static load.

Several studies focus on the numerical evaluation of the circulation of fluid through the brain and spinal cord which is essential for maintaining fluid homoeostasis, and for clearing metabolites and other substances from the CNS [88]. The glymphatic system (a functional waste clearance pathway for the vertebrate CNS) is proposed to be a unidirectional fluid and solute circulation pathway in the brain involving transport through perivascular spaces, brain interstitium and glial cells. The glymphatic hypothesis includes controversial issues, especially concerning the outflow pathway, and limited knowledge is available on the forces governing fluid transport at each stage. Mathematical and computational modeling approaches are valuable tools for hypotheses assessment and provide supplementary information to experimental analysis in this area. Various modeling approaches have been applied to study solute and fluid transport from purely analytical models to hydraulic resistance networks and CFD models. The modeling interest has mainly focused on transport through the parenchyma and periarterial inflow. The main outcome of these studies is that arterial pulsation is unlikely to be the sole inflow driving force, and diffusion is most likely the dominant mode of transport in the parenchymal extracellular spaces. Models of efflux are limited and have not been able to provide new knowledge on the driving forces for fluid outflow from the CNS.

Studies on peripheral nerves present computational models of one or more axons. The models' variability mainly deals with the specific ionic currents as well as their conductances through the axon membrane, while another significant factor is the approach used to model myelination. Typically, peripheral nerve models are nonlinear in nature and require the solution of systems of differential equations, although there are fast approximation techniques that can provide insight into the tendency of axonal response to varying stimulation. Peripheral nerve models are used to study single neurons in the research domain of neurosciences, while in the areas of neural and biomedical engineering, they are used in large population models to evaluate the global response to electrical nerve stimulation or for nerve recording, mainly for application in functional electrical stimulation (FES) neuromodulation or neuroprosthetic systems [89]. More specifically, the distribution of potential (voltage) within a volume conductor must be assessed when the response of an axon to extracellular electrical stimulation is the parameter of interest. The peripheral axons can be discriminated into two categories, known as myelinated and unmyelinated, according to the occurrence of periodically spaced sheaths of myelin enrobing the axon. A small portion of the axon membrane remains exposed between adjacent myelin sheaths and is called the node of Ranvier. The simplest representation of a volume conductor is that of a homogeneous, isotropic medium in which the source of stimulation is at a sufficiently far distance from the axon. In this case, the source can be simulated as a point source and the voltage at each node of Ranvier and along each internode can be calculated by the equation:

$$V = \frac{I}{4\pi\sigma R}, \tag{6.21}$$

in which $\sigma$ denotes the conductivity of the medium, $R$ is the distance from the point source to the location of interest along the axon and $I$ denotes the current (a first analysis of the extracellular electric field deals with point monopole sources, i.e. local points in the space emitting a current $I$). Alternative equations can be applied to represent the voltage associated with a disc electrode as well as when the medium does not satisfy the assumptions of homogeneity and isotropy (e.g. when an electrode is on the surface of the skin). Nevertheless, as the models become more complex integrating tissues with dissimilar conductances, the research interest turns to the application of FE models. The aforementioned approaches can be further extended to simulate populations of axons (e.g. when electrically stimulating or recording from a peripheral nerve) considering that each axon is treated independently with the assumption that no axon directly affects the reaction of any other axon to electrical stimulation. As the access to computer clusters and the computing power have improved in the current era, researchers have managed to establish full-nerve, 3D, multifasciculated population models. Although the increasing complexity of the models leads to the consequent increase in the time required to simulate the axon(s), this time cost can be tolerated for a small number of axons thanks to the continuously improving computing power. Nevertheless, the conduction of sensitivity analyses with a fine resolution of the swept variable or the simulation of large populations of axons can increase significantly the simulation time, even if parallel computing is used, and the establishment of the appropriate model depends on the nature of the study.

## 6.11 The Reproductive System

The reproductive system or genital system includes all the sex organs of an organism which collaborate to enable reproduction. Additionally, several nonliving substances like hormones, fluids, and pheromones play a significant role to support the reproductive system's functions. The main difference between the reproductive system compared to other organ systems is that the sexes of different species may have significant differences allowing for a combination of genetic material between two individuals.

The reproductive system of males consists of a series of organs outside of the body, around the pelvic region contributing toward the reproduction function. The reproductive organs of the male can be grouped into three categories having as main function the generation of the male sperm for fertilization of the ovum: (i) the first category is sperm production and storage. Production takes place in the testes which are housed in the temperature-regulating scrotum. Immature sperm then travels to the epididymis for development and storage. (ii) The second category refers to the ejaculatory fluid-producing glands with the seminal vesicles, the vas deferens, and the prostate. (iii) The third category is used for copulation, and deposition of the spermatozoa (sperm) within the male including the penis, vas deferens, urethra, and Cowper's gland.

The female reproductive system includes one pair of ovaries as well as one pair of uterine tubes, the uterus, the breasts, and the vagina. The main function of this system is to generate the ovum, which is the female reproductive cell, and hormones and provides a cavity for the development of the zygote [90].

### 6.11.1 Computational Modeling of the Reproductive System

Although modeling in most areas of reproductive biology can still be considered to be in its infancy compared to more established fields such as cardiac or tumor physiology, an increasing number of new and sophisticated models have appeared, which focus on the reproductive process [91]. Computational models have been used successfully to test existing hypotheses regarding the mechanisms of female infertility and pathological fetal development, and also to provide new experimentally testable hypotheses regarding the process of development.

During pregnancy, traumas can threaten maternal and fetal health [92, 93]. A biomechanical model of the female reproductive system and the fetus for the realization of a childbirth virtual simulator was presented in [92]. This study was based on the FEM and mechanical laws to simulate the delivery and to recover the different forces applied to the different organs concerned by childbirth. The model took into account the principal obstetrical organs. The different simplifications that were applied (elastic behavior law, homogeneous elastic mechanical properties for the fetus, simplification of the force fields applied) did not result in erroneous behaviors. Dynamic simulations were performed and different mechanical parameters were estimated in order to find the right boundary conditions for each organ. Two models of abdominal efforts were developed and compared. The first model assumed unidirectional forces field, from the top of the uterus toward the center of the cervix. The second model was based on the principle that abdominal muscles cover a more important surface of the uterus and that forces are applied perpendicularly to the surface of the uterus. The results indicated that the distribution is slightly more homogeneous in the second case, with efforts following the normal direction.

Additionally, various trauma effects on a pregnant uterus have been investigated to a limited extent. In [93], an FE model was presented of a uterus along with a fetus, placenta, and amniotic fluid, and two effective ligament sets were developed. This model evaluated numerically various loadings on a pregnant uterus. The geometry of the model was established using CT images and validated based on anthropometric data. The material properties of the placenta and uterine wall were assigned using the Ogden hyperelastic theory. After simulating the "rigid-bar" abdominal loading, the impact force and abdominal penetration were investigated. It was found that at low abdominal penetrations (less than 45 mm), the pregnant abdomen response is highly compatible with the nonpregnant case, while at large penetrations, the pregnant abdomen reveals stiffer behavior. It was concluded that the developed model enables the investigation of the response of a pregnant abdomen under different environmental loading conditions. This can be used as a basis to study the injury risk to the uterus and fetus.

The authors of [94] studied numerically preterm birth which is the leading cause of childhood mortality and may cause health risks in survivors. The mechanical function of the uterus, cervix, and fetal membranes plays a significant role in the protection of the fetus during gestation. To this end, a 3D parameterized FE model of pregnancy was presented in [94] to better understand this mechanical function and impact on preterm birth. This model was developed based on an automatic procedure using maternal ultrasound measurements. More specifically, a baseline model at 25 weeks of gestation was presented. In order to visualize the impact of cervical structural parameters on tissue stretch, the model sensitivity was assessed to: (i) anterior uterocervical angle, (ii) cervical length, (iii) posterior cervical offset, and (iv) cervical stiffness. It was found that cervical tissue stretching is minimal when the cervical canal is aligned with the longitudinal uterine axis and a softer cervix is more sensitive to changes in the geometric variables.

Fewer numerical studies have been presented in the literature dealing with the male reproductive system. In [95], the consequences of the ciliary motion on the transport of seminal liquid through the ductus efferentes of the male reproductive tract were examined. By assuming the seminal liquid as a couple stress fluid, a mathematical model was developed for a 2D flow through an axially symmetric tube whose inner surface is ciliated in the form of a metachronal wave. The governing system consisted of nonlinear coupled partial differential equations which can be reduced to a system of ordinary differential equations by utilizing the long wavelength approximation in an environment of inertia-free flow. Exact solutions for the pressure gradient, the velocity distribution and the stream function were derived according to the ciliary metachronism and couple stress parameters. Also, the trapping and the pumping characteristics due to the cilia motility were of primary importance. This study revealed that ciliary pumping has to be more efficient to transport a couple of stress fluids as compared to Newtonian fluid. As the fluid behavior turns from Newtonian to the couple stress, the overall magnitude of the velocity decreases.

## 6.12 Conclusion

This chapter presented computational modeling approaches at the organ level focusing on each one of the eleven distinct organ systems in human beings. Although computational modeling at the macroscale level can provide supplementary information to experimental and clinical research, the application of multiscale-level approaches is the main challenge of the current era in biomedical engineering, as they provide information on how changes occurring at one level of simulation can propagate to higher levels. In this direction, two significant issues must be addressed: (i) How can parameters be propagated across scales? (ii) How can simulations at multiple levels be run concurrently?

## References


1 Bartsch, R.P., Liu, K.K.L., Bashan, A., and Ivanov, P.C. (2015). Network physiology: how organ systems dynamically interact. *PLoS One* 10 (11): e0142143. https://doi.org/10.1371/journal.pone.0142143.

2 Cerrolaza, J.J., Picazo, M.L., Humbert, L. et al. Computational anatomy for multi-organ analysis in medical imaging: a review. *Med. Image Anal.* 56: 44–67. https://doi.org/10.1016/j.media.2019.04.002.

3 Empey, D.W. (1978). Diseases of the respiratory system. Introduction: structure and function of the lungs. *Br. Med. J.* 1 (6113): 631–633. https://doi.org/10.1136/bmj.1.6113.631.

4 Roth, C.J., Yoshihara, L., Ismail, M., and Wall, W.A. (2017). Computational modelling of the respiratory system: discussion of coupled modelling approaches and two recent extensions. *Comput. Meth. Appl. Mech. Eng.* 314: 473–493. https://doi.org/10.1016/j.cma.2016.08.010.

5 Oakes, J.M., Marsden, A.L., Grandmont, C. et al. (2014). Airflow and particle deposition simulations in health and emphysema: from in vivo to in silico animal experiments. *Ann. Biomed. Eng.* 42: 899–914.

6 Ismail, M., Comerford, A., and Wall, W.A. (2013). Coupled and reduced dimensional modelling of respiratory mechanics during spontaneous breathing. *Int. J. Numer. Methods Biomed. Eng.* 29: 1285–1305.

7 Yoshihara, L., Roth, C.J., and Wall, W.A. (2016). Fluid–structure interaction including volumetric coupling with homogenised subdomains for modelling respiratory mechanics. *Int. J. Numer. Methods Biomed. Eng.* 33 (4): e2812. https://doi.org/10.1002/cnm.2812.

8 Ma, B. and Bates, J.H.T. (2010). Modelling the complex dynamics of derecruitment in the lung. *Ann. Biomed. Eng.* 38: 3466–3477.

9 Roth, C.J., Ismail, M., Yoshihara, L., and Wall, W.A. (2017). A comprehensive computational human lung model incorporating interacinar dependencies: application to spontaneous breathing and mechanical ventilation. *Int. J. Numer. Methods Biomed. Eng.* 1–24.

10 Tsega, E.G. (2018). Computational fluid dynamics modelling of respiratory airflow in tracheobronchial airways of infant, child, and adult. *Comput. Math. Methods Med.* 1–9. https://doi.org/10.1155/2018/9603451.

11 Hofmann, W. (1982). Mathematical model for the postnatal growth of the human lung. *Respir. Physiol.* 49: 115–129.

12 Thiriet, M. (2014). *Biomathematical and Biomechanical Modelling of the Circulatory and Ventilatory Systems.* New York, NY: Springer.

13 Brandstaeter, S., Fuchs, S.L., Aydin, R.C., and Cyron, C.J. (2019). Mechanics of the stomach: a review of an emerging field of biomechanics. *GAMM-Mitteilungen* 42 (3): e201900001. https://doi.org/10.1002/gamm.201900001.

14 Ferrua, M.J. and Singh, R.P. (2010). Modelling the fluid dynamics in a human stomach to gain insight of food digestion. *J. Food Sci.* 75 (7): R151–R162. https://doi.org/10.1111/j.1750-3841.2010.01748.x.

15 Imai, Y., Kobayashi, I., Ishida, S. et al. (2013). Antral recirculation in the stomach during gastric mixing. *Am. J. Physiol. Gastrointest. Liver Physiol.* 304 (5): G536.

16 Cleary, P., Sinnott, M., Hari, B. et al. (2015). Modelling food digestion, Chapter 10. In: *Woodhead Publishing Series in Food Science, Technology and Nutrition, Modelling Food Processing Operations* (ed. S. Bakalis, K. Knoerzer and P. Fryer), 255–s. Woodhead Publishing.

17 Harrison, S.M., Cleary, P.W., and Sinnott, M.D. (2018). Investigating mixing and emptying for aqueous liquid content from the stomach using a coupled biomechanical-SPH model. *Food Funct.* 9 (6): 3202–3219.

18 Niederer, S.A., Lumens, J., and Trayanova, N.A. (2019). Computational models in cardiology. *Nat. Rev. Cardiol.* 16 (2): 100–111.

19 Suinesiaputra, A., McCulloch, A.D., Nash, M.P. et al. (2016). Cardiac image modelling: breadth and depth in heart disease. *Med. Image Anal.* 33: 38–43.

**20** Zhang, X., Ambale-Venkatesh, B., Bluemke, D.A. et al. (2015). Information maximizing component analysis of left ventricular remodeling due to myocardial infarction. *J. Transl. Med.* 13 (343): 1–9. https://doi.org/10.1186/s12967-015-0709-4.

**21** Humphrey, J.D. and McCulloch, A.D. (2003). The cardiovascular system – anatomy, physiology and cell biology. In: *Biomechanics of Soft Tissue in Cardiovascular Systems. International Centre for Mechanical Sciences (Courses and Lectures)*, vol. 441 (ed. G.A. Holzapfel and R.W. Ogden). Vienna: Springer https://doi.org/10.1007/978-3-7091-2736-0_1.

**22** Armentano, R.L., Cabrera Fischer, E.I., and Cymberknop, L.J. (2019). *Biomechanical Modelling of the Cardiovascular System*, 1–18. IOP Publishing Ltd.

**23** Gray, R.A. and Pathmanathan, P. (2018). Patient-specific cardiovascular computational modeling: diversity of personalization and challenges. *J. Cardiovasc. Transl. Res.* 11 (2): 80–88. https://doi.org/10.1007/s12265-018-9792-2.

**24** HeartFlow. (2016). https://www.accessdata.fda.gov/cdrh_docs/pdf15/K152733.pdf (accessed 29 November 2022).

**25** CardioInsight. (2016) https://www.accessdata.fda.gov/cdrh_docs/pdf16/k162440.pdf (accessed 29 November 2022).

**26** Sarvas, J. (1987). Basic mathematical and electromagnetic concepts of the biomagnetic inverse problem. *Phys. Med. Biol.* 32 (1): 11–22.

**27** Rudy, Y. (2013). Noninvasive electrocardiographic imaging of arrhythmogenic substrates in humans. *Circ. Res.* 112 (5): 863–874.

**28** Marieb, E.N. (2015). *Essentials of Human Anatomy and Physiology*, Chapter 15, 10e. Pearson Education, Inc. publishing as Benjamin Cummings.

**29** Fry, C.H., Sadananda, P., Wood, D.N. et al. (2011). Modeling the urinary tract-computational, physical, and biological methods. *Neurourol. Urodyn.* 30 (5): 692–699. https://doi.org/10.1002/nau.21131.

**30** Griffiths, D. (1985). The pressure within a collapsed tube, with special reference to urethral pressure. *Phys. Med. Biol.* 30: 951–963.

**31** Griffiths, D.J. (1987). Dynamics of the upper urinary tract: I. Peristaltic flow through a distensible tube of limited length. *Phys. Med. Biol.* 32: 813–822.

**32** van Mastrigt, R. and Kranse, M. (1993). Automated evaluation of urethral obstruction. *Urology* 42: 216–224.

**33** Postnov, D.D., Marsh, D.J., Postnov, D.E. et al. (2016). Modeling of kidney hemodynamics: probability-based topology of an arterial network. *PLoS Comput. Biol.* 12 (7): e1004922. https://doi.org/10.1371/journal.pcbi.1004922.

**34** Moss, R., EdKazmierczak, Kirley, M., and Harris, P. (2009). A computational model for emergent dynamics in the kidney. *Philos. Trans. R. Soc.* 367: 2125–2140. https://doi.org/10.1098/rsta.2008.0313.

**35** Kolarsick, P., Kolarsick, M.A., and Goodwin, C. (2011). Anatomy and physiology of the skin. *J. Dermatol. Nurses Assoc.* 3 (4): 203–221.

**36** Limbert, G. (2017). Mathematical and computational modelling of skin biophysics: a review. *Proc. Math. Phys. Eng. Sci.* 473 (2203): 20170257. https://doi.org/10.1098/rspa.2017.0257.

**37** Chan, L.S. (1997). Human skin basement membrane in health and autoimmune diseases. *Front. Biosci.* 2: 343–352. https://doi.org/10.2741/A196.

**38** J. F. Ribeiro, E.H.M. dos Anjos, M.L.S. Mello, B. de Campos Vidal. Skin collagen fiber molecular order: a pattern of distributional fiber orientation as assessed by optical anisotropy and image analysis. PLoS One 8, e54724. 2013. (https://doi.org/10.1371/journal.pone.0054724).

**39** Ottenio, M., Tran, D., Annaidh, A.N. et al. (2015). Strain rate and anisotropy effects on the tensile failure characteristics of human skin. *J. Mech. Behav. Biomed. Mater.* 41: 241–250. https://doi.org/10.1016/j.jmbbm.2014.10.006.

**40** Kvistedal, Y.A. and P.M.F. Nielsen PMF. (2009). Estimating material parameters of human skin in vivo. *Biomech. Model. Mechanobiol.* 8: 1–8. https://doi.org/10.1007/s10237-007-0112-.

**41** Batisse, D., Bazin, R., Baldeweck, T. et al. (2022). Influence of age on the wrinkling capacities of skin. *Skin Res. Technol.* 8: 148–154. https://doi.org/10.1034/j.1600-0846. 2002.10308.x.

**42** Delalleau, A., Josse, G., Lagarde, J.M. et al. (2006). Characterization of the mechanical properties of skin by inverse analysis combined with the indentation test. *J. Biomech.* 39: 1603–1610. https://doi.org/10.1016/j.jbiomech.2005.05.001.

**43** Weickenmeier, J., Jabareen, M., and Mazza, E. (2015). Suction based mechanical characterization of superficial facial soft tissues. *J. Biomech.* 48: 4279–4286. https://doi.org/10.1016/j.jbiomech.2015. 10.039.

**44** Tonge, T.K., Atlan, L.S., Voo, L.M., and Nguyen, T.D. (2013). Full-field bulge test for planar anisotropic tissues: part I – experimental methods applied to human skin tissue. *Acta Biomater.* 9: 5913–5925. https://doi.org/10.1016/j.actbio.2012.11.035.

**45** Geerligs, M., van Breemen, L.C.A., Peters, G.W.M. et al. (2011). *In vitro* indentation to determine the mechanical properties of epidermis. *J. Biomech.* 44: 1176–1181. https://doi.org/10.1016/j.jbiomech.2011.01.015.

**46** Lamers E, van Kempen THS, Baaijens FPT, Peters GWM, Oomens CWJ. 2013 Large amplitude oscillatory shear properties of human skin. J. Mech. Behav. Biomed. Mater. 28, 462–470. (https://doi.org/10.1016/j.jmbbm.2013.01.024).

**47** Veronda, D.R. and Westmann, R. (1970). Mechanical characterization of skin – finite deformations. *J. Biomech.* 3: 111–124. https://doi.org/10.1016/0021-9290(70)90055-2.

**48** Fung, Y.C. (1981). *Biomechanics: Mechanical Properties of Living Tissues.* New York, NY: Springer.

**49** Marino, M. (2016). Molecular and intermolecular effects in collagen fibril mechanics: a multiscale analytical model compared with atomistic and experimental studies. *Biomech. Model. Mechanobiol.* 15: 133–154. https://doi.org/10.1007/s10237-015-0707-8.

**50** Spencer, A.J.M. (1984). Constitutive theory for strongly anisotropic solids. In: *Continuum Theory of the Mechanics of Fibre-Reinforced Composites* (ed. A.J.M. Spencer), 1–32. Vienna, Austria: Springer.

**51** Tepole, A.B., Joseph Ploch, C.P., Wong, J. et al. (2011). Growing skin: a computational model for skin expansion in reconstructive surgery. *J. Mech. Phys. Solids* 59 (10): 2177–2190. https://doi.org/10.1016/j.jmps.2011.05.004.

**52** Tepole, A.B., Wong, J., Gosain, A.K., and Kuhl, E. (2014). Computational modeling of skin: using stress profiles as predictor for tissue necrosis in reconstructive surgery. *Comput. Struct.* 143: 32–39. https://doi.org/10.1016/j.compstruc.2014.07.004.

**53** Walker, J. (2020). Skeletal system 1: the anatomy and physiology of bones. *Nursing Times* 116 (2): 38–42.

**54** Potsika, V.T., Vavva, M.G., Protopappas, V.C. et al. (2013). *Computational Modeling of Ultrasound Wave Propagation in Bone*. New York: Springer Science+Business Media.

**55** Muscolino, J.E. (2016). *The Muscular System Manual: The Skeletal Muscles of the Human Body* 4th revised edition. St Louis: Elsevier - Health Sciences Division.

**56** Jung, G.S. and Buehler, M.J. (2017). Multiscale modeling of muscular-skeletal systems. *Annu. Rev. Biomed. Eng.* 19 (1): 435–457. https://doi.org/10.1146/annurev-bioeng-071516-044555.

**57** Griffith, A.A. (1921). The phenomena of rupture and flow in solids. *Philos. Trans. R. Soc. A* 221: 163–198.

**58** Irwin, G.R. (1957). Analysis of stresses and strains near the end of a crack traversing a plate. *J. Appl. Mech.* 24: 361–364.

**59** Jung, G.S., Qin, Z., and Buehler, M.J. (2015). Molecular mechanics of polycrystalline graphene with enhanced fracture toughness. *Extreme Mech. Lett.* 2: 52–59.

**60** Phillips, A.T.M., Villette, C.C., and Modenese, L. (2015). Femoral bone mesoscale structural architecture prediction using musculoskeletal and finite element modelling. *Int. Biomech.* 2 (1): 43–61. https://doi.org/10.1080/23335432.2015.1017609.

**61** Siebert, T. and Rode, C. (2014). Computational modeling of muscle biomechanics. In: *Computational Modelling of Biomechanics and Biotribology in the Musculoskeletal System*, 173–204. https://doi.org/10.1533/9780857096739.2.173.

**62** Podshivalov, L., Fischer, A., and Bar-Yoseph, P.Z. (2014). On the road to personalized medicine: multiscale computational modeling of bone tissue. *Arch. Comput. Meth. Eng.* 21 (4): 399–479. https://doi.org/10.1007/s11831-014-9120-1.

**63** Hiller-Sturmhöfel, S. and Bartke, A. (1998). The endocrine system: an overview. *Alcohol Health Res. World* 22 (3): 153–164.
**64** Zavala, E., Wedgwood, K.C.A., Voliotis, M. et al. (2019). Mathematical modelling of endocrine systems. *Trends Endocrinol. Metab.* 30 (4): 244–257.
**65** Boron, W.F. and Boulpaep, E.L. (2012). *Medical Physiology*, 2e, 1052. ISBN 978-1-4377-1753-2,. Philadelphia, PA: Saunders.
**66** Santulli, G. (2015). *Adrenal Glands: From Pathophysiology to Clinical Evidence.* New York, NY. ISBN 978-1-63483-570-1.: Nova Science Publishers.
**67** Riggs, D.S. (1952). Quantitative aspects of iodine metabolism in man. *Pharmacol. Rev.* 4 (3): 284–370.
**68** Degroot, L.J., Decostre, P., and Phair, R. (1971). A mathematical model of human iodine metabolism. *Clin. Endocrinol. Metab.* 32 (6): 757–765.
**69** Merrill, E.A., Clewell, R., Robinson, P. et al. (2005). PBPK model for radioactive iodide and perchlorate kinetics and perchlorate-induced inhibition of iodide uptake in humans. *Toxicol. Sci.* 83 (1): 25–43.
**70** Distefano, J.J. and Mak, P.H. (1978). Optimal control policies for the prescription of thyroid hormones. *Math. Biosci.* 42 (3, 4): 159–189.
**71** Yamada, H., Distefano, J.J., Yen, Y.-M., and Nguyen, T.T. (1996). Steady-state regulation of whole-body thyroid hormone pool sizes and interconversion rates in hypothyroid and moderately T3-stimulated rates. *Endocrinology* 137: 5624–5633.
**72** Li, G., Liu, B., and Liu, Y. (1995). A dynamical model of the pulsatile secretion of the hypothalomo-pituitary-thyroid axis. *Biosystems* 35 (1): 83–92.
**73** Dietrich, J.W., Tesche, A., Pickardt, C.R., and Mitzdorf, U. (2002). Fractal properties of the thyrotropic feedback control implications of a nonlinear model compared with empirical data. In: *Cybernetics and Systems 2002* (ed. R. Trappl), 329–334. Austrian Society for Cybernetic Studies.
**74** Degon, M., Chipkin, S.R., Hollot, C.V. et al. (2008). A computational model of the human thyroid. *Math. Biosci.* 212: 22–53.
**75** Jina, C., He, Z., and Liu, J. (2014). MRI-based finite element simulation on radiofrequency ablation of thyroid cancer. *Comput. Meth. Programs Biomed.* 113: 529–538.
**76** Helmy, A.W., Rizkalla, M.E., Holdmann, M.M., and Salama, P. (2000). Finite element analysis for simulating a hot thyroid nodule. In: *Proceedings of the 43rd IEEE Midwest Symposium on Circuits and Systems* (Cat.No.CH37144), Lansing, MI, vol. 3, 1064–1067. https://doi.org/10.1109/MWSCAS.2000.951399.
**77** Fallahi, H., Clausing, D., Shahzad, A. et al. (2019). Microwave antennas for thermal ablation of benign adrenal adenomas. *Biomed. Phys. Eng. Express* 5 (2): https://doi.org/10.1088/2057-1976/ab068b.
**78** Cueni, L.N. and Detmar, M.D. (2008). The lymphatic system in health and disease. *Lymphat Res. Biol.* 6 (3, 4): 109–122.

**79** Watson, D.J., Sazonov, I., Zawieja, D. et al. (2017). Integrated geometric and mechanical analysis of an image-based lymphatic valve. *J. Biomech.* 7 (64): 172–179.

**80** Macdonald, A.J., Arkill, K.P., Tabor, G.R. et al. (2008). Modelling flow in collecting lymphatic vessels: one-dimensional flow through a series of contractile elements. *Am. J. Physiol. Heart Circ. Physiol.* 295 (1): H305–H313. https://doi.org/10.1152/ajpheart.00004.2008.

**81** Reddy, N.P., Krouskop, T.A., and Newell, P.H. (1977). A computer model of the lymphatic system. *Comput. Biol. Med.* 7: 181–197.

**82** Margaris, K.N. and Black, R.A. (2012). Modelling the lymphatic system: challenges and opportunities. *J. R. Soc. Interface* 9: 601–612. https://doi.org/10.1098/rsif.2011.0751.

**83** Kandel, E.R., Schwartz, J.H., and Jesse, T.M. (2000). The anatomical organization of the central nervous system, Chapter 17. In: *Principles of Neural Science*. McGraw-Hill Professional. ISBN 978-0-8385-7701-1,.

**84** D. Purves, G. J. Augustine, D. Fitzpatrick, W. C. Hall, A. S. LaMantia, J. O. McNamara, L. E. White. Neuroscience. 4th ed. Sinauer Associates. 15–16, 2008.

**85** Pan, Y., Sullivan, D., Shreiber, D.I., and Pelegri, A.A. (2013). Finite element modeling of CNS white matter kinematics: use of a 3D RVE to determine material properties. *Front. Bioeng. Biotechnol.* 1: 19. https://doi.org/10.3389/fbioe.2013.00019.

**86** Song, X., Wang, C., Hu, H. et al. (2015). A finite element study of the dynamic response of brain based on two parasagittal slice models. hindawi publishing corporation. *Comput. Math. Methods Med.* 816405. http://dx.doi.org/10.1155/2015/816405.

**87** Greaves, C.Y., Gadala, M.S., and Oxland, T.R. (2008). A three-dimensional finite element model of the cervical spine with spinal cord: an investigation of three injury mechanisms. *Ann. Biomed. Eng.* 36 (3): 396–405.

**88** Martinac, A.D. and Bilston, L.E. (2020). Computational modelling of fluid and solute transport in the brain. *Biomech. Model. Mechanobiol.* 19: 781–800. https://doi.org/10.1007/s10237-019-01253-y.

**89** Schiefer, M. (2014). Peripheral nerve models. In: *Encyclopedia of Computational Neuroscience* (ed. D. Jaeger and R. Jung). New York, NY: Springer. https://doi.org/10.1007/978-1-4614-6675-8_213.

**90** Evans, K.D. (2009). Reproductive systems, Chapter 10. In: *Radiographic Pathology for Technologists*, 5e, 284–303. Elsevier Publishing.

**91** Clark, A.R. and Kruger, J.A. (2016). Mathematical modeling of the female reproductive system: from oocyte to delivery. *WIREs, Wiley Interdisciplinary Rev.* https://doi.org/10.1002/wsbm.1353.

**92** Buttin, R., Zara, F., Shariat, B., and Redarce, T. (2009). A biomechanical model of the female reproductive system and the fetus for the realization of a childbirth virtual simulator. In: *Conference proceedings: International Conference of the IEEE Engineering in Medicine and Biology Society*. IEEE Engineering in Medicine and Biology Society https://doi.org/10.1109/IEMBS.2009.5334085.

**93** Irannejad Parizi, M., Ahmadian, M.T., and Mohammadi, H. (2020). Rigid-bar loading on pregnant uterus and development of pregnant abdominal response corridor based on finite element biomechanical model. *Int. J. Numer. Methods Biomed. Eng.* 36 (1): e3284.

**94** Westervelt, A.R., Fernandez, M., House, M. et al. (2017). A parameterized ultrasound-based finite element analysis of the mechanical environment of pregnancy. *ASME J. Biomech. Eng.* 139 (5): 051004. https://doi.org/10.1115/1.4036259.

**95** Farooq, A.A. and Siddiqui, A.M. (2017). Mathematical model for the ciliary-induced transport of seminal liquids through the ductuli efferentes. *Int. J. Biomath.* 10 (3): 1750031.

# 7

# Mechanotransduction Perspective, Recent Progress and Future Challenges

## 7.1 Introduction

The process through which organisms understand and act to physical forces and respond with biochemical signals is known as mechanotransduction. Cell proliferation, migration, differentiation, and death have all been demonstrated to be controlled by such signals [1, 2]. The enormous effect of mechanotransduction on cells implies that modifying, changing, and directing the process might be of therapeutic value. Internal and external mechanical forces on the human body, such as shear forces in blood vessels, stretching, actomyosin contraction, gravity, acoustic vibration, and pressure, can all influence Internal and external mechanical forces on the human body can all impact cellular development and other activities, such as shear forces in blood vessels, stretching, actomyosin contraction, gravity, acoustic vibration, and pressure [3, 4].

For example, exposure to the microgravity environment of space travel reduces bone and muscle development and alters immune reaction [5]. Mechanical environments affects the differentiation and proliferation of cells [6]. Human epithelial cell stiffness increases significantly with age, impacting progenitor cell differentiation [7]. Numerous expressed genes were revealed by DNA sequencing to be activated by signals generated by mechanotransduction in response to mechanical inputs in diverse biological processes [8]. However, in mechanotransduction research, a complete understanding of how forces are sensed, transmitted, and transduced into gene expression is still absent.

*Multiscale Modelling in Biomedical Engineering*, First Edition. Dimitrios I. Fotiadis, Antonis I. Sakellarios, and Vassiliki T. Potsika.

Published 2023 by John Wiley & Sons, Inc.

## 7.2 Methods for Studying Mechanotransduction

In order to comprehend a biological system, scientists often apply "input" into the system referred as mechanical forces applied to cells or tissues in mechanotransduction and read "output." There are various methods for applying a controlled force input. Tissue culture cells, for instance, may be repeatedly stretched in multiple directions and rates on an elastic substrate to simulate muscle, blood arteries, and the lungs [9]. Channel proteins can be activated by hydrostatic pressure or membrane stretching [10]. A dynamic compressive bioreactor can be used to compress cells to imitate periodontal cells and create articular cartilage tissue [11]. A pump can create shear stress to simulate blood flow [12]. Microgravity may also be tested on satellites and space stations, as well as in a rotating wall vessel bioreactor on Earth [13]. Mechanical stimuli like as infrasound (0–20 Hz) and low-frequency noise (20–500 Hz) can be employed in research [14]. Internal forces, in addition to external pressures, can be controlled by inhibiting myosin. The tension on adhesion molecules that link to the cytoskeleton and the nucleus is influenced by the stiffness of the substrate [15]. As these procedures are well-established, "output" in the experimental methods above comprises gene expression, alterations of the morphology, protein transformation, and other.

### 7.2.1 How Mechanical Forces Are Detected

Research indicates that mechanical forces cause protein structural changes in order to activate them. To operate as a mechanotransduction channel, piezo cation channels, for instance, utilize a lever-like mechanogating mechanism [16]. When actomyosin contraction diffuses the domain–domain pair of filamin A, a cryptic binding site for integrins is exposed [17, p. 21]. When stress is applied to the R8 domain, talin unfolding occurs to trigger downstream signaling [18, p. 1]. Blood shear stress can cause von Willebrand factor's A2 domain to unfold, exposing the binding sites for the glycoprotein I receptor in the A1 domain, cryptic A disintegrin and metalloproteinase with thrombospondin motifs 13 (ADAMTS13) binding sites, and the cleavage site in the A2 domain [19]. Acoustic forces deflect the stereocilia of hair cells, opening a calcium channel and activating the current via the channel [20].

All of these demonstrations rely on extensive structural knowledge because there is no reliable mechanism for identifying a mechanosensing molecule. To increase our understanding of force sensing systems, a nanotechnology-based technique that selectively detects mechanosensitive changes without identifying non-mechanosensitive changes that may occur concurrently with mechanosensitive changes is required. These modifications might involve not just protein conformational changes, but also alterations in other biological components such as membrane lipids and carbohydrates.

### 7.2.2 Transmission of Mechanical Forces

Force transmission can be mediated in principle via cell–matrix interaction, cell-to-cell contact, cytoskeleton, pressure, fluidic flow, and vibration [21]. When compared to molecular diffusion in cells, cytoskeleton-mediated transmission is faster even across great distances, owing to the sensor's direct link to a target [22]. However, pressure change, stretching, and vibration may all convey information quicker than molecule diffusion. To control gene expression, the cytoskeleton is directly coupled to the linker of nucleoskeleton and cytoskeleton (LINC) complex [15]. Furthermore, osmotic shocks expand the nucleus and nuclear pores, allowing active trafficking of the transcriptional cofactor Yes-associated protein 1 (YAP) into the nucleus [23].

The disturbance of a candidate is the sole way to identify a force transmitter. Depolymerization drugs such as latrunculin and nocodazole, for example, can disrupt the cytoskeleton [24]. The LINC complex can be damaged by utilizing siRNA and genome editing to target the complex's components [23]. To identify and monitor force transmission at micro or nanoscales, another specialized approach is required. Clustered regularly interspaced short palindromic repeats (CRISPR) screening, for example, may aid in the finding of important mechanotransmitters, and a rationally designed fluorescent sensor may monitor force transfer in real time [25].

### 7.2.3 Conversion of Mechanical Forces to Signals

By integrating ions into the cytosol, mechanotransduction channels can directly transform mechanical stresses into biochemical signals [16]. Filamin A may link to integrin, smoothelin, and fimbacin during mechanical stimulation and dissociate FilGAP, a Rac-specific GTPase-activating protein [26]. The link between filamin A and integrin modulates cell adhesion and migration, although the roles of force-dependent interactions with smoothelin and fimbacin remain unknown [27]. FilGAP binding to filamin A directs FilGAP to membrane protrusion sites, where it antagonizes Rac to govern actin remodeling [28]. Mechanical forces, which expose a cleavage site, also cause proteolysis. The transport of YAP from the cytosol to the nucleus is facilitated by mechanical forces such as substrate stiffness, indentation of the plasma membrane by atomic force microscopy (AFM), and osmotic shocks [29]. Actin polymerization, which is activated by mechanical forces via unknown processes, is another method for converting mechanical forces into biochemical signals [30]. Actin polymerization separates myocardin-related transcription factor A from unpolymerized actin in the cytoplasm, allowing mitochondrial transcription factor A to connect with multiple serum response factor target promoters [31]. Although numerous additional molecules, including cadherin, catenin, and merlin, have been found to be involved in mechanotransduction pathways, it is unknown how mechanical stresses are translated to biochemical signals via these molecules [32]. The main impediment is a lack of nanoscale structure information before and after mechanical force activation.

## 7.3 Mathematical Models of Mechanotransduction

Numerous mathematical models of mechanotransduction have been developed to explain the differentiation of cells in response to mechanical stimuli [33–35]. These include analyses of the involvement of YAP/transcriptional co-activator (TAZ), the transcription factors YAP and transcriptional co-activator with PDZ-binding motif (TAZ), in mechanosensing [36], as well as models aimed at predicting cell differentiation during bone healing [37]. Mousavi et al. created a 3D mechanosensing computational model to show how matrix stiffness affects Mesenchymal stem cell (MSC) fates. Their modeling results for MSC differentiation in response to substrate stiffness agree with previously published experimental data [35]. Burke and Kelly created a computer model to evaluate whether substrate stiffness and blood oxygen tension impact stem cell differentiation during fracture healing [34]. Their model predicted the occurrence of key processes involved in fracture healing, such as cartilaginous bridging, endosteal and periosteal bony bridging, and bone remodeling, using features related to cell proliferation, oxygen tension, and substrate stiffness. However, these models are constrained in that they do not account for the impact of regulatory variables. Additionally, these investigations employed several of models to describe a variety of experimental findings. As a result, describing the whole cell state space and studying the transitions between cell fates is challenging. Thus, a dynamic mathematical model is required that can stimulate a continuous range of stiffness values and their corresponding cell fates. Concluding there are several different approaches to model the multi-scale mechanics of mechanotransduction.

### 7.3.1 ODE Based Computational Model

In this section, a mathematical model of MSC differentiation is presented, which is driven by collection of fundamental events. The model of this section was previously presented by Peng et al. [33] and distributed under the terms of the Creative Commons Attribution 4.0 International License (http://creativecommons.org/licenses/by/4.0). MSCs immediately detect the stiffness of their surroundings by their adhesion to the substrate. The transcription factors YAP and TAZ mediate the signal via their interaction with downstream genes involved in cell development. When MSCs are activated in a soft stiffness environment (1 kPa), the tubulin beta-3 chain gene TUBB3 is expressed. PPARG, or peroxisome proliferator-activated receptor gamma, is a gene that encodes an adipogenic marker that has been shown to be active in low-stiffness environments (1 kPa). MYOD1, or myogenic differentiation 1, is a myogenic gene which is activated under moderately stiff (10 kPa) conditions and produces essential proteins regulating muscle growth. RUNX2, also known as runt-related transcription factor 2, is an osteogenic gene that is upregulated in environments with a high stiffness (40 kPa) and plays a vital role in osteoblast formation. The collection of four lineage-specific genes in our model is used to summarize the transcriptional changes seen during MSC differentiation into four distinct cell fates in response to mechanical stimuli mediated by YAP/TAZ signaling.

A simplified gene regulatory network that underpins MSC fate with ordinary differential equations (ODEs) has been developed:

$$\begin{aligned}\frac{\mathrm{d}[\mathrm{SAA}]}{\mathrm{d}t} &= k_1 \frac{\left(\frac{s}{K_1}\right)^{n_1} + \left(\frac{[\mathrm{TUBB3}]}{K_2}\right)^{n_2}}{1+\left(\frac{s}{K_1}\right)^{n_1} + \left(\frac{[\mathrm{TUBB3}]}{K_2}\right)^{n_2}} + k_2 \frac{\left(\frac{s}{K_3}\right)^{n_3} + \left(\frac{[\mathrm{PPARG}]}{K_4}\right)^{n_4}}{1+\left(\frac{s}{K_3}\right)^{n_3} + \left(\frac{[\mathrm{PPARG}]}{K_4}\right)^{n_4}} \\ &\quad + k_3 \frac{\left(\frac{s}{K_5}\right)^{n_5} + \left(\frac{[\mathrm{MYOD1}]}{K_6}\right)^{n_6}}{1+\left(\frac{s}{K_5}\right)^{n_5} + \left(\frac{[\mathrm{MYOD1}]}{K_6}\right)^{n_6}} + k_4 \frac{\left(s/K_7\right)^{n_7} + \left([\mathrm{RUNX2}]/K_8\right)^{n_8}}{1+\left(s/K_7\right)^{n_7} + \left([\mathrm{RUNX2}]/K_8\right)^{n_8}} \\ &\quad - d_1[\mathrm{SAA}],\end{aligned} \tag{7.1}$$

$$\frac{\mathrm{d}[\mathrm{YAPTAZ}]}{\mathrm{d}t} = k_5[\mathrm{SAA}] - d_2[\mathrm{YAPTAZ}], \tag{7.2}$$

$$\frac{\mathrm{d}[\mathrm{TUBB3}]}{\mathrm{d}t} = k_6 \frac{\left(\mathrm{SAA}/K_9\right)^{n_9}}{1+\left(\mathrm{SAA}/K_9\right)^{n_9} + \left([\mathrm{YAPTAZ}]/K_{10}\right)^{n_{10}}} - d_3[\mathrm{TUBB3}], \tag{7.3}$$

$$\frac{\mathrm{d}[\mathrm{PPARG}]}{\mathrm{d}t} = k_7 \frac{\left(\mathrm{SAA}/K_{11}\right)^{n_{11}}}{1+\left(\mathrm{SAA}/K_{11}\right)^{n_{11}} + \left([\mathrm{YAPTAZ}]/K_{12}\right)^{n_{12}}} - d_4[\mathrm{PPARG}], \tag{7.4}$$

$$\frac{d[\text{MYOD1}]}{dt} = k_8 \frac{(\text{YAPTAZ}/K_{13})^{n_{13}}}{1+(\text{SAA}/K_{14})^{n_{14}} + ([\text{YAPTAZ}]/K_{13})^{n_{13}}} - d_5[\text{MYOD1}], \tag{7.5}$$

$$\frac{d[\text{RUNX2}]}{dt} = k_9 \frac{(\text{YAPTAZ}/K_{15})^{n_{15}}}{1+(\text{SAA}/K_{16})^{n_{16}} + ([\text{YAPTAZ}]/K_{15})^{n_{15}}} - d_6[\text{RUNX2}], \tag{7.6}$$

where $S$ and [SAA], are the relative levels of the stiffness (input to the system) and of the effective stiffness adhesion area, respectively. [YAPTAZ], [TUBB3], [PPARG], [MYOD1], and [RUNX2] denote the concentrations of YAP/TAZ, TUBB3, PPARG, MYOD1, and RUNX2. Since concentration and time in the model are given in relative units, i.e. are dimensionless, then all parameters in the above equations are also dimensionless. $d_i (i = 1, 2, \ldots, 6)$ are the degradation rates of the corresponding genes/factors.

The constructed model exhibits various stable states throughout the behavioral areas examined (with first seeding stiffness values ranging from 0.1 kPa to larger than 100 kPa; Figure 7.1) which was determined experimentally. Multiple stable states of the YAP/TAZ expression are shown in Figure 7.1a,b spanning the stiffness range investigated, and variations in the YAP/TAZ state may be viewed when stiffness rises (right directed grey arrows) or decreases (right directed grey arrows). Along the blue lines, the nonlinear connection between YAP/TAZ and the stiffness of the substrate is consistent with prior studies.

Bias in the relative gene expression of TUBB3 (a neurogenic differentiation driver) downstream of YAP/TAZ is seen in Figure 7.1c. When the stiffness is less than 0.2 kPa, TUBB3 is turned off. It will be activated when the stiffness reaches 0.25 kPa. It reverts to "OFF" mode as the stiffness rises. Meanwhile, when the stiffness falls below 0.2 kPa, TUBB3 remains "ON", emphasizing the mechanical memory exhibited during neurogenic differentiation. Notably, when the stiffness drops from 0.6 kPa, TUBB3 remains "OFF." The stiffness range of 0.25–0.55 kPa is defined as a "differentiation memory zone" for TUBB3. This indicates that if the stiffness of the first seeding substrate is within this range, the cell will "remember" it and differentiate accordingly (toward a neurogenic destiny) upon reseeding. Additionally, our model predicts unique memory areas for differentiation for PPARG (0.6–3 kPa; Figure 7.1d) and MYOD1 (10–15 kPa; Figure 7.1e). Of the four lineage-specific marker genes investigated, RUNX2 exhibited the biggest differential memory area.

Each of the four lineage-specific genes investigated had a bistable region, as seen in Figure 7.1c–f. This is a remarkable prediction: that an area of

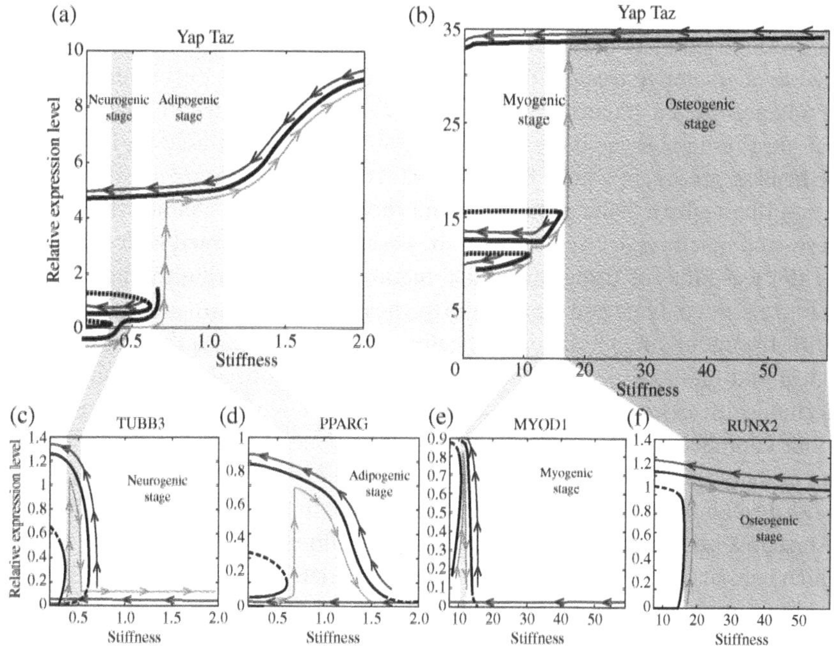


**Figure 7.1** (a)–(f) Multistability in the MSC differentiation network. The relative expression level of YAP/TAZ in a stiffness range from 0.1 kPa to 60 kPa is shown. *Source:* [33]/Springer Nature/CC BY 4.0.

mechanical memory exists for all cell fates, not only osteogenic differentiation, as previously described [8]. While the memory areas of neurogenic and adipogenic differentiation are smaller than those of osteoblasts, they may still be critical for stem cell destiny control. The true contribution of each will require additional investigation, as a variety of interacting factors influence neurogenic and adipogenic cell fate decisions, including those not currently included in our model, such as the role of substrate-induced stemness and epithelial to mesenchymal transition.

### 7.3.2 PDE Based Computational Model

Numerous computational models have been presented to mimic bone healing in order to better understand the fundamental principles governing cell activity and angiogenesis [38]. Typically, a set of partial differential equations is used, with the unknown variables being cell types, growth factors, and tissues densities. The spatiotemporal evolution of endothelial cell concentration and vascular density were used to simulate the angiogenesis process. A lattice-based model was developed to characterize tissue differentiation and angiogenesis during shear stress on a bone/implant fracture [39]. Also, a lot of effort was given to demonstrate that ultrasound plays a vital role in angiogenesis and can be used for improving the healing process of a bone fracture [40, 41]. These models are mostly utilized to describe realistic bone geometries and their mechanical characteristics on a meso- and macro-scale. Additionally, they are mainly concerned with determining the feasibility of using innovative methods for measuring the mechanical characteristics of bone and monitoring the healing process without attempting to define the underlying physiological healing processes. On contrary, mechanobiological and mathematical models have been widely utilized to (i) explain how mechanical stresses govern biological processes through signals to cells, and (ii) mimic complicated biological processes such as bone regeneration that are difficult to study experimentally or clinically. As a result, such models may contribute to the development of unique insights and a basic understanding of the effect of ultrasound on bone repair. A detailed and multi-scale computational model which consists of partial differential equations (PDEs) was previously presented by Vavva et al. [42] and re-presented here for the educational purposes of our textbook. The model predicts bone healing and angiogenesis by employing ultrasound stimulation. The model includes 11 differential equations describing the spatiotemporal variation of mesenchymal stem cells ($c_{\mathrm{m}}$), fibroblasts ($c_{\mathrm{f}}$), chondrocytes ($c_{\mathrm{c}}$), osteoblasts ($c_{\mathrm{b}}$), fibrous extracellular matrix ($m_{\mathrm{f}}$), cartilaginous extracellular matrix ($m_{\mathrm{c}}$), bone extracellular matrix ($m_{\mathrm{b}}$), generic osteogenic ($g_{\mathrm{b}}$), chondrogenic ($g_{\mathrm{c}}$), and vascular growth factors ($g_{\mathrm{v}}$), as well as the concentration of oxygen and nutrients ($n$) [42, 43]:

$$\begin{aligned}\frac{\vartheta c_{\mathrm{m}}}{\partial t} &= \nabla\cdot\left(D_{\mathrm{m}}\nabla c_{\mathrm{m}} - C_{\mathrm{mCT}}c_{\mathrm{m}}\nabla\left(g_{\mathrm{b}}+g_{\mathrm{v}}\right) - C_{\mathrm{mHT}}c_{\mathrm{m}}\nabla_{\mathrm{m}}\right)\\ &\quad + A\_m\,c\_m\left(1-\left[\left(a\_m\,c\right)\right]\_m\right) - F\_1\,c\_m - F\_2\,c\_m - F\_4\,c\_m,\end{aligned} \tag{7.7}$$

$$\frac{\vartheta c_{\mathrm{f}}}{\partial t} = \nabla\cdot\left(D_{\mathrm{f}}\nabla c_{\mathrm{f}} - C_{\mathrm{f}}c_{\mathrm{f}}\nabla g_{\mathrm{b}}\right) + A_{\mathrm{f}}c_{\mathrm{f}}\left(1-a_{\mathrm{f}}c_{\mathrm{f}}\right) + F_4 c_{\mathrm{m}} - F_3 d_{\mathrm{f}} c_{\mathrm{f}}, \tag{7.8}$$

$$\frac{\vartheta c_{\mathrm{c}}}{\partial t} = A_{\mathrm{c}}c_{\mathrm{c}}\left(1-a_{\mathrm{c}}c_{\mathrm{c}}\right) + F_2 c_{\mathrm{m}} - F_3 c_{\mathrm{c}}, \tag{7.9}$$

$$\frac{\vartheta c_{\mathrm{b}}}{\partial t} = A_{\mathrm{b}}c_{\mathrm{b}}\left(1-a_{\mathrm{b}}c_{\mathrm{b}}\right) + F_1 c_{\mathrm{m}} + F_3 c_{\mathrm{c}} - d_{\mathrm{b}}c_{\mathrm{b}}, \tag{7.10}$$

$$\frac{\vartheta m_{\mathrm{f}}}{\partial t} = P_{\mathrm{fs}}\left(1 - k_{\mathrm{f}} m_{\mathrm{f}}\right) c_{\mathrm{f}} - Q_{\mathrm{f}} m_{\mathrm{f}} m_{\mathrm{c}} c_{\mathrm{b}}, \tag{7.11}$$

$$\frac{\vartheta m_{\mathrm{c}}}{\partial t} = P_{\mathrm{cs}}\left(1 - k_{\mathrm{c}} m_{\mathrm{c}}\right) c_{\mathrm{c}} - Q_{\mathrm{c}} m_{\mathrm{c}} c_{\mathrm{b}}, \tag{7.12}$$

$$\frac{\vartheta m_{\mathrm{b}}}{\partial t} = P_{\mathrm{bs}}\left(1 - k_{\mathrm{b}} m_{\mathrm{b}}\right) c_{\mathrm{b}}, \tag{7.13}$$

$$\frac{\vartheta g_{\mathrm{c}}}{\partial t} = \nabla \cdot \left(D_{\mathrm{gc}} \nabla_{\mathrm{gc}}\right) + E_{\mathrm{gc}} c_{\mathrm{c}} - d_{\mathrm{gc}} g_{\mathrm{c}}, \tag{7.14}$$

$$\frac{\vartheta g_{\mathrm{b}}}{\partial t} = \nabla \cdot \left(D_{\mathrm{gb}} \nabla_{\mathrm{gb}}\right) + E_{\mathrm{gb}} c_{\mathrm{b}} - d_{\mathrm{gb}} g_{\mathrm{c}}, \tag{7.15}$$

$$\frac{\vartheta g_{\mathrm{v}}}{\partial t} = \nabla \cdot \left(D_{\mathrm{gv}} \nabla_{\mathrm{gv}}\right) + E_{\mathrm{gvb}} c_{\mathrm{b}} + d_{\mathrm{gvb}} c_{\mathrm{c}} - g_{\mathrm{v}}\left(d_{\mathrm{gv}} + d_{\mathrm{gvc}} c_{\mathrm{v}}\right), \tag{7.16}$$

$$\frac{\vartheta n}{\partial t} = \nabla \cdot \left(D_n \nabla_n\right) + E_n c_n - d_n n, \tag{7.17}$$

$m = m_{\mathrm{f}} + m_{\mathrm{c}} + m_{\mathrm{b}}$ is the total tissue density. The processes described by these equations as well as the parameters used in this model are extensively discussed in Geris et al. [44].

The ultrasound effect is modeled based on the rational that for a fluid-saturated medium subjected to a small amplitude oscillatory pressure gradient, such as in the case of ultrasound presence, the pressure fluctuation causes micro fluid flow through the sample to release the differential pressure. This phenomenon can be described by dynamic diffusion as:

$$\frac{\vartheta p}{\partial t} = \nabla \cdot \left(D_{\mathrm{p}} \nabla_{\mathrm{p}}\right), \tag{7.18}$$

where $D_{\mathrm{p}}$ is the diffusivity of the ultrasound acoustic pressure.

The Darcy's law is employed to associate pressure with the fluid velocity:

$$\boldsymbol{u} = -K \nabla_{\mathrm{p}}, \tag{7.19}$$

where $K$ is the permeability coefficient assuming that the interstitium is a porous medium.

The interstitial fluid velocity is given by:

$$\frac{\vartheta g_{\mathrm{v}}}{\partial t} = \nabla \cdot \left(D_{\mathrm{gv}} \nabla_{\mathrm{gv}}\right) + E_{\mathrm{gvb}} c_{\mathrm{b}} + d_{\mathrm{gvb}} c_{\mathrm{c}} - g_{\mathrm{v}}\left(d_{\mathrm{gv}} + d_{\mathrm{gvc}} c_{\mathrm{v}}\right) - \nabla \cdot \left(\boldsymbol{u} g_{\mathrm{v}}\right). \tag{7.20}$$

Regarding blood vessel formation when a grid volume of the spatial discretization contains a vessel, the variable $c_v$ is set to 1, otherwise $c_v = 0$. The evolution of $c_v$ is determined by blood vessel growth, branching and anastomosis. Blood vessel growth is modeled by solving tip velocity equations that describe the movement of the corresponding tip cell. Branching occurs for high vascular endothelial growth factor (VEGF) concentrations and anastomosis when a tip cell meets a blood vessel, i.e. when the tip cell reaches a grid volume i.e. when the tip cell reaches a grid volume with $c_v = 1$.

The initial and boundary conditions' values are based on earlier work. Except for the variables presented in Figure 7.2, no-flux border conditions are applied for all variables. Due to symmetry difficulties, only one-fourth of the domain is taken into account. Initially, it is considered that the callus is made of fibrous tissue. During the first three post-fracture (PF) days, MSCs and fibroblasts are thought to be released into the callus tissue from the periosteum, surrounding tissues, and bone marrow. During the first 5 PF days, an initial amount of chondrogenic growth factors is assumed at the deteriorating bone ends. During the first 10 PF days, the cortical factors also promote osteogenic development. Figure 7.2 depicts the initial placements of the tip cells. Endothelial cells are able to depart the callus area. To imitate transducer application during axial ultrasound transmission, the periosteal area, which is in close contact with the soft tissues, acts as a source of

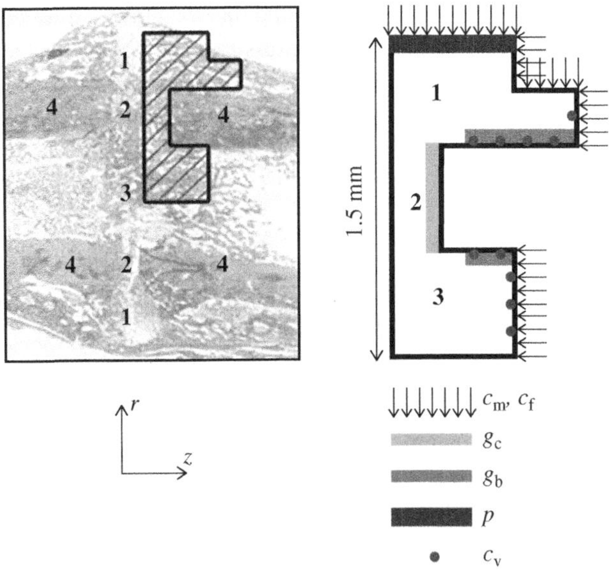


**Figure 7.2** The geometry used in the presented model. *Source:* Reproduced with permission from Vavva et al. [42]/ELSEVIER.

ultrasound acoustic pressure. We look at four distinct situations of acoustic pressure boundary conditions that imitate ultrasound application at various intensities.

The results of the provided model correctly characterized the most significant elements of bone healing, which begin with the migration of mesenchymal cells, fibroblasts, and the release of growth factors into the callus from surrounding tissues. MSCs develop into osteoblasts toward the cortex and away from the fracture gap, whereas chondrocytes differentiate throughout the callus. When chondrocytes hypertrophy, the angiogenic and osteogenic process begins with the production of vascular growth factors. Angiogenesis begins in the periosteal callus on day 3 PF and in the endosteal callus on day 8, almost a week sooner than without the ultrasonic impact. Following that, the arteries provide oxygen and nutrients, resulting in endochondral ossification. The gap is subsequently gradually filled with bone, while cartilage and fibrous tissue densities diminish. Meanwhile, blood vessels sprout and form a network that covers the entire callus region and gives oxygen and nutrients to the entire fracture. It is clear that ultrasound causes increased branching and anastomosis, resulting in a quicker vascular network inside the callus area. In such situation, bone healing is complete by day 26 PF. However, without the ultrasound, bone repair takes around four to five weeks (Figure 7.3).

#### 7.3.2.1 Mechanical Factors that Affect Cell Differentiation and Proliferation

Cell differentiation, proliferation, apoptosis, and migration all play important roles in the early stages of tissue regeneration. The ability of a stem cell to differentiate into many cell types allows it to produce separate tissues. MSCs, for example, can differentiate into fibroblasts, chondrocytes, osteoblasts, neuronal precursors, adipocytes, and a variety of other cell types [45]. Although stem cells' capacity to differentiate into several lineages is beneficial, it may be fatal if they differentiate at the wrong time, in the wrong place, or to the incorrect cell type. This can lead to a pathological state or the creation of non-functional tissue. To overcome such aberrations, stem cells have been designed to differentiate only in response to certain biological cues. As a result, while other signals such as chemotaxis can cause cells to differentiate, proliferate, and/or die, our purpose here is to investigate it from a mechanotactic approach.

Cell differentiation and proliferation are governed by a combination of chemical [46] and mechanical [47] cues, with scientists claiming that other signals such as growth factors and cytokines may be involved in stem cell differentiation. Recent research indicates that mechanical cues can have a significant influence on cell differentiation and proliferation [48]. Mechanical characteristics such substrate stiffness, adhesion surface nanotopography, mechanical stresses, fluid flow, and cell colony sizes have been shown in tests to drive stem cell fate even in the

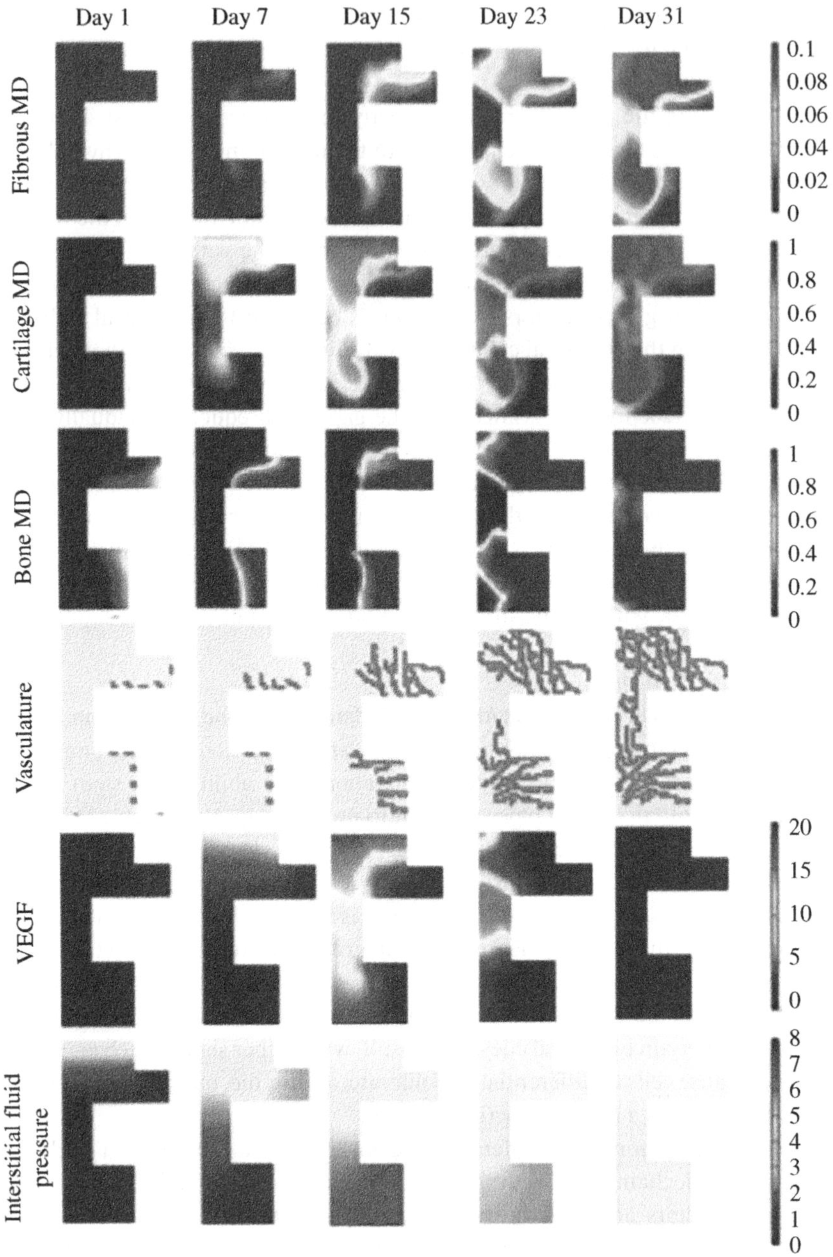


**Figure 7.3** Predicted spatiotemporal evolution of fibrous tissue, cartilage bone matrix density, vasculature and VEGF (×100 ng/ml) under the presence of Ultrasound. *Source:* Reproduced with permission from Vavva et al. [42]/ELSEVIER.

absence of biochemical inputs [47]. Many studies have been carried out to study the effect of mechanical stimulation on cell differentiation and proliferation in tissue regeneration. Pauwels [49], for example, stated that distortional shear stress is a specific trigger for MSCs to differentiate into fibroblasts for the formation of fibrous tissue. MSCs differentiate into chondrocytes in cartilage synthesis in response to hydrostatic compression, but MSCs evolve into the osteogenic pathway (ossification) only when the strain felt by the cell is less than a particular threshold.

Cells actively monitor and respond to mechanical changes in their microenvironment (mechano-sensing) via their focal adhesions [50]. It has been established, for example, that altering the stiffness of the matrix from soft to somewhat rigid may impact MSC fate.

The signaling mechanisms that govern cell lineage by microenvironment stiffness are unclear. Many mechano-biological models have been developed to define cell differentiation during fracture healing [37]. Mechanical strain and perfusive fluid flow were employed by Stops et al. [37] to mimic cell differentiation and proliferation in a collagen-glycosaminoglycan scaffold. They predicted that the scaffold strain and the velocity of the entering fluid would influence the responses of the representative cells. They showed that certain scaffold strains and intake fluid flows, as defined by MSC differentiation patterns, result in phenotypic assemblies dominated by single cell types.

#### 7.3.2.2 A Case Example of Multi-Scale Modeling Cell Differentiation and Proliferation

A discrete finite element approach has been presented to formulate cell migration, differentiation, proliferation and apoptosis in defined substrates [35] distributed under the terms of the Creative Commons Attribution 4.0 International License (http://creativecommons.org/licenses/by/4.0). The primary goal of this study is to expand the prior model to investigate the effect of substrate mechanical conditions on cell differentiation, proliferation, and apoptosis during migration.

***7.3.2.2.1 Cell Migration*** During cell translocation, the actin cytoskeleton (CSK) controls the driving forces at the cell front while the microtubule network regulates the rear retraction of the cell [51]. Active stress generated by actin filaments and myosin II, active cellular elements, depends on the maximum, $\epsilon_{max}$, and the minimum, $\epsilon_{min}$, internal cell strains. Besides, passive stress is related to the microtubules and the cell membrane, passive cellular elements. The cell stress which is transmitted to the extracellular matrix (ECM) can be defined as the sum of the passive and active stresses

*7.3.2.2.2* ***Effective Mechanical Force*** During cell migration two main mechanical forces affect the cell body, the traction force and the drag force. The former, which is transmitted to the substrate through integrins, is generated due to the contraction of the actin-myosin apparatus. This force drives the cell body forward and is directly proportional to the cell stress, $\sigma_{\text{cell}}$. Representing the cell by a connected group of finite elements, the nodal traction force of the cell can be represented as:

$$\boldsymbol{F}_i^{\text{trac}} = \sigma_{\text{cell}} S \zeta \boldsymbol{e}_i, \tag{7.21}$$

where $\boldsymbol{e}_i$ denotes a unit vector passing through the $i$th node of the cell membrane toward the cell centroid (Figure 7.4a). $S$ represents the cell membrane area and $\zeta$ is a dimensionless parameter named "adhesivity" which is directly proportional to the concentration of the ligands at the leading edge of the cell, $\psi$, the total number of available receptors, $n_{\text{r}}$, and the binding constant of the cell integrins, $k$:

$$\zeta = k n_{\text{r}} \psi. \tag{7.22}$$

*7.3.2.2.3* ***Protrusion Force*** To migrate, cells extend local protrusions by exerting a protrusion force to evaluate their surrounding substrate. This refers to the actin polymerization and differs from the cytoskeletal contractile force transmitted to the substrate. It is a random force that causes cells to move along a directed

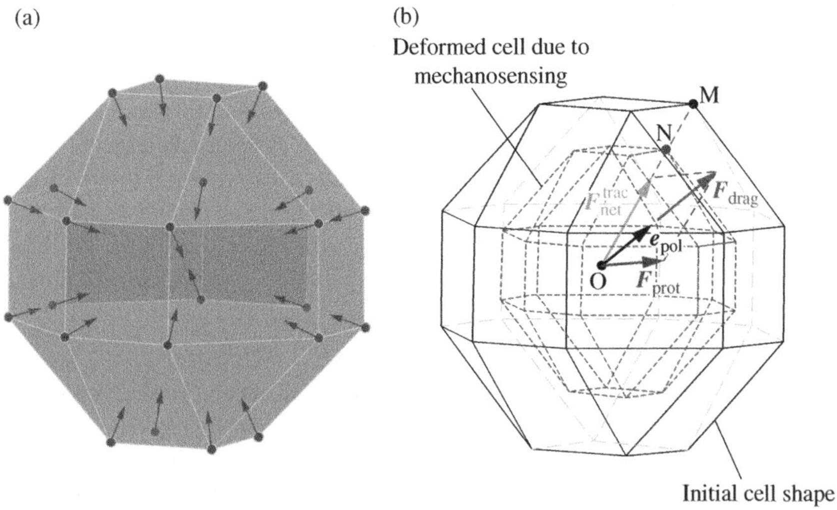


**Figure 7.4** (a) Spherical configuration of the cell in which sensing forces are exerted at each membrane node toward the cell centroid (mechano-sensing process). (b) Calculation of the cell internal deformation due to cell mechano-sensing. *Source:* Mousavi and Doweidar [35]/PLOS/CC BY 4.0.

random path toward the effective cue. It is remarkable that the order of the protrusion force magnitude is the same as that of the traction force but with lower amplitude.

*7.3.2.2.4* ***Cell Deformation and Reorientation*** For the sake of simplicity, a spherical cell shape is considered here (solid line in Figure 7.4b). However, any cell shape can be considered using the present model. In the mechano-sensing stage, a cell firstly exerts sensing forces at each finite element node located on the cell membrane toward the cell centroid to probe its surrounding environment. The cell deformation resulting from the mechano-sensing step is shown by dashed lines in Figure 7.4b. Therefore, the internal deformation of the cell at each finite element node of the cell membrane can be defined as:

$$\varepsilon_{\text{cell}} = \frac{\text{MN}}{\text{OM}}. \tag{7.23}$$

*7.3.2.2.5* ***Cell Differentiation, Proliferation, and Apoptosis*** Cells may respond to the mechanical signals received from their micro-environment by differentiation or apoptosis [52]. A specific deformation range experienced by a cell is shown to lead to a specific differentiation. This diversity may arise from differences in their tissue origin, in the magnitude and duration of the mechanical signal sensed by the cell and in the degree of preconditioning. Although, the precise effect of mechanical cues on cell apoptosis is still poorly understood, there are experimental works reporting that cell death may occur due to the deformation which a typical cell can bear. Experimental data are used to define the cell differentiation considering the mechano-biological scale as well as the time dependent nature. For example, MSCs and chondrocyte need to be at a certain level of maturity before they undergo differentiation or proliferation. A maturation time is defined for this purpose for the cells. Specific rules are applied to check the mechanisms of migration, proliferation and apoptosis which may be performed in parallel and one affects each other.

Cell proliferation is the process of producing two daughter cells from a mother. In normal tissues, this, in general, refers to cells that replenish the tissue by cell growth followed by cell division. Cell proliferation occurs in defined steps including the first growth phase, the synthesis phase, the second growth phase and the mitosis phase. During the first growth phase, known as the G1 phase, the cell synthesizes a huge content of biological material. As soon as the G1 phase is completed the cell enters the synthesis phase, the S phase, to replicate its DNA. At the end of the S phase it starts the second growth phase, G2, that finally leads to the mitosis phase, the M phase. Subsequently, reorganization of the cell chromosomes is followed by the cell division so that a mother cell is divided into two daughter cells. This is a critical instant because some cells temporarily stop proliferation by entering into the quiescence state which is called the G0 phase.

Computationally speaking, it is hypothesized that there are no concerns about shortage of oxygen or nutrients for the cells in culture. Therefore, we intend to model the dominant cell division cycle through two main steps. It is assumed that during the G1, S, and G2 phases the cell grows and matures such that when the cell maturation is achieved, depending on the strength of the mechanical signal received by the cell, one mature mother cell may enter into the mitosis phase and divide into two non-mature daughters. Thus, in the present model, the cell is either under maturation or in the proliferation phase. In other words, each cell is in the quiescence phase unless it delivers two daughter cells.

***7.3.2.2.6 Finite Element Implementation*** We have applied the model for several numerical cases where the cell is embedded within a 400 × 200 × 200 μm substrate with different ranges of stiffness. It is assumed that there is no external force acting on the substrate and all of the boundary surfaces are considered free. The substrate is meshed by 16,000 regular hexahedral elements and 18,081 nodes while the cell has a constant spherical shape with 24 nodes on its membrane. The calculation time is about one minute for each time step in which each step corresponds to approximately six hours of real cell–substrate interaction. The concept of the proposed algorithm is shown in Figure 7.5.

The developed model can be used for several biological scenarios such as to predict cell fate within soft (0.1–1 kPa), intermediate (20–25 kPa), and hard

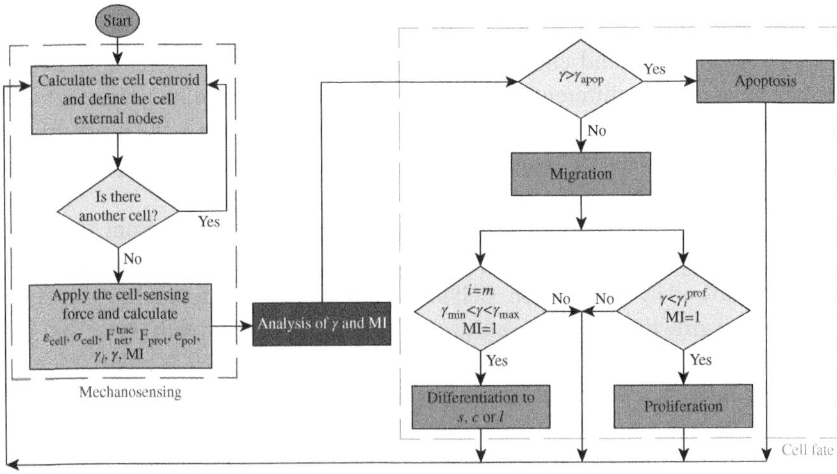


**Figure 7.5** Computational algorithm of cell mechano-sensing and consequent cell fate due to mechanotaxis. *Source:* Mousavi and Doweidar [35]/PLOS/CC BY 4.0.

(30–45 kPa) substrates which are comparable with neurogenic, chondrogenic and osteogenic tissue, respectively. This model can be considered as a first step toward the interpretation of existing knowledge about the effect of cellular micro-environment on cell fate. However, further research on mechanical and physical factors, such as cell shape, topographic changes, external mechanical forces and colony size, can provide a better understanding of the determination of the cell fate. It was investigated how a 3D substrate with soft, intermediate and hard stiffnesses affects the differentiation of MSCs toward neurogenic, chondrogenic or osteogenic lineage, respectively. The present 3D numerical model can successfully predict essential aspects of cell maturation, differentiation, proliferation and apoptosis during regenerative events.

### 7.3.3 Methodology of a Hybrid Multi-Scale Approach

To that yet, a thorough understanding of the interaction between active mechano-sensing, mechanotransduction, and ECM material property regulation remains elusive. This section is separated into two sections to discuss the hybrid model, also known as the mechano-agent-based model (Mech-ABM) [53] distributed under the terms of the Creative Commons Attribution 4.0 International License (http://creativecommons.org/licenses/by/4.0). The first section discusses the ABM, which simulates events at the cellular and molecular levels, while the second section covers the mechanical model, which simulates events at the tissue level. Communication between the two models occurs in discrete time increments. The mechanical model mimics tissue biomechanics and calculates numerical force values at each integrin site. The ABM replicates mechanotransduction processes in a 3D osteoblasts (OB) environment in an ECM (osteoid).

The Mech-ABM investigated the impact of changing the mechanical characteristics of integrins on mechanotransduction and mechanoreciprocity. It specifically looked at the effect of integrin mechanosensitivity and heterogeneity among the integrin population on (i) mechanotransduction dynamics, (ii) tissue material property regulation, and (iii) OB response to mechanical stimulation. The mechanosensitivity threshold (MT) is a numerical value of mechanical force to which an integrin is exposed. Integrins were activated if the applied tissue load (AFT) value was equal to or larger than the MT threshold.

#### 7.3.3.1 The Agent-Based Model (ABM)

A complete summary of the ABM may be found here [53] distributed under the terms of the Creative Commons Attribution 4.0 International License (http://creativecommons.org/licenses/by/4.0). The latter was customized to the standard procedure for defining ABMs: the Overview, Design principles, and Description (ODD) [54].The ABM was founded on chemical interactions among intracellular molecules. In summary, mechanotransduction occurs downstream of integrins and is activated by mechanical stimulation of the ECM. The extracellular signal-regulated kinase (ERK) cascade (Raf–MEK–ERK) is the primary mechanotransduction route that connects cytoplasmic events to the nucleus and the production of osteogenic genes via the interaction with RUNX2 TF. As a result, genes encoding osteogenic indicators such alkaline phosphatase (ALP), osteopontin (OPN), osteocalcin (OCN), and bone sialoprotein (BSP) are transcribed and translated into the relevant proteins (ECMp). Following that, ECMps were placed in the surrounding ECM to stiffen it. All of these molecules were modeled as agents, and each ECMp deposited increased ECM elasticity (ECMelasticmod) by 0.001. The agents were believed to be well-mixed and evenly distributed inside the cell membrane, cytoplasm, and nucleus. The number of molecules was calculated using information from the literature. The ABM was ran in real time for the equivalent of four days.

The ABM was implemented using the generic ABM framework Flexible Large Agent-Based Modeling Environment (FLAME). It schedules agents interactions in discrete time-steps. A time-step was defined as one second. After the time-step execution, FLAME communicates with the mechanical model. Every molecule-agent is an autonomous communicating X-machine governed by transition functions which are executed serially by the agents.

Transition functions determine molecular behaviors and interactions, controlling molecules' state transitions from dormant to active states and vice versa as modeled previously [55]. The agents were heterogeneous (HM) and their heterogeneity arises from: agents' occupancy of different states; spatial separation and compartmentalization into either the cytoplasm or the nucleus; and stochastic updates of every agent's global and local variable over time; specifically the Activation Cycle Switch (ACS) variable [55]. This stochasticity was implemented by random selection of numerical values from either uniform or Gaussian distributions. The distribution selected depended on the modeled parameter, and the agent. For instance, numerical values of parameters ACS and agent's interaction radius were extracted from uniform distributions, while rate of protein syntheses was extracted from a Gaussian distribution.

Integrins mechanosensitivity threshold is a pivotal state variable and it was examined using four models. In models one and two, the integrin population was homogeneous with respect to the mechanosensitivity threshold; while the third and fourth models were heterogeneous (HM). The first mechanosensitivity threshold was set to 10% of $AF_T$ (sensitive model [SM]), while the second threshold was set to 1% of $AF_T$ (ultrasensitive model [USM]). In the third model, the integrin population was divided into ultrasensitive and sensitive agents with a ratio of 1 : 10 respectively (10%-HM), while in the fourth model the heterogeneity was changed to 1 : 100 ratios (1%-HM).

Emergent behavior related to activation dynamics variables of an agent population were monitored and analyzed. These were: maximal magnitude of agents' active state ($E_{max}$), time to achieve $E_{max}(t{-}E_{max})$, half $E_{max}(EC_{50})$, time to achieve $EC_{50}(t\text{-}EC_{50})$ and magnitude of deposited ECMp levels individually and collectively. The emergence of molecular memory was also of interest. These variables were sampled at every 100th time-step to minimize the signal noise without compromising on detail of the simulation output.

Molecular events occur within 3D space. The distance between two pixels was calibrated to 1 nm. Adopting the OB's physiological volume would have led to agent numbers to be in magnitude of millions; substantially increasing computational costs and model run time. Hence, to minimize this drawback, total cell volume and cellular and nuclear radii were attuned to 1% of the average OB volume. It was previously shown that such an approach had an insignificant influence on altering interaction dynamics in intracellular ABMs. The spatiality was partitioned to extracellular environment (ECM), plasma membrane, cytoplasm, nuclear membrane and the nucleus.

#### 7.3.3.2 Mechanical Model

The normal stress at the interface of the cell and the ECM can be determined from the expression:

$$\sigma_{rr} = \frac{5\eta\kappa}{[6(2\eta+\kappa)+\dfrac{4\eta\kappa(71\eta+16\kappa)}{152\eta+19\eta\kappa+16\kappa)}}\sigma_{xy,\infty}\sin^2\theta\sin 2\varphi. \tag{7.24}$$

Here, $\eta = \mu C/\mu M$ is the ratio of shear moduli of the cell ($\mu C$) and the ECM ($\mu M$) and $k = \alpha R/\mu M$ is non-dimensional ratio relating the stiffness of the interface to the elasticity of the ECM. Specifically, $\alpha$ is the spring constant in both normal and tangential directions and the interface is considered to behave as a linear spring. Finally, $R$ is the cell radius, $\sigma_{xy,\infty}$ is the uniform shear stress applied at an infinitely large distance from the cell and $\theta$ and $\phi$ are spherical polar angles. It is assumed here that this stress is mediated through integrins, which are of equal size. Each of $N$ equal-sized integrins will cover an area equal to $4\pi R2/N$ and a circle possessing the same area has radius $a = 2R/N1/2$. If the number of integrins is large, this circular patch is small, and the variation in the normal stress $\sigma_{rr}$ over this area can be neglected. For example, with $N = 500$ integrins the term $\sin^2\theta\sin 2\varphi$ varies, on average, less than 7.5% of its maximum value over the cell's surface. Thus, the force on the integrin is $\pi a2\sigma_{rr}$ where $a$ is the integrin radius. The total number of ECMp deposited regulates the elastic moduli of the ECM. Integrin 3D coordinates are taken from the ABM and used to determine numerical values of force exerted on individual integrin agents.

Modulation of integrins' mechanosensitivity to $AF_T$ and its impact on mechanotransduction, and consequently, synthesis of osteogenic proteins was investigated via a hybrid model which coupled a mechanical model and an Agent-Based Model (ABM), hence referred to as Mech-ABM (Figure 7.6).

The key principles of Mech-ABM are outlined here: The applied tissue load ($AF_T$), which was modeled as a unidirectional constant shear stress, propagates through the ECM to reach the cell's plasma membrane through the anchorage proteins and mechanoreceptors at the interface with the ECM. Integrins are the primary anchorage proteins and mechanoreceptors at the plasma membrane. $AF_T$ stretches the integrins, modeled as linear springs, and therefore change their confirmation leading to their priming for full activation. Mechanical excitation drives OB differentiation to OCy and the expression of osteogenic markers such as ALP, osteopontin (OCP), osteocalcin (OCN) and BSP, which are also extracellular matrix proteins (ECMp). Each ECMp modulates the material properties of the surrounding ECM by increasing its stiffness. Subsequently transfer of $AF_T$ changes due to reduced ECM stretching and therefore mechanosensation via integrins is altered. Integrin activation initiates intracellular mechanotransduction and provides the link between the mechanical and ABM models.

This study presents a novel hybrid 3D model of mechanosensation and mechanotransduction (Mech-ABM) that integrates mechanoreciprocity and links mechanosensation and mechanotransduction. It demonstrated that integrins' mechanosensing properties were important for maintaining response to applied mechanical load. The study also highlighted that heterogeneity of integrin mechanical properties are determinants of mechanotransduction dynamics. Moreover, the Mech-ABM illustrated how Integrins' mechanosensation in conjunction with ERK pathway contributed to the emergence of mechanical memory. The Mech-ABM model simulated mechanotransduction events downstream of integrins, triggering the ERK pathway and ERK-dependent osteogenic gene expression events. These involved Runx2 activation and TFs recruitment. The Mech-ABM illustrated that heterogeneity of integrin's mechanosensitivity contributed to the emergence of mechanical memory. It pointed to pERK as a candidate factor with the establishment of new pERK baseline that was sustainable for longer than four days. *In vivo* and *in vitro* observations in neuronal cells support this, where the establishment of new pERK baseline is vital for development of neuronal memory to previous excitations, termed long-term potentiation (LTP).

## 7.4 Challenges

Utilizing computer models to mimic biological events at all scales incurs personnel and computational expenses. To begin, every computational platform depends on well-developed tools that assist model definition and enable simulations to be

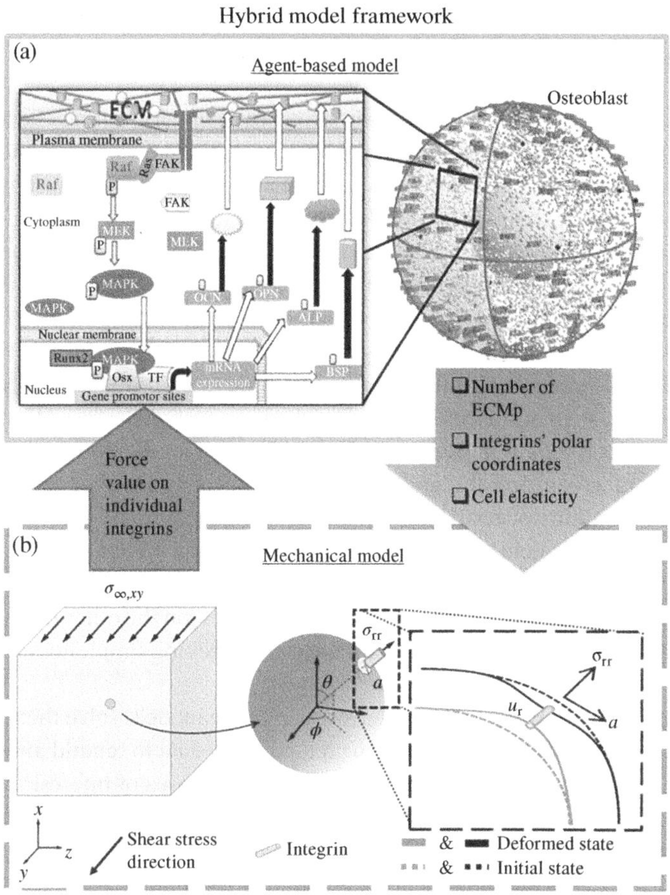


**Figure 7.6** A hybrid multi-scale model of mechanotransduction combining an agent based model (ABM) and a mechanical model. (a) The ABM simulated molecular events downstream of integrin mechanoreceptors in a spherical osteoblast responding to mechanical stimulation imposed at the ECM. The enlarged box shows the cascade simulated by the ABM which focused on biochemical mechanotransduction. This recruited the ERK pathway and its induced osteogenic gene expression events. These involved TF Runx2 recruitment, and its ensemble to express osteogenic ECM genes. Their corresponding mRNA and proteins were produced and the latter were deposited in the surrounding ECM. (b) A representation of the mechanical model; at the tissue level a unidirectional and constant shear stress was applied and was propagated through the ECM. The mechanical force was transferred to the interface between the cell and the ECM, resulting in cell and plasma membrane compression. Integrins resided at the interface, and the opposing stretch forces at the ECM and the plasma membrane were triggers for integrin activation. The values of the forces were computed and quantified by the mechanical model. The enlarged box illustrates the required parameters to calculate the forces generated on individual integrin-agents. The two models communicated every iteration which was equivalent to one seconds. The arrows between the two models represent these communications and the input/output parameters. ECM, ECMp FAK, Ras, Raf, mitogen-activated protein kinase (MEK), MAPK, Runx2, Osx, TF, ALP, OCN, OPN, BSP, $U_r$, $r$, $\theta$, and $\phi$ stand for: extracellular matrix, extracellular matrix protein, focal adhesion kinase, Rat sarcoma, Rapidly Accelerated Fibrosarcoma protein, Mitogen-activated protein kinase, Mitogen-activated protein kinase, Runt-related transcription factor 2, Ostrix, transcription factor, alkaline phosphatase, osteocalcin, osteopontin, bone sialoprotein, deformation of cell and ECM surfaces, and spherical polar coordinates locations respectively. *Source:* Shuaib et al. [53]/Springer Nature/CC BY 4.0.

run in a reasonable amount of time. Developing strong tools involves much effort and knowledge of computer science and computational mathematics. Given the model building and numerical solution tools, another time-consuming phase is the construction of a model that is directly relevant to the study subject. This procedure includes compiling appropriate model input parameters and expressing anatomical structures (e.g. by constructing complicated geometry meshes). This is frequently an unreported expense that is not included in the study's findings. After developing the model, the next stage is to utilize it to simulate settings in order to solve clinical or research problems. Simulations are computationally demanding, even more so in multiscale modeling, when connectivity across physical domains and scales is required. While solutions may be achieved, interpreting them may be difficult, requiring the assistance of an expert and maybe hours spent confirming the validity of results and then extracting meaningful information related to the research field. All of these broad issues may be handled in a research context if the development cost of tools and models is balanced against the loss of research productivity. The urgency of clinical issues, on the other hand, raises the strain on model developers, who are increasingly seeking simple-to-use, robust models coupled with rapid answers.

Tissue mechanics modeling needs robust software to create models, solve them, and then handle the findings. Models are often created using data to rebuild anatomical geometry, to represent properties unique to a physical area of interest, to characterize simulation settings, and to validate. For model creation and display of outcomes, graphics processors capable of processing massive datasets are often required. By and large, the simulation process requires technology capable of high-speed numerical calculation and substantial memory. It is not commonplace to use high-performance computers equipped with several processors; parallel processing through shared or distributed memory architectures.

Multiscale modeling need increased computational requirements. For example, for each dimension of the musculoskeletal system, separate and specialized computational techniques are utilized. Multibody dynamics at the body level, continuum finite element techniques at the organ level, and specialized algorithms and solvers at the tissue/cell level are the key computational tools for musculoskeletal modeling. While the multi-body technique is computationally efficient, it lacks the granularity necessary to adequately represent tissue activity. While finite element models of organs may be used to predict tissue deformation, they are often time and resource costly in terms of model development and processing. Electromyographic activity, ground response forces, kinematics, musculotendon dynamics models, muscle activation optimization techniques, and forward or inverse dynamics may all be used to estimate net joint loading and individual muscle forces. Joint loading and muscle force data may then be used to generate finite element models that compute tissue deformation. For example, boundary conditions for finite element prediction of cartilage stress have been established using projected quasi-static muscle forces and relative bone displacements [36]. The muscle control approach in this system excludes factors at the organ and tissue level. Although the finite element approach and optimization-based force prediction may be combined [37], the computing cost of solving the finite element model repeatedly is too expensive. Additionally, viscoelastic behavior and contact friction cannot be reproduced unless time history is also transferred across several computing domains.

Predictive modeling of musculoskeletal motions at the body level is achievable via the use of forward dynamics and effective management of muscle activation [38]. These models are capable of predicting muscle control patterns for performance-related activities such as maximal height leaping [39], as well as for energy-efficient motions such as walking [40]. These simulations are already expensive owing to the frequent integration of motion equations to find the ideal muscle control pattern. Regardless, adding another degree of complexity via the use of tissue- or even cell-scale models has practical ramifications. For instance, one may develop rehabilitation procedures that optimize system-level loads in order to encourage cell-level remodeling and repair. Additionally, safe movement methods may be anticipated to avoid failures at the tissue level. Simultaneous simulations are achievable by linking the forward dynamics of a musculoskeletal multi-body system to tissue-level finite element models (Figure 7.1). However, this raises the computational cost of prediction from seconds to hours when utilizing a single forward dynamics solution [41]. Bear in mind that the ideal control process may require hundreds if not thousands of these simulations.

Another key problem is the validation of multiscale musculoskeletal models. Muscle activations, body segment motion, and ground reaction forces represent the current level of *in vivo* data collecting for the musculoskeletal system. Additionally, magnetic resonance imaging (MRI) may be used to get musculoskeletal organ geometry, however high resolution pictures need *in vitro* MRI or computed tomography (CT) scans. Muscle forces, with the exception of a few rare exceptions, cannot be quantified *in vivo*. Models that require the calculation of muscular force are often validated using electromyography (EMG) [42]. Typically, data on organs, tissues, and subtissues are collected *in vitro*. This may include cortical bone's mechanical characteristics, collagen fiber orientation, and protein content. The scarcity of *in vivo* subject-specific data and the difficulty and expense of experimental measures underscore the need of musculoskeletal biomechanics data banks. Sensitivity analyses of parameters in conjunction with statistical populations of *in vivo* and mostly *in vitro* data may give viable validation avenues.

The primary obstacle in studying the mechanics of lung tissue remains the same as it was over 25 years ago: "First, stress-strain relationships must be established, either by independent material testing or microstructural modeling with some conformational testing." [43]. The first computer investigations of lung tissue mechanics relied on linear elastic theory and linear material constants, modeling the lung as a uniformly inflated structure subjected to incremental deformation [44]. By solving for repeated tiny displacement increments and using elastic moduli that were dependent on the related incremental change in transmural pressure, the technique was expanded [45]. While the lung undergoes considerable strain during normal breathing, necessitating a large number of these small increments and potentially accumulating numerical error with each increment, the appeal of the linear elastic approach is clear: the governing equations are simpler than equations valid for large deformations, allowing for analytic solutions for simple shapes and loading [46], and the elastic moduli are relatively easily obtained [17].

To conduct an appropriate analysis, however, governing equations must be valid for large deformations, i.e. finite deformation elasticity. Only a few studies have made use of deformation theory, and all of them have made certain concessions to their constitutive law [47]. The lung parenchyma's elasticity is a combination of tissue and surface forces generated by the alveolar air-liquid contact. Surface forces vary in amplitude according to surface area, and surface tension modifies alveolar geometry as the tissue expands or recoils [48]. A mathematical model of the tissue microstructure has been constructed [49], but it has not been expanded to incorporate interactions between alveolar structure and surface tension. This connection might be formed using the technique described in [50], in which moduli were generated from an examination of the elastic response of a lung microstructure model. Testing lung tissue experimentally to discover the link between stress and strain is technically challenging since it is conducted on tissue that lacks its usual surface forces. When there is no surface tension, the tissue component of parenchymal elasticity is different than it is in the intact, *in vivo* lung, because surface tension distorts the tissue, allowing for the storage of greater elastic energy. As a result, amassing sufficient precise facts on which to construct a constitutive rule is incredibly difficult. In an ideal world, a study of lung tissue mechanics would include viscoelasticity and tissue hysteresis, as well as the distinct contributions – and interactions – of tissue elasticity and surface forces.

## References

**1** S. Urner, M. Kelly-Goss, S. M. Peirce, and E. Lammert, "Chapter 6 – Mechanotransduction in blood and lymphatic vascular development and disease," in *Advances in Pharmacology*, vol. 81, R. A. Khalil, Ed. Academic Press, 2018, pp. 155–208. doi: https://doi.org/10.1016/bs.apha.2017.08.009.

**2** Broders-Bondon, F., Nguyen Ho-Bouldoires, T.H., Fernandez-Sanchez, M.-E., and Farge, E. (2018). Mechanotransduction in tumor progression: the dark side of the force. *J. Cell Biol.* 217 (5): 1571–1587. https://doi.org/10.1083/jcb.201701039.

**3** N. I. Petridou, Z. Spiró, and C.-P. Heisenberg, "Multiscale force sensing in development," *Nat. Cell Biol.*, vol. 19, no. 6, Art. no. 6, 2017, https://doi.org/10.1038/ncb3524.

**4** M. A. Wozniak and C. S. Chen, "Mechanotransduction in development: a growing role for contractility," *Nat. Rev. Mol. Cell Biol.*, vol. 10, no. 1, Art. no. 1, 2009, https://doi.org/10.1038/nrm2592.

**5** Bradamante, S., Barenghi, L., and Maier, J.A.M. (2014). Stem cells toward the future: the space challenge. *Life (Basel)* 4 (2): 267–280. https://doi.org/10.3390/life4020267.

**6** Wei, S.C. and Yang, J. (2016). Forcing through tumor metastasis: the interplay between tissue rigidity and epithelial–mesenchymal transition. *Trends Cell Biol.* 26 (2): 111–120. https://doi.org/10.1016/j.tcb.2015.09.009.

**7** Pelissier, F.A., Garbe, J.C., Ananthanarayanan, B. et al. (2014). Age-related dysfunction in mechanotransduction impairs differentiation of human mammary epithelial progenitors. *Cell Reports* 7 (6): 1926–1939.

**8** C. Uhler and G. V. Shivashankar, "Regulation of genome organization and gene expression by nuclear mechanotransduction," *Nat. Rev. Mol. Cell Biol.*, vol. 18, no. 12, Art. no. 12, 2017, https://doi.org/10.1038/nrm.2017.101.

**9** Yamamoto, K. and Ando, J. (2015). Vascular endothelial cell membranes differentiate between stretch and shear stress through transitions in their lipid phases. *Am. J. Physiol. Heart Circ. Physiol.* 309 (7): H1178–H1185. https://doi.org/10.1152/ajpheart.00241.2015.

**10** Vitzthum, C., Clauss, W.G., and Fronius, M. (2015). Mechanosensitive activation of CFTR by increased cell volume and hydrostatic pressure but not shear stress. *Biochim. Biophys. Acta, Biomembr.* 1848 (11, Part A): 2942–2951. https://doi.org/10.1016/j.bbamem.2015.09.009.

**11** Chukkapalli, S.S. and Lele, T.P. Periodontal cell mechanotransduction. *Open Biol.* 8 (9): 180053. https://doi.org/10.1098/rsob.180053.

**12** Shear stress regulation of miR-93 and miR-484 maturation through nucleolin, PNAS. https://www.pnas.org/doi/abs/10.1073/pnas.1902844116 (accessed 3 March 2022).

**13** Baio, J., Martinez, A.F., Silva, I. et al. (2018). Cardiovascular progenitor cells cultured aboard the International Space Station exhibit altered developmental and functional properties. *NPJ Microgravity* 4 (1): 1. https://doi.org/10.1038/s41526-018-0048-x.

**14** Lim, K., Kim, J., Seonwoo, H. et al. (2013). In vitro effects of low-intensity pulsed ultrasound stimulation on the osteogenic differentiation of human alveolar bone-derived mesenchymal stem cells for tooth tissue engineering. *BioMed. Res. Int.* 2013: 269724. https://doi.org/10.1155/2013/269724.

**15** Hieda, M. (2019). Signal transduction across the nuclear envelope: role of the LINC complex in bidirectional signaling. *Cells* 8 (2): 2. https://doi.org/10.3390/cells8020124.

**16** Zhao, Q., Zhou, H., Li, X., and Xiao, B. (2019). The mechanosensitive Piezo1 channel: a three-bladed propeller-like structure and a lever-like mechanogating mechanism. *The FEBS Journal* 286 (13): 2461–2470.

**17** Nakamura, F. and Wang, L. (2019). Identification of Filamin A Mechanobinding partner I: Smoothelin specifically interacts with the Filamin a mechanosensitive domain 21. *Biochemistry* 58: https://doi.org/10.1021/acs.biochem.9b00100.

**18** Haining, A.W.M., Rahikainen, R., Cortes, E. et al. (2018). Mechanotransduction in Talin through the interaction of the R8 domain with DLC1. *PLoS Biol.* 16 (7): e2005599. https://doi.org/10.1371/journal.pbio.2005599.

**19** Zhang, X., Halvorsen, K., Zhang, C.-Z. et al. (2009). Mechanoenzymatic cleavage of the Ultralarge vascular protein von Willebrand factor. *Science* 324 (5932): 1330–1334. https://doi.org/10.1126/science.1170905.

**20** Goutman, J.D., Elgoyhen, A.B., and Gómez-Casati, M.E. (2015). Cochlear hair cells: the sound-sensing machines. *FEBS Lett.* 589 (22): 3354–3361. https://doi.org/10.1016/j.febslet.2015.08.030.

**21** Mathieu, S. and Manneville, J.-B. (2019). Intracellular mechanics: connecting rheology and mechanotransduction. *Curr. Opin. Cell Biol.* 56: 34–44. https://doi.org/10.1016/j.ceb.2018.08.007.

**22** Wang, N., Tytell, J., and Ingber, D. (2009). Mechanotransduction at a distance: mechanically coupling the extracellular matrix with the nucleus. *Nat. Rev. Molecular Cell Biol.* 10: 75–82.

**23** Elosegui-Artola, A., Andreu, I., Beedle, A.E.M. et al. (2017). Force triggers YAP nuclear entry by regulating transport across nuclear pores. *Cell* 171 (6): 1397–1410.e14. https://doi.org/10.1016/j.cell.2017.10.008.

**24** Thomasy, S.M., Morgan, J.T., Wood, J.A. et al. (2013). Substratum stiffness and Latrunculin B modulate the gene expression of the mechanotransducers YAP and TAZ in human trabecular meshwork cells. *Exp. Eye Res.* 113: 66–73. https://doi.org/10.1016/j.exer.2013.05.014.

**25** Yang, B. and McJunkin, K. (2020). CRISPR screening strategies for microRNA target identification. *FEBS J.* 287 (14): 2914–2922. https://doi.org/10.1111/febs.15218.

**26** Ehrlicher, A.J., Nakamura, F., Hartwig, J.H. et al. (2011). Mechanical strain in actin networks regulates FilGAP and integrin binding to Filamin A. *Nature* 478: 7368. https://doi.org/10.1038/nature10430.

**27** D. A. Calderwood, A. Huttenlocher, W. B. Kiosses et al., "Increased filamin binding to β-integrin cytoplasmic domains inhibits cell migration," *Nat. Cell Biol.*, vol. 3, no. 12, Art. no. 12, 2001, https://doi.org/10.1038/ncb1201-1060.

**28** Y. Ohta, J. H. Hartwig, and T. P. Stossel, "FilGAP, a rho- and ROCK-regulated GAP for Rac binds filamin a to control actin remodelling," *Nat. Cell Biol.*, vol. 8, no. 8, Art. no. 8, 2006, https://doi.org/10.1038/ncb1437.

**29** Fal, K., Asnacios, A., Chabouté, M.-E., and Hamant, O. (2017). Nuclear envelope: a new frontier in plant mechanosensing? *Biophys. Rev.* 9 (4): 389–403. https://doi.org/10.1007/s12551-017-0302-6.

**30** Zhang, X., Hu, X., Lei, H. et al. (2016). Mechanical force-induced polymerization and depolymerization of F-actin at water/solid interfaces. *Nanoscale* 8 (11): 6008–6013. https://doi.org/10.1039/C5NR08713A.

**31** X. Hu Z. Z. Liu, X. Chen et al., "MKL1-actin pathway restricts chromatin accessibility and prevents mature pluripotency activation," *Nat. Commun.*, vol. 10, no. 1, Art. no. 1, 2019, https://doi.org/10.1038/s41467-019-09636-6.

**32** Chen, T., Saw, T.B., Mège, R.-M., and Ladoux, B. (2018). Mechanical forces in cell monolayers. *J. Cell Sci.* 131 (24): jcs218156. https://doi.org/10.1242/jcs.218156.

**33** Peng, T., Liu, L., MacLean, A.L. et al. (2017). A mathematical model of mechanotransduction reveals how mechanical memory regulates mesenchymal stem cell fate decisions. *BMC Syst. Biol.* 11 (1): 55. https://doi.org/10.1186/s12918-017-0429-x.

**34** Burke, D.P. and Kelly, D.J. (2012). Substrate stiffness and oxygen as regulators of stem cell differentiation during skeletal tissue regeneration: a mechanobiological model. *PLoS One* 7 (7): e40737. https://doi.org/10.1371/journal.pone.0040737.

**35** Mousavi, S.J. and Doweidar, M.H. (2015). Role of mechanical cues in cell differentiation and proliferation: a 3D numerical model. *PLoS One* 10 (5): e0124529. https://doi.org/10.1371/journal.pone.0124529.

**36** Sun, M., Spill, F., and Zaman, M.H. (2016). A computational model of YAP/TAZ Mechanosensing. *Biophys. J.* 110 (11): 2540–2550. https://doi.org/10.1016/j.bpj.2016.04.040.

**37** Stops, A.J.F., Heraty, K.B., Browne, M. et al. (2010). A prediction of cell differentiation and proliferation within a collagen-glycosaminoglycan scaffold subjected to mechanical strain and perfusive fluid flow. *J. Biomech.* 43 (4): 618–626. https://doi.org/10.1016/j.jbiomech.2009.10.037.

**38** Ghiasi, M.S., Chen, J., Vaziri, A. et al. (2017). Bone fracture healing in mechanobiological modeling: a review of principles and methods. *Bone Rep.* 6: 87–100. https://doi.org/10.1016/j.bonr.2017.03.002.

**39** Checa, S. and Prendergast, P.J. (2009). A mechanobiological model for tissue differentiation that includes angiogenesis: a lattice-based modeling approach. *Ann. Biomed. Eng.* 37 (1): 129–145. https://doi.org/10.1007/s10439-008-9594-9.

**40** Protopappas, V.C., Vavva, M.G., Fotiadis, D.I., and Malizos, K.N. (2008). Ultrasonic monitoring of bone fracture healing. *IEEE Trans. Ultrason. Ferroelectr. Freq. Control* 55 (6): 1243–1255. https://doi.org/10.1109/TUFFC.2008.787.

**41** Claes, L. and Willie, B. (2007). The enhancement of bone regeneration by ultrasound. *Prog. Biophys. Mol. Biol.* 93 (1): 384–398. https://doi.org/10.1016/j.pbiomolbio.2006.07.021.

**42** Vavva, M.G., Grivas, K.N., Carlier, A. et al. (2018). Effect of ultrasound on bone fracture healing: a computational bioregulatory model. *Comput. Biol. Med.* 100: 74–85. https://doi.org/10.1016/j.compbiomed.2018.06.024.

**43** Peiffer, V., Gerisch, A., Vandepitte, D. et al. (2011). A hybrid bioregulatory model of angiogenesis during bone fracture healing. *Biomech. Model. Mechanobiol.* 10 (3): 383–395. https://doi.org/10.1007/s10237-010-0241-7.

**44** Geris, L., Gerisch, A., Sloten, J.V. et al. (2008). Angiogenesis in bone fracture healing: a bioregulatory model. *J. Theor. Biol.* 251 (1): 137–158. https://doi.org/10.1016/j.jtbi.2007.11.008.

**45** Li, D., Zhou, J., Chowdhury, F. et al. (2011). Role of mechanical factors in fate decisions of stem cells. *NPJ Regener. Med.* 6 (2): 229–240. https://doi.org/10.2217/rme.11.2.

**46** Pittenger, M.F., Mackay, A.M., Beck, S.C. et al. (1999). Multilineage potential of adult human mesenchymal stem cells. *Science* 284 (5411): 143–147. https://doi.org/10.1126/science.284.5411.143.

**47** Subramony, S.D., Dargis, B.R., Castillo, M. et al. (2013). The guidance of stem cell differentiation by substrate alignment and mechanical stimulation. *Biomaterials* 34 (8): 1942–1953. https://doi.org/10.1016/j.biomaterials.2012.11.012.

**48** Lee, D.A., Knight, M.M., Campbell, J.J., and Bader, D.L. (2011). Stem cell mechanobiology. *J. Cell. Biochem.* 112 (1): 1–9. https://doi.org/10.1002/jcb.22758.

**49** Pauwels, F. (1980). A new theory concerning the influence of mechanical stimuli on the differentiation of the supporting tissues. In: *Biomechanics of the Locomotor Apparatus: Contributions on the Functional Anatomy of the Locomotor Apparatus* (ed. F. Pauwels), 375–407. Berlin, Heidelberg: Springer https://doi.org/10.1007/978-3-642-67138-8_14.

**50** Delaine-Smith, R.M. and Reilly, G.C. (2012). Mesenchymal stem cell responses to mechanical stimuli. *Muscles Ligaments Tendons J.* 2 (3): 169–180.

**51** Allena, R. (2013). Cell migration with multiple pseudopodia: temporal and spatial sensing models. *Bull. Math. Biol.* 75 (2): 288–316. https://doi.org/10.1007/s11538-012-9806-1.

**52** Harrison, N.C., Diez del Corral, R., and Vasiev, B. (2011). Coordination of cell differentiation and migration in mathematical models of caudal embryonic axis extension. *PLoS One* 6 (7): e22700. https://doi.org/10.1371/journal.pone.0022700.

**53** A. Shuaib, D. Motan, P. Bhattacharya, A. McNabb, T. M. Skerry, and D. Lacroix, “Heterogeneity in the mechanical properties of Integrins determines Mechanotransduction dynamics in bone osteoblasts,” *Sci. Rep.*, vol. 9, no. 1, Art. no. 1, 2019, https://doi.org/10.1038/s41598-019-47958-z.

**54** Grimm, V., Berger, U., DeAngelis, D.L. et al. (2010). The ODD protocol: a review and first update. *Ecol. Modell.* 221 (23): 2760–2768. https://doi.org/10.1016/j.ecolmodel.2010.08.019.

**55** Shuaib, A., Hartwell, A., Kiss-Toth, E., and Holcombe, M. (2016). Multi-compartmentalisation in the MAPK Signalling pathway contributes to the emergence of oscillatory behaviour and to Ultrasensitivity. *PLoS One* 11 (5): e0156139. https://doi.org/10.1371/journal.pone.0156139.

# 8

# Multiscale Modeling of the Musculoskeletal System

## 8.1 Introduction

The musculoskeletal system is composed of bones, muscles, tendons, ligaments, cartilage, and other connective tissues. The most fundamental and crucial function of this system is to support body structures, protect vital organs, and facilitate mobility. More specifically, the muscular system includes the skeletal and smooth muscles which allow body movement, while the skeletal system consists of bones and joints which provide body shape and form, store vital minerals, and protect soft organs. The skeletal system also includes tendons and ligaments which are flexible and tough bands of fibrous connective tissues. The main role of tendons is to connect bone and muscle, while the main function of ligaments is to connect bones. Cartilage is a smooth and resilient elastic tissue which bridges the joints between bones, preventing wear by providing lubrication. Cartilage is not as rigid and hard as bone, but it is much stiffer and less flexible than muscle. The aforementioned components of the musculoskeletal system consist of similar materials such as collagen, mineral, and elastin [1]. This chapter is structured as follows: (i) initially the structure of the musculoskeletal system is presented and then (ii) computational modeling approaches are discussed for the main components of the musculoskeletal system based on the current state of the art.

## 8.2 Structure of the Musculoskeletal System

### 8.2.1 Structure of the Skeletal System Components

The skeletal system includes the bones, ligaments, and cartilage which support the body and form the shape of the body structures. The skeleton consists of the

*Multiscale Modelling in Biomedical Engineering*, First Edition. Dimitrios I. Fotiadis, Antonis I. Sakellarios, and Vassiliki T. Potsika.

Published 2023 by John Wiley & Sons, Inc.

bones of the body and for adults, there are around 206–210 bones, depending on the counting of sternum, while for newborns, there are 270 bones later decreasing to 80 bones in the axial skeleton and 126 bones in the appendicular skeleton.

Bone is a heterogeneous, porous, and anisotropic material with a complex structure. In order to fully evaluate bone's mechanical properties, it is important to provide insight into the mechanical properties of its components and the structural relationship between them at the various levels of hierarchical structural organization [2–5]. At the macrostructure level, bone consists of the periosteum, bone tissue, blood vessels, nerves, and bone marrow. Bone tissue is composed of the cortical or compact bone and the cancellous or trabecular bone. The main differentiation between these two types of bone is their degree of porosity or density. Trabecular bone is found in the inner parts of bones and has a significantly porous structure. Its porosity ranges from 50% to 95%, usually found in cuboidal bones, flat bones, and at the ends of long bones. At the microstructure level, trabecular bone consists of three-dimensional cylindrical structures, called trabeculae, with a thickness of about 100 μm and a variable arrangement form. This porous network of trabecular bone includes pores filled with marrow which produces the blood cells and consists of blood vessels, nerves, and various types of cells. Cortical bone covers the external surface of all bones and its porosity ranges from 5% to 10%. From a microscopic point of view, it consists of cylindrical structures called osteons or Haversian systems with a diameter of about 10–500 μm formed by cylindrical lamellae surrounding the Haversian canal (size of 3–7 μm). The boundary between the osteon and the surrounding bone is called the cement line. The most significant type of porosity, known as vascular porosity, is formed by the Haversian canals (aligned with the long axis of the bone) and the Volkmann's canals (transverse canals connecting Haversian canals) with capillaries and nerves. Experimental studies have shown that the shear modulus of wet single femoral osteons is higher than that of the whole bone which suggests that microstructural effects are significant. The basic components of bone at the nanostructure level are hydroxyapatite and collagen which are made of collagen molecules organized in fibrils. Inorganic components (hydroxyapatite crystals) are mainly responsible for compression strength and stiffness, while organic components (collagen fibers, proteoglycans, and osteocalcin) are responsible for tension properties. The hydroxyapatite crystals are located in the interfibrillar spaces. Mineralized fibers are aligned to form bone lamellae of the typical thickness of a few micrometers. The orientation of the fibers depends on the lamellae and may vary within lamellar sublayers. Figure 8.1 depicts the structure of a long bone and Figure 8.2 depicts the hierarchical structure of a typical long bone at various length scales.

Bone's material and structural properties differ according to the examined hierarchical level implying that the development of realistic models necessitates consideration of bone's microstructural effects as well as porosity and anisotropy properties.

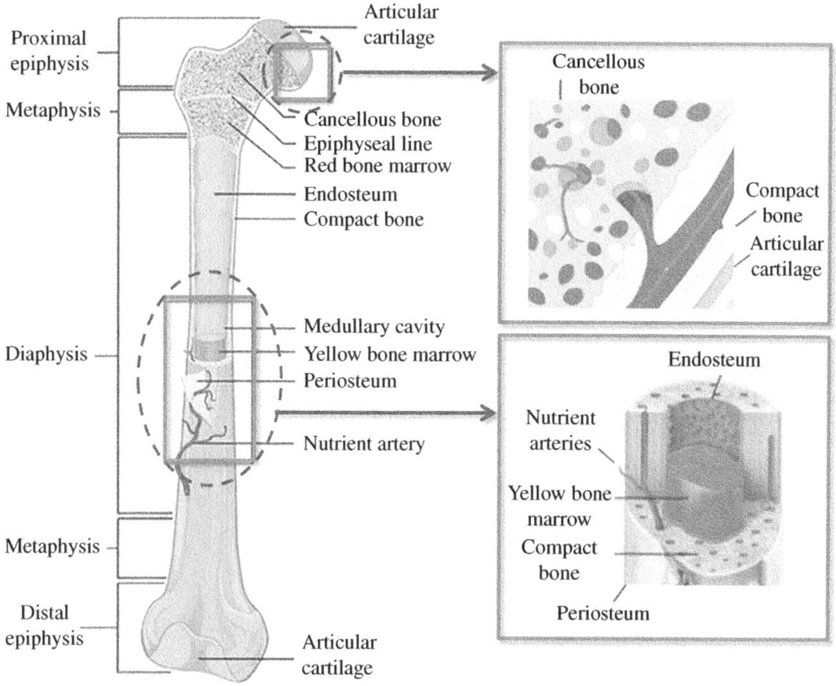


**Figure 8.1** The structure of a long bone.

Articular cartilage is located at opposing bone surfaces in joints and is considered a remarkable tissue, characterized by mechanical function and extremes of physiological structure. It lacks vasculature, lymphatics, and nerves yet exhibiting tribological properties that surpass engineering standards [6]. Articular cartilage is mechanically complex and prone to degeneration and pathology. The mechanical performance of cartilage is underpinned by its hierarchical structure. The main cartilage components include an anisotropic and heterogeneous matrix of predominantly type II collagen, intermeshed with high molecular weight proteoglycans, mainly aggrecans, immersed in an interstitial fluid containing numerous physiological electrolytes. Near the interface with bone, the collagen matrix is oriented predominantly perpendicular to the bone when averaged at the mesoscale, with average fibril orientation rotating on moving up through the cartilage until the fibrils are mainly parallel to the articulating surface, forming the classic "Benninghoff arcades." According to theoretical studies, this mesoscale architecture not only improves tissue stresses under load but it also enables a smooth transition of functionalities from microstructural effects. At the microscale level, the collagen orientation is pseudorandomly leading to the creation of an interconnected network which provides an effective entanglement of aggrecans. This interaction

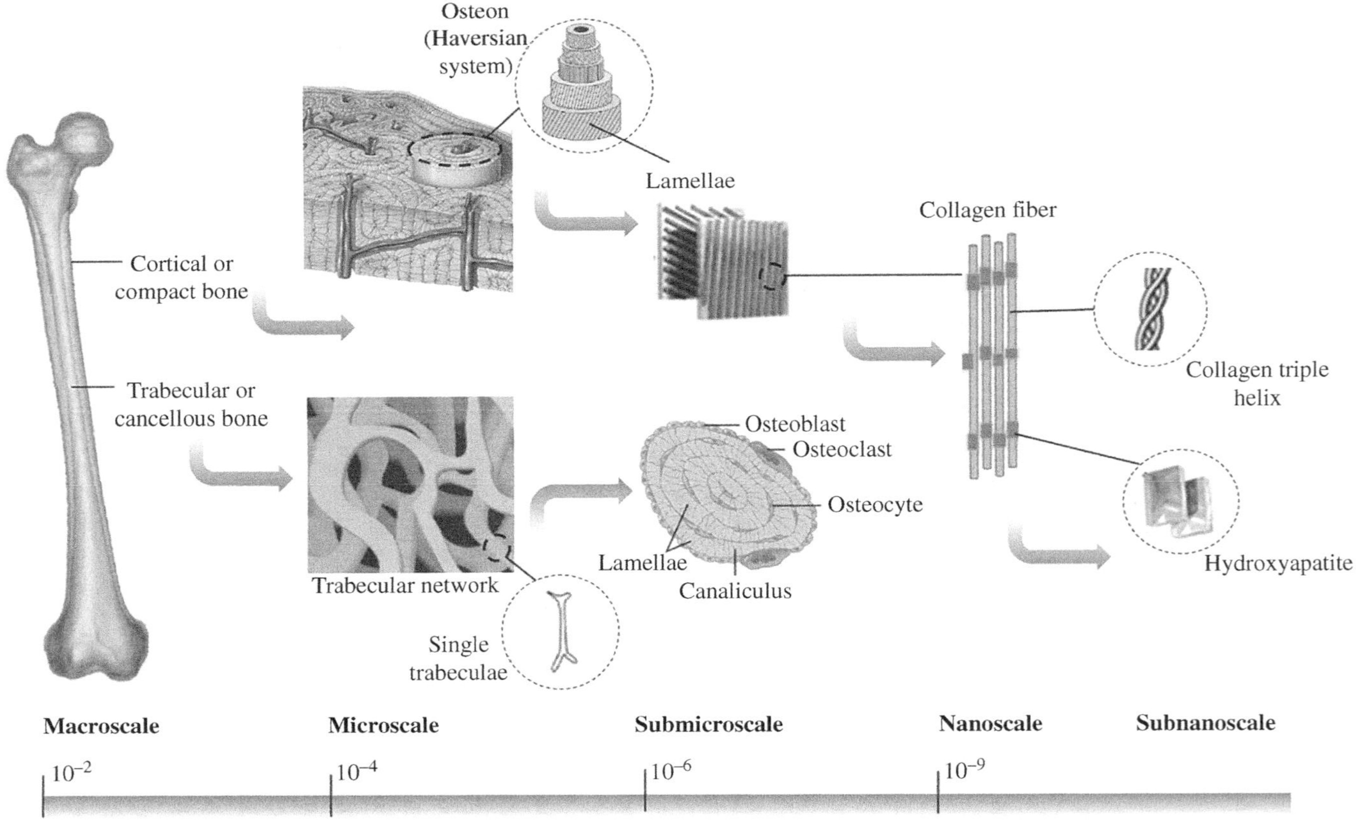


**Figure 8.2** The hierarchical structure of a typical long bone at various length scales.

with aggrecans is of fundamental importance for the load-bearing properties of the joints. From a mechanical point of view, the large aggrecan macromolecules are deformed and compressed in the collagen network, forming a low-permeability composite structure that serves to retain fluid pressure and provides compressive stiffness for cartilage. Concerning the electrochemical perspective, the aggrecans are relatively immobile, with a bottlebrush structure consisting of numerous glycosaminoglycan sidechains possessing electric charge, in turn, inducing a shielding layer of electrophysiological ions, that is a Debye layer, and setting up ion concentration gradients and thus gradients of chemical potential. In turn, osmotic pressures act to reduce these ionic gradients, hydrating and thus swelling the solid cartilage constituent, placing the collagen under tensile stress. This overall tendency of the solid constituent to imbibe fluid is considered to have a mechanical impact due to the induction of a prestress and assisting in the prevention of cartilage being wrung out under load, especially in the context of recovery and repeated loading [6].

The main components of ligaments are elastin, collagen, protein polysaccharides, glycolipids, glycoproteins, water, and cells [7]. More specifically, collagen and ground substances are the largest quantities of ligament components. Collagen constitutes a ratio from 70% to 80% of the dry weight of ligament, the majority being type I collagen which is also found in bone, skin, and tendon. Collagen has a relatively long turnover rate (the average half-life is around 300 and 500 days), which is slightly longer compared to bone, implying that several months may be required after injury for a ligament to alter its structure to meet changes in physical loading conditions or to repair itself. The wet weight of ligaments mainly consists of water which is its main component with a ratio ranging from 60% to 80%. A significant amount of water is associated with the ground substance. The ground substance comprises only about 1% of the total mass of the tissue on a dry weight basis. The ground substance likely provides lubrication and spacing that enhance the sliding of fibers. In addition, the ligament's viscoelastic behavior depends on the ground substance. Ligaments have a hierarchical structure, while the basic structural element is the tropocollagen molecule. Closely packed, parallel collagen fiber bundles are oriented to provide motion and stability for the musculoskeletal system. Properties can change according to strain rate, temperature, hydration, maturation, aging, immobilization, exercise, and healing. The structural properties of isolated ligaments and bone-ligament-bone preparations are normally determined via tensile tests. In such a test, a ligament, tendon, or bone-ligament-bone complex is subjected to a tensile load applied at a constant rate. The force-elongation curve represents structural properties of the ligament. The force-elongation curve is initially upwardly concave, but the slope becomes nearly linear in the prefailure phase of tensile loading. The shape of the curve depends on the geometry of the specimen tested (e.g. tissue length and cross-sectional area) [7].

### 8.2.2 Structure of the Muscular System Components

The muscular system, as an organ system, consists of three different types of muscles which are skeletal, smooth, and cardiac muscles. It enables the body movement and blood circulation and maintains its posture. The muscular system in vertebrates is controlled through the nervous system, although some muscles (e.g. the cardiac muscle) can be completely autonomous. The term musculoskeletal system describes the interaction among the muscular and skeletal systems which is responsible for the movement of the human body.

From a structural point of view, the skeletal muscle is characterized by a hierarchical organization (Figure 8.3). Skeletal muscles can be found at the openings of internal tracts to control the movement of various substances. These muscles allow several functions (e.g. swallowing and urination) to be under voluntary control. Additionally, they contribute to the protection of internal organs (particularly abdominal and pelvic organs) by acting as an external "shield" against potential trauma and by supporting the organs' weight. Skeletal muscles contribute to the maintenance of homeostasis in the body by generating heat. Muscle contraction requires energy, and when adenosine triphosphate (ATP) is broken down, heat is produced. The heat is significant during exercise, as body temperature rises due to sustained muscle movement, and in cases of extreme cold, when shivering produces random skeletal muscle contractions to generate heat. Each skeletal muscle is an organ which consists of various integrated tissues such as the skeletal muscle fibers, nerve fibers, blood vessels, and connective tissue. Skeletal muscles have three layers of connective tissue, known as "mysia," which enclose it, provide structure to the muscle, and compartmentalize the muscle fibers within the muscle. Each muscle is wrapped in a sheath of dense, irregular connective tissue, known as the epimysium, which enables the contraction and powerful movement of a muscle while maintaining its structural integrity. The epimysium also separates muscle from other tissues and organs in the area, allowing the muscle to move independently. Concerning muscle fibers, they are organized into individual bundles, known as fascicles (fascicular organization), by a middle layer of connective tissue called the perimysium. This organization of the muscle fibers is common in the limbs allowing the nervous system to trigger a specific movement of a muscle by activating a subset of muscle fibers within a fascicle.

The muscle fiber in the fascicle is surrounded by a thin connective tissue layer of collagen and reticular fibers, known as the endomysium. The endomysium consists of the nutrients and extracellular fluid which support the muscle fiber. The blood transfers the nutrients to the muscle tissue. The collagen in the three tissue layers (the mysia) coexists with the collagen of tendons in skeletal muscles which work with tendons to pull on bones. At the other end of the tendon, it fuses with the periosteum covering the bone's surface. The mysia transfers to the

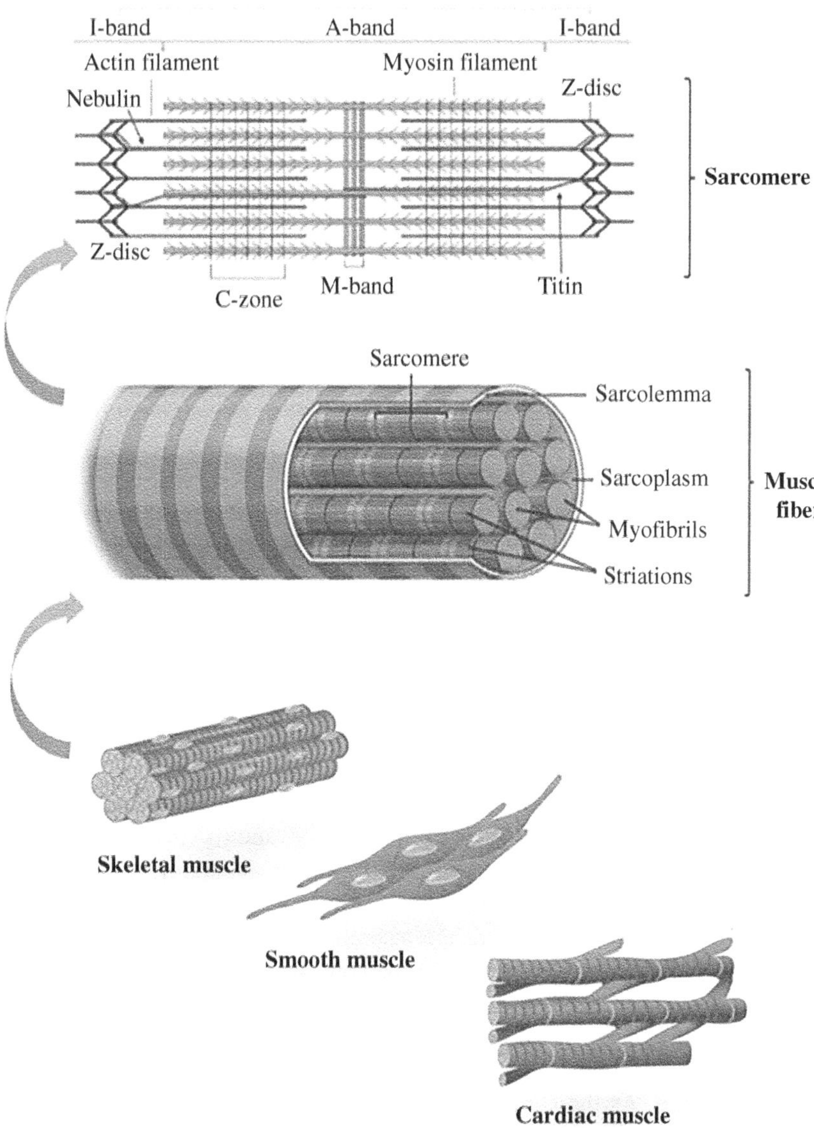


**Figure 8.3** The hierarchical structure of skeletal muscle.

tendon the tension created by the contraction of the muscle fibers and then to the periosteum to pull on the bone to enable the movement of the skeleton. In other regions, the mysia may fuse with a broad, tendon-like sheet known as aponeurosis, or to fascia, the connective tissue between bones and skin. The broad sheet of connective tissue in the lower back that the latissimus dorsi muscles fuse into is an example of aponeurosis. Blood vessels supply to skeletal muscles their nourishment, deliver oxygen, and remove waste. In addition, each muscle fiber of a skeletal muscle is supplied by the axon branch of a somatic motor neuron, which signals the fiber to contract. In comparison with smooth and cardiac muscles, the contraction of skeletal muscles occurs through signaling from the nervous system [8].

Smooth muscles are controlled directly by the autonomic nervous system and are involuntary, implying that they are not able to move under conscious thought. It should be noted that functions such as heartbeat and lungs are involuntary muscles but are not smooth muscles. Smooth muscles can be found in all viscera and vessels, and are a major component in the wall of all tubular organs (one of the few exceptions is the bile duct of some species, such as the rat, which has no muscle). The appearance of many smooth muscles under a microscope varies to an extent which is surprisingly small compared to the wide range of functional properties found in different muscles. The structure of the muscle has features which account for its different mechanical properties. Some structural properties are common to all smooth muscle cells: (i) they are single (uninuclear) cells, (ii) they have no transverse striation, (iii) they have no T-tubules but abundant caveolae, and (iv) they have numerous cell-to-cell and cell-to-stroma junctions. Other aspects of the muscle structure vary from organ to organ and are related to the characteristic functions of each muscle. Specific features are found at all levels of the muscle structure: (i) in the three-dimensional arrangement of muscle cells, layers, bundles, and cords, (ii) in the packing density of muscle cells, the amount of intercellular material, and the occurrence and distribution of non-muscle cells, (iii) in the cytological characteristics of muscle cells, and (iv) in the nature and distribution of proteins associated with the contractile apparatus, the cytoskeleton, the cell membrane, and the stroma [9].

Although the cardiac muscle has a few similar properties with both smooth muscle and skeletal muscle, it also has unique features. The contractions of the heart, known as heartbeats, are controlled by specialized cardiac muscle cells called pacemaker cells. The property of the pacemaker cells to control heart rate is known as autorhythmicity. Neither the smooth nor the skeletal muscle reveals similar properties. As the cardiac muscle cannot be consciously regulated, the pacemaker cells interact with the signals from the autonomic nervous system to slow down or speed up the heart rate. Also, the pacemaker cells are able to interact with various hormones which regulate heart rate to control blood pressure. There are two main types of cardiac muscle cells which are called the myocardial conducting cells and the myocardial contractile cells. The myocardial contractile cells, which constitute the bulk (99%) of the cells in the ventricles and atria, conduct impulses and are responsible for contractions pumping blood through the body. The myocardial conducting cells (1% of the cells) are the autorhythmic cells which form the conduction system of the heart. These cells initiate and propagate the action potential (the electrical impulse) which travels throughout the heart muscle and triggers the contractions that promote blood circulation. Compared to the size of the skeletal muscle cells, the cardiac muscle cells (cardiomyocytes) are considerably shorter with much smaller diameters. Cardiac muscle also demonstrates striations, the alternating pattern of dark A bands, and light I bands attributed to the precise arrangement of the myofilaments and fibrils that are organized in sarcomeres along the length of the cell [10]. These contractile elements are virtually identical to skeletal muscle. T (transverse) tubules penetrate from the surface plasma membrane, the sarcolemma, to the interior of the cell, allowing the electrical impulse to reach the interior. The T tubules are only found at the Z discs, whereas in skeletal muscle, they are found at the junction of the A and I bands. Therefore, there are one-half as many T tubules in cardiac muscle as in skeletal muscle. In addition, the sarcoplasmic reticulum stores few calcium ions. So, most of the calcium ions must come from outside the cells. The result is a slower onset of contraction. The presence of mitochondria is significant in number, providing the necessary energy for the contractions of the heart. Although, in most cases, cardiac muscle cells include a single and central nucleus, two or more nuclei can be identified in some cells. Cardiac muscle cells branch freely. A junction between two adjoining cells is marked by a critical structure called an intercalated disc supporting the synchronized contraction of the muscle. The sarcolemmas from adjacent cells bind together at the intercalated discs. They consist of desmosomes, specialized linking proteoglycans, tight junctions, and large numbers of gap junctions that allow the passage of ions between the cells and help synchronize the contraction. Intercellular connective tissue also helps bind the cells together. The significance of binding these cells together is required by the forces exerted by contraction [10].

## 8.3 Elasticity

Bones are physical objects that obey the laws of mechanics. The primary laws of mechanics that concern deformable objects like bone are the three laws of motion by Sir Isaac Newton in 1687 and the law of elasticity of solid materials described by Robert Hooke in 1678 [11]. The theory of elasticity deals with the systematic study of the stress, strain, and displacement fields in an elastic body under the influence of external forces. The response of a solid body to external forces is influenced by the geometric configuration of the body, as well as by the mechanical properties of the material. The strain and stress fields in a medium are related by laws called constitutive or material laws, characterizing the mechanical behavior of the medium. These are derived from experimental procedures and allow us to have the best description of the experimental results. Although living tissues are not elastic in nature, in limited ranges of stress, strain, and temperature, they may follow an elastic behavior. Bone mechanics deals with elastic materials, i.e. materials in which the deformation and stress disappear with the removal of the external forces. A deeper understanding of the tissue elastic properties can provide invaluable information about the normal functioning of various organs, predict changes due to pathophysiological alterations, and help propose methods of surgical treatment and artificial intervention.

In daily life, the skeletal system is subject to a variety of loads that alternate within the bone according to the type of movement and joint. Biomechanical experiments can be performed using compression, tension, shear, torsion, or bending loading to evaluate the rigidity and ultimate load of a long bone. A compressive strength causes shortening and extension, a tensile strength causes narrowing and elongation, torsional strength creates an angular distortion and the bending strength includes all changes seen in compression, tension, and shear. Shear load is characterized by two forces acting parallel to each other but in opposite directions, so that one part of the object is moved or displaced relative to another part.

Considering the upper extremity of the femur loaded in a single stance phase configuration, the monopodial loading is mimicked. This loading condition can be reproduced *ex vivo* on a testing machine. From this specific experiment, the load applied to simulate a single stance phase configuration and the corresponding displacement can be measured. In Figure 8.4, the load-displacement curve is depicted.

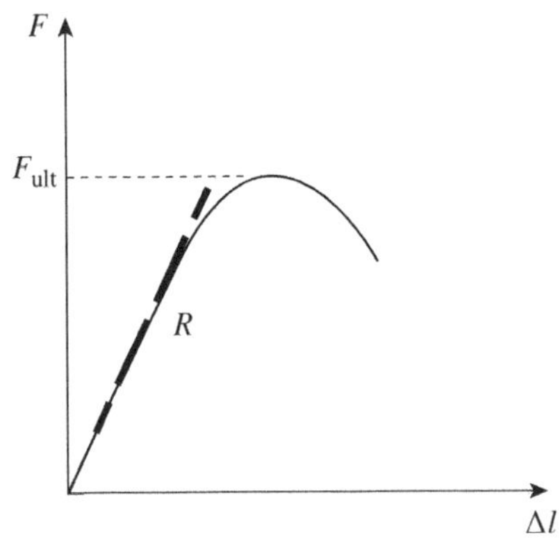


**Figure 8.4** Load-displacement curve, ultimate load, and rigidity.

The rigidity ($R$) can be evaluated from the linear part of the curve from the following equation:

$$R = \frac{F}{\Delta l}, \tag{8.1}$$

where $F$ is the load (in $N$) and $\Delta l$ the displacement (in m).

The rigidity reflects the capability of the bone to withstand a load. This parameter evaluates the elasticity of a complex shape (such as bone or a portion of bone, i.e. the proximal femur). The ultimate response of the structure is defined by the ultimate load (failure load) which corresponds to the maximum of the load-displacement curve. These parameters depend on the geometry of the bone. More specifically, with increasing size of the bone, the rigidity and the failure load are higher. Figure 8.3 presents the load-displacement curve, ultimate load ($F_{\mathrm{ult}}$) and rigidity ($R$).

The stress is the central physical concept of continuum mechanics and elasticity theory and corresponds to the transmission of force through a continuum medium. The stress ($\sigma$) is determined from the measurement of the load ($F$) applied to a given area ($A$) as described by the following equation:

$$\sigma = \frac{F}{A}. \tag{8.2}$$

The strain ($\varepsilon$) can be computed from the ratio of the measured displacement ($\Delta L$) and the initial specimen length ($L$) (Figure 8.4):

$$\varepsilon = \frac{\Delta L}{L}. \tag{8.3}$$

Figure 8.5 presents the stress–strain curve for cortical and trabecular bone. If the material after loading returns to the initial position, the material presents an elastic behavior. The yield point corresponds to the end of the elastic domain and is defined by the yield stress ($\sigma_{\mathrm{y}}$). The most elementary description of material behavior is the well-known Hooke's law which refers to a one-dimensional extensional test:

$$E = \frac{\sigma}{\varepsilon}, \tag{8.4}$$

in which $E$ denotes Young's modulus or modulus of elasticity.

The material elasticity is defined by Young's modulus which can be assessed as the slope of the linear part of the stress–strain curve. The material that is loaded only to a level below the yield strain and unloads along the same path is called elastic.

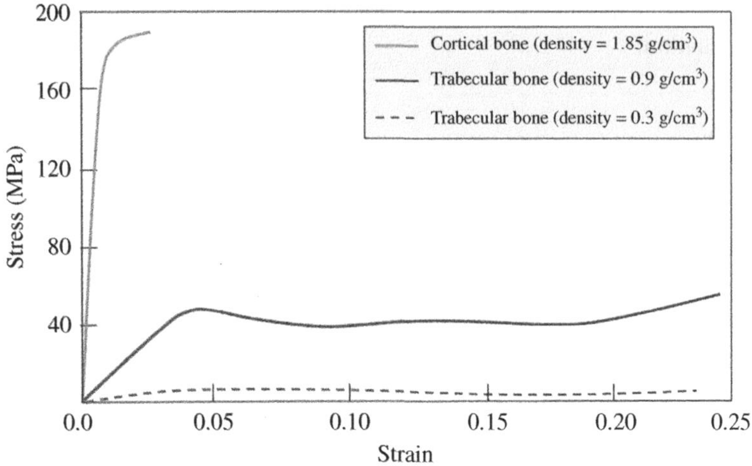


**Figure 8.5** Stress–strain curve for cortical and trabecular bone.

If the path is linear as well, the material is said to be linearly elastic. In order to postulate a linear relationship between each component of stress and strain, it is necessary to establish the existence of a strain energy density $W$ that is a homogeneous quadratic function of strain components $W(\varepsilon_{ij})$. For a body which is slightly strained by gradual application of the loading, while the temperature remains constant, this will produce stress components derivable as follows:

$$\sigma_{ij} = \frac{\partial W}{\partial \varepsilon_{ij}}. \tag{8.5}$$

It should be noted that the function should have coefficients such that in order to insure the stability of the material, corresponding to a natural or zero energy state.

The generalized Hooke's law is expressed by the equation:

$$\sigma_{ij} = C_{ijkl}\varepsilon_{kl}, \tag{8.6}$$

where $C_{ijkl}$ is a fourth-order tensor (34 = 81 components).

The coefficients $\sigma_{11}, \sigma_{12}, \ldots, \sigma_{33}$ are the components of stress. The entire array consists of the stress tensor based on the transformation rule, $\sigma_{kl} = \lambda_{ki}\lambda_{lj}\sigma_{ij}$, $i,j,k,l = 1,2,3$, where $\lambda_{ki}$ are the direction cosines.

Since $W$ is continuous, the order of differentiation in Eq. (8.5) is immaterial and symmetric. Thus, the number of independent components (elastic constants) can be reduced to 21.

The corresponding strain energy function is:

$$\begin{aligned}2W = {} & C_{11}\varepsilon_{11}^2 + 2C_{12}\varepsilon_{11}\varepsilon_{22} + 2C_{13}\varepsilon_{11}\varepsilon_{33} + 2C_{14}\varepsilon_{11}\varepsilon_{12} + 2C_{15}\varepsilon_{11}\varepsilon_{23} \\ & + 2C_{16}\varepsilon_{11}\varepsilon_{31} + C_{22}\varepsilon_{22}^2 + 2C_{23}\varepsilon_{22}\varepsilon_{33} + 2C_{24}\varepsilon_{22}\varepsilon_{12} + 2C_{25}\varepsilon_{22}\varepsilon_{23} \\ & + 2C_{26}\varepsilon_{22}\varepsilon_{31} + C_{33}\varepsilon_{33}^2 + 2C_{34}\varepsilon_{33}\varepsilon_{12} + 2C_{35}\varepsilon_{33}\varepsilon_{23} + 2C_{36}\varepsilon_{33}\varepsilon_{31} \\ & + C_{44}\varepsilon_{12}^2 + 2C_{45}\varepsilon_{12}\varepsilon_{23} + 2C_{46}\varepsilon_{12}\varepsilon_{31} \\ & + C_{55}\varepsilon_{23}^2 + 2C_{56}\varepsilon_{23}\varepsilon_{31} \\ & C_{66}\varepsilon_{31}^2 .\end{aligned} \tag{8.7}$$

The generalized Hooke's law in matrix form is:

$$\begin{bmatrix}\sigma_{11}\\ \sigma_{22}\\ \sigma_{33}\\ \sigma_{12}\\ \sigma_{23}\\ \sigma_{31}\end{bmatrix} = \begin{bmatrix} C_{11} & C_{12} & C_{13} & C_{14} & C_{15} & C_{16}\\ C_{21} & C_{22} & C_{23} & C_{24} & C_{25} & C_{26}\\ C_{13} & C_{23} & C_{33} & C_{34} & C_{35} & C_{36}\\ C_{41} & C_{42} & C_{43} & C_{44} & C_{45} & C_{46}\\ C_{51} & C_{52} & C_{53} & C_{54} & C_{55} & C_{56}\\ C_{61} & C_{62} & C_{63} & C_{64} & C_{65} & C_{66}\end{bmatrix} \begin{bmatrix}\varepsilon_{11}\\ \varepsilon_{22}\\ \varepsilon_{33}\\ \varepsilon_{12}\\ \varepsilon_{23}\\ \varepsilon_{31}\end{bmatrix}, \tag{8.8}$$

where $C_{IJ} = C_{JI}$.

Equation (8.6) can be written in the inverse form as follows:

$$\varepsilon_{ij} = S_{ijkl}\sigma_{kl}, \tag{8.9}$$

or in the analogous to (8.8) form as:

$$\mathbf{S} = \mathbf{C}^{-1}. \tag{8.10}$$

The coefficients are called compliance constants.

The transformation of (8.6) and (8.9) follows the rules of the Cartesian tensors.

The preceding characterization of a medium is the most general and refers to an anisotropic material. More specifically, a material is anisotropic when its mechanical properties vary according to the direction of analysis. Several factors contribute to the mechanical anisotropy of bone. These include the orientation of the Haversian porous network, the orientation of the lamellae and the alignment of collagen fibers and hydroxyapatite crystals. In hierarchical structures like bone, the anisotropy depends on the observational level. Most of the engineering materials possess properties of symmetry about one or more of the planes or axes, which allows the number of independent constants to be reduced. This reduction is a function of the symmetry exhibited by the material under consideration.

An orthotropic material has three planes of symmetry which are mutually orthogonal. The existence of the two orthogonal symmetry planes implies the existence of the third. Therefore, is derived by adding a symmetry plane orthogonal to the preceding one of the monoclinic material. The invariance of under a change of reference system carried out by symmetry about the second plane leads to:

$$\begin{bmatrix} C_{11} & C_{12} & C_{13} & 0 & 0 & 0 \\ C_{12} & C_{22} & C_{23} & 0 & 0 & 0 \\ C_{13} & C_{23} & C_{33} & 0 & 0 & 0 \\ 0 & 0 & 0 & C_{44} & 0 & 0 \\ 0 & 0 & 0 & 0 & C_{55} & 0 \\ 0 & 0 & 0 & 0 & 0 & C_{66} \end{bmatrix} \tag{8.11}$$

Thus, the number of independent elastic constants is reduced to nine.

In the first approximation, cortical and cancellous bones can be considered orthotropic, which means that the properties differ according to orthogonal directions. A composite medium such as bone can be considered as being consisted of a fiber embedded in a cylinder of matrix. The material behaves as an orthotropic material having in addition one axis of revolution. The material is then called a transversely isotropic material. A change of reference system produced by an arbitrary rotation about this axis must leave unchanged. Considering this property, the following equations are satisfied:

$$\begin{gathered} C_{13} = C_{12}, \quad C_{33} = C_{22}, \quad C_{55} = C_{66} \\ C_{44} = \frac{1}{2}\left(C_{22} - C_{23}\right). \end{gathered} \tag{8.12}$$

Thus, the properties of a transverse isotropic material can be determined by five independent elastic constants.

A material is isotropic if its properties are independent of the choice of its reference axes. The usual materials, with the exception of wood, usually satisfy this model at a macroscopic level. In this case, **C** must be invariant under any change of the orthonormal basis. The application of this property to a unidirectional material leads to:

$$C_{22} = C_{11}, \quad C_{23} = C_{12}, \quad C_{66} = \frac{1}{2}\left(C_{11} - C_{12}\right). \tag{8.13}$$

Thus, the number of independent constants is reduced to two.

Introducing the Lamé coefficients $\lambda$ and $\mu$, the following equations are derived:

$$C_{12} = \lambda, \quad \mu = \frac{1}{2}\left(C_{11} - C_{12}\right), \tag{8.14}$$

and

$$C_{11} = \lambda + 2\mu. \tag{8.15}$$

After (8.14) and (8.15), Hooke's law (8.8) can be written as:

$$\sigma_{ij} = \lambda\delta_{ij}\varepsilon_{kk} + 2\mu\varepsilon_{ij}, \tag{8.16}$$

where $\varepsilon_{kk} = \varepsilon_{11} + \varepsilon_{22} + \varepsilon_{33}$ is the dilatation of the material and $\delta_{ij}$ is Kronecker's delta.

The normal stresses ($i = j$) are written as:

$$\sigma_{ii} = \lambda\delta_{kk} + 2\mu\varepsilon_{ii}, \tag{8.17}$$

and the shearing stresses ($i \neq j$) as:

$$\sigma_{ij} = 2\mu\varepsilon_{ij}. \tag{8.18}$$

The deformations as functions of stresses are easily obtained as:

$$\varepsilon_{ij} = -\frac{\lambda}{2\mu(3\lambda + 2\mu)}\delta_{ij}\sigma_{kk} + \frac{1}{2\mu}\sigma_{ij}. \tag{8.19}$$

In the case of uniaxial tension or compression, the only non-vanishing stress is $\sigma_{11} \neq 0$.

It follows from this that $\sigma_{ij} = 0$ for $i \neq j$. The normal deformation in the test direction is, by (8.19):

$$\varepsilon_{11} = -\frac{\lambda}{2\mu(3\lambda + 2\mu)}\sigma_{11} + \frac{1}{2\mu}\sigma_{11} \tag{8.20}$$

and

$$\sigma_{11} = E\varepsilon_{11}, \tag{8.21}$$

where:

$$E = \frac{\mu(3\lambda + 2\mu)}{\lambda + \mu}, \tag{8.22}$$

is Young's modulus.

The transverse strains are:

$$\varepsilon_{22} = \varepsilon_{33} = -\frac{\lambda}{2\mu(3\lambda + 2\mu)}\sigma_{11} = -\frac{\lambda}{2(\lambda + \mu)}\varepsilon_{11}, \tag{8.23}$$

or

$$\varepsilon_{22} = \varepsilon_{33} = -\frac{v}{E}\sigma_{11} = -v\varepsilon_{11}, \tag{8.24}$$

where

$$v = \frac{\lambda}{2(\lambda + \mu)}, \tag{8.25}$$

is called Poisson's ratio.

In a simple shear test, the only non-vanishing stress is $\sigma_{12} = \tau \neq 0$.

The stress-strain relations are then reduced to:

$$\sigma_{12} = 2\mu\varepsilon_{12} = \mu\gamma_{12}, \tag{8.26}$$

where $\gamma_{12}$ denotes the shear strains between the two orthogonal directions $x_1$, $x_2$. The coefficient $\mu = G$ is called the shear modulus.

The shear modulus is related to the Young's modulus and the Poisson's ratio as:

$$G = \frac{E}{2(1+v)}. \tag{8.27}$$

The bulk modulus ($K$) of an isotropic material measures its resistance to uniform compression (i.e. uniform load applied in all directions). It is linked to the Young's modulus ($E$) and the Poisson's ratio ($\nu$) by the following relationship:

$$K = \frac{E}{3(1-2v)}. \tag{8.28}$$

Only two constants are independent since $G = \mu$ and $E$, $\nu$ are defined in terms of $\lambda$ and $\mu$. The Lamé coefficients in terms of $E$ and $\nu$ are given as:

$$\mu = \frac{E}{2(1+v)} \text{ and } \lambda = \frac{vE}{(1+v)(1-2v)}. \tag{8.29}$$

In order for $\lambda$ to remain finite, the Poisson's ratio should be in the range:

$$-1 < v < 0.5. \tag{8.30}$$

It is difficult to derive $\nu < 0$ as this implies that an elongational strain in the direction $\mathbf{e}_1$ would result in an expansion in the direction $\mathbf{e}_3$ as well. For $\nu = 0.5$ the material is said to be incompressible.

The elasticity equations are usually written as functions of the engineering constants. For an isotropic material, these constants are: (i) Young's modulus, (ii) Poisson's ratio, and (iii) shear modulus. These engineering constants are measured employing simple mechanical tests.

In conclusion, in Section 6.5, we will present the numerical approaches which are currently used in the literature to deal with multiscale modeling of the musculoskeletal system. Nevertheless, in Section 6.3, we focused on fundamental principles of mechanics and elasticity which should be used as basic knowledge by biomedical engineers in the area of computational modeling of the musculoskeletal system.

## 8.4 Mechanical Characteristics of Muscles

The activated muscles pull about equally on all their attachments to generate forces. This tensile force has two sources: active and passive tension. The term active tension corresponds to the forces created between myosin and actin fibers in the sarcomeres of activated motor units. So, active tension is the force created by the contractile proteins (actin and myosin) using chemical energy stored in ATP. This ability of muscles to create active tensile forces is unique compared to the connective tissue components (ligaments, tendons, and bone) of the musculoskeletal system. The shape of this active tension potential of skeletal muscle is called the force–velocity relationship (Figure 8.6) [12].

The term passive tension corresponds to the force that comes from an elongation of the connective tissue components of the muscle-tendon unit. When a person does a stretching exercise, the tension in the muscles is the internal resistance of the muscle-tendon unit to the elongation of the stretch. The passive tension in stretching exercises can be significant and may cause muscular weakness in muscles following stretching. In the midranges of joint motion, passive tension does not significantly contribute to muscle forces in normal movement. Nevertheless, it is an important factor in low-force movements, in various neuromuscular

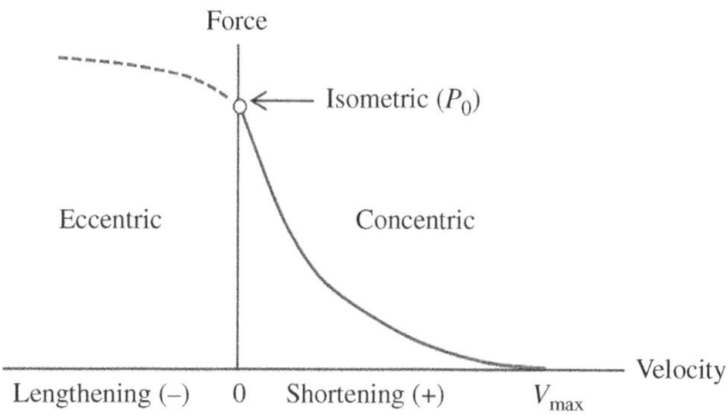


**Figure 8.6** Force–velocity relationship of muscle.

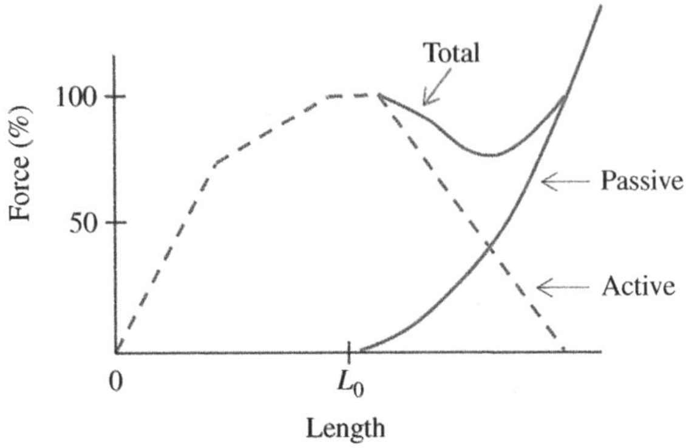


**Figure 8.7** Force–length relationship of human skeletal muscle.

disorders and affects significantly movement at the extremities of joint range of motion. The increase in passive tension limiting the range of joint motion is quite apparent in multiarticular muscles and is called passive insufficiency. Passive tension is an important component of the force–length relationship of muscle. The tension a muscle can create has both active and passive sources. So, the length–tension graph of muscle will have both of these components as shown in Figure 8.7. The force–length relationship reflects the variation of muscle tension at different muscle lengths and has a significant impact on how movement is created.

From a mechanical point of view, another important characteristic of the muscle is the temporal delay in the development of tension. The force–time relationship refers to the delay in the development of muscle tension of the whole muscle–tendon unit and can be expressed as the time from the motor action potential (electrical signal of depolarization of the fiber that makes of the electromyographic signal) to the rise or peak in muscle tension [12].

One of the most widely used mechanical models of muscle that takes into account both the active and passive components of muscle tension is the Hill muscle model. This model of muscle force consists of three components including a series of elastic component, a parallel elastic component and a contractile component. The relationship between the velocity of shortening $v$ and a constant load $w$ on an excited striated muscle is given by the Hill's equation:

$$(P-w)b=(w+a)v, \tag{8.31}$$

where $P$ denotes the isometric tension and $a$ and $b$ are constants. This relation describes the force–velocity relation for several types of muscles.

## 8.5 Multiscale Modeling Approaches of the Musculoskeletal System

This subsection deals with computational modeling approaches based on the current literature focusing on the components of the musculoskeletal system individually as presented in Section 6.2.

### 8.5.1 Multiscale Modeling of Bones

Computational modeling is a valuable tool to better understand bone mechanical properties and the relationship between its structural, mechanical, and biological environment. Bone is a living organ characterized by the unique ability to adapt to mechanical usage and biophysical stimuli in terms of its architecture and mass. This phenomenon is known as functional adaptation in the forms of modeling and remodeling [13]. Within the last decades, several computational modeling approaches have been presented focusing on different bone types, hierarchical levels, numerical methods, or even for different bone pathologies.

Among the existing numerical approaches, the most popular one is the finite element method (FEM) (Figure 8.8). The FEM was first used in bone biomechanics for the analysis of the mechanical behavior of skeletal parts in 1972 [14].

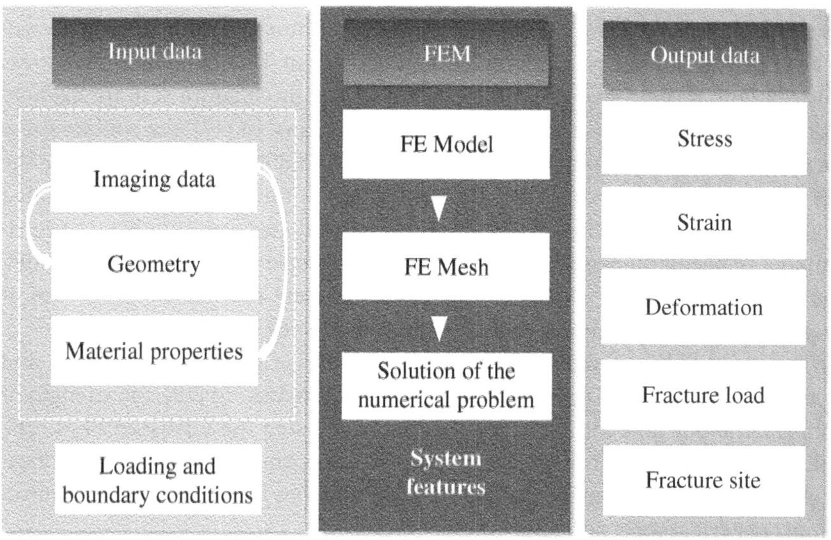


**Figure 8.8** General scheme applied in computational modeling of bone using the finite element method. This can be further modified and extended for different hierarchical scales and various numerical methods.

In [15], FEM applications were presented during the last decades in the four main areas of bone biomechanics: (i) analysis of stress and strain, (ii) determination of mechanical properties, (iii) fracture load prediction, and (iv) fracture fixation design.

The application of the FEM includes three successive stages for the analysis of human bones which are the stage of preprocessing followed by the level of numerical solution, and finally the postprocessing stage. In the preprocessing stage, a computer-aided design (CAD) model is required to be generated. The geometry and material properties of bone can be acquired from computed tomography (CT) or other imaging modalities depending also on the examined hierarchical level. The geometry of fracture fixation (implant) is usually developed on CAD commercial software like CATIA® or Solidworks®, while several researchers develop in-house 3D reconstruction software. After the establishment of the bone model, the material properties are assigned to all the included parts. The next step is the implementation of the proper boundary conditions followed by mesh generation (Figure 8.9). A similar approach is followed by other computational methods such as the boundary element method or the finite differences method (see Chapter 1, Section 2.3).

It is crucial to implement the proper boundary conditions to get accurate results after the numerical solution of the problem. The stress analysis of bones using FEM can provide insights into a better understanding of bone remodeling mechanism and assessment of fracture risk and contribute to the improvement of fracture fixation. In [16], the stresses in the adult human femur were analyzed. The geometry of bone was obtained from the CT scan. Due to the irregular stress distribution at the contact region (joint) of bones, the fracture is more likely to occur at the joints in case of impact falling of humans. In [17], the contact stress between the bones of human knee joints (femur and tibia) was assessed using the FEM software NASTRAN®. The cartilage stresses in the hip joint were evaluated in [18] assuming the material of the bone to be rigid. In several studies, the cortical and trabecular bones are considered to be isotropic and homogenous media, but in real conditions, both bone types have different mechanical properties. It has been shown that human bone toughness is affected by the strain rate.

The study of the mechanical properties of bones (e.g. stiffness, strength, and toughness) is crucial in order to estimate the fracture risk. In [19], the strength of proximal femur and surface strain was assessed using the CT-based FEM. The FEM analysis results of bones were observed to depend on the anatomical location of the specimen due to the fact that the bone has heterogeneous and anisotropic material properties.

In humans, usually after thirty years of age, the mass of bone starts to decrease, a phenomenon known as osteoporosis. Osteoporosis is a disease characterized by significant bone loss, leading to an increased risk of bone fracture. Due to

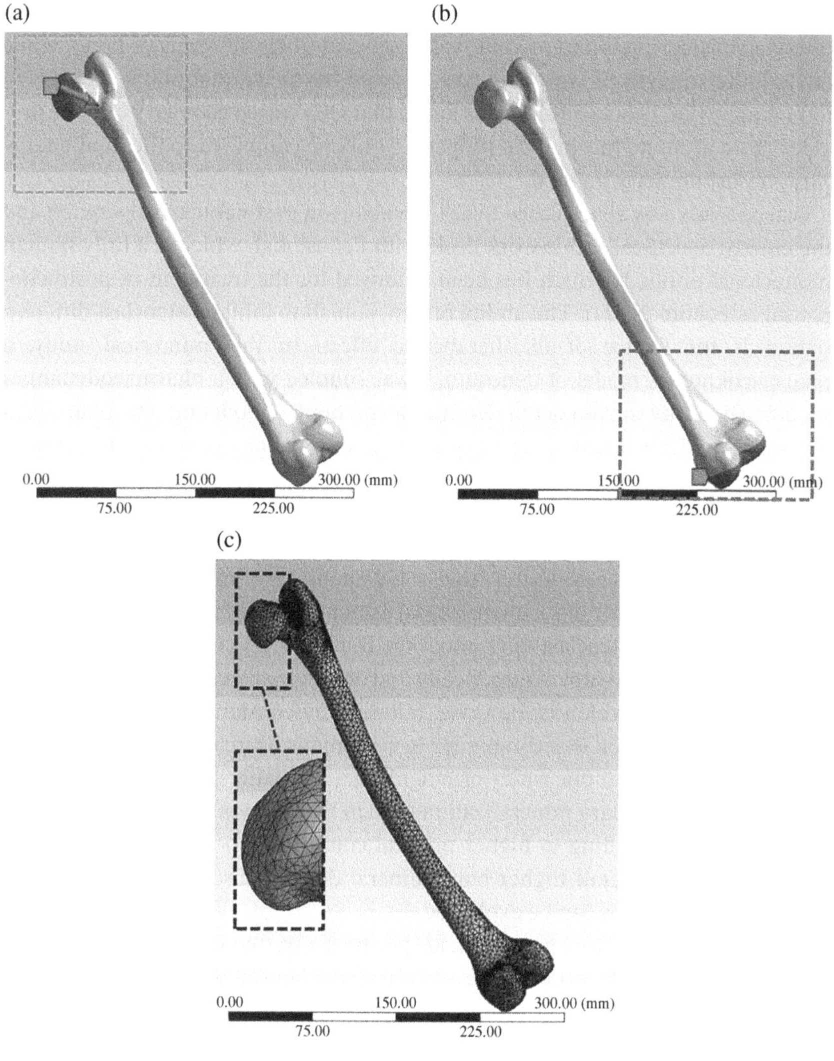


**Figure 8.9** Computational model of the femur: (a and b) boundary conditions including the application of a gait cycle at the region of the head of the trochanter and a fixed constraint at the region of the medial condyle and lateral condyle, and (c) generation of the mesh using triangular elements.

osteoporosis, the mechanical properties of bones are negatively affected. In [20], the strength of bone was evaluated at the three major osteoporotic fracture sites (vertebral sections, distal radii, and in side loading of proximal femora), while in [21], additionally to the strength, the stiffness of the femoral bone was also determined using FEM. Another study [22] estimated the mechanical properties of bone considering different volume fractions at different levels of bone, while in [23], the strength of vertebrae was assessed based on quantitative computed tomography (QCT) and FEM. It was found that QCT-based geometry of bone provides more accurate mechanical properties of bone compared to the dual-energy X-ray absorptiometry method.

Osteoporosis was also studied in [24] focusing on postmenopausal women and the elderly. Specifically, this work deals with denosumab which is a fully human monoclonal antibody which has been approved for the treatment of postmenopausal osteoporosis [24]. The ability of denosumab to inhibit osteoclast differentiation is the source of its therapeutic effect. In this numerical study, a pharmacokinetics model of denosumab was coupled with a pharmacodynamics model to quantify the impact of denosumab on bone remodeling. The pharmacodynamics model is based on an integrated systems biology-continuum micromechanics approach. More specifically, a bone cell population model is included, considering the governing biochemical factors of bone remodeling (including the action of denosumab), as well as a multiscale micromechanics-based bone mechanics model considering the mechanobiology of bone remodeling. Computational studies of postmenopausal osteoporosis have indicated that denosumab suppresses osteoclast differentiation, thus strongly curtailing bone resorption. Also, it has been shown numerically that denosumab may trigger a short-term bone volume gain, which is, however, followed by constant or decreasing bone volume. This evolution is accompanied by a dramatic decrease in the bone turnover rate by more than one order of magnitude. The latter proposes a dominant occurrence of secondary mineralization (which is not anymore impeded through cellular activity), leading to higher mineral concentration per bone volume. In this manner, the overall higher bone mineral density observed in denosumab-related clinical studies can be explained.

Osteoporosis was also examined in [25] by presenting computational models of the pathology for both cortical bone and trabecular bone at the basic multicellular unit (BMU) level according to bone functional adaptation and Frost's mechanostat theory. The term "mechanostat" refers to the influence of mechanical loading on bone structure when its mass and architecture are modified to provide a structure which resists habitual loads with an economical amount of material. For cortical bone, cortical thickness was taken as the controlling variable, and for trabecular bone, porosity was the main parameter of interest. The coupling relationship between the biological and mechanical factors was fulfilled by the osteoblastic formation threshold and BMU activation threshold. For cortical bone remodeling model, the analysis was performed on a representative rectangular slice of the cross section of cortical bone volume. Peripheral QCT (pQCT) data of femur and tibia were used as an example of mechanical disuse to validate the model. For the trabecular bone remodeling model, the analysis was conducted using a representative cross section of trabecular bone in the lumbar spine of a postmenopausal woman ($100\,mm^2$). Clinical data were used for the validation of the numerical findings. For the femur cortical model, the simulated steady-state value of cortical thickness was 2.04 mm with a relative difference of 5.99% compared to the clinical data. For the tibia cortical model, the simulated steady-state value was 4.13 mm with a relative difference of 4.82% compared to the clinical data. Concerning the trabecular bone, it was found that changes in osteoblastic formation threshold and/or changes in BMU activation threshold could provide results consistent with the bone loss patterns identified at menopause when disuse was maintained. These computational models for osteoporosis for both trabecular and cortical bone can be further used to predict and quantify the effects of biological and mechanical factors on trabecular architecture and cortical thickness, provide new insights into the relationship between the mechanical and biological environment of bone with its morphology.

Additionally, computational models play a key role not only for basic science but also for the analysis of clinical interventions and the support of clinicians prior to intervention. In [26], vertebroplasty has been used to stabilize compression fractures in the spine for years, yet there are still diverging ideas on the ideal deposition location, volume, and augmentation material. More specifically, the long-term effects of the intervention on the surrounding biological tissue have been examined to a limited extent. Computational studies in the area of vertebroplasty have been presented covering a variety of topics ranging from the augmentation procedure to strength prediction and long-term *in silico* bone biology in augmented human vertebrae. Although the augmentation procedure and strength prediction have been examined widely with numerical methods, this is not the case for the prediction of long-term effects as studies in this domain are lacking. Recent numerical approaches in this area have the potential to show adaptation to cement augmentation and, thus, cover this gap [26].

In [27], the contribution of FEM and CT (CT/FEM) were assessed as means to evaluate accuracy and reliability in human bones, as well as fracture risk and effects of osteoporosis medication. For clinicians, fracture load, fracture site, and fracture pattern are very useful data. For the hip, validation studies indicated that CT/FEM might be able to predict femoral strength better than quantitative CT (QCT) or dual-energy X-ray absorptiometry and was able to assess fracture site of the proximal femur. Nonlinear CT/FEE showed more accurate predictions of fracture load at proximal femur. Nonlinear CT/FEM applied shell-solid meshing to analyze fracture loads of the proximal femurs as well as principal strains of the surface of the femoral bones with accuracy. The Drucker–Prager yield criterion was used which is a pressure-dependent model for determining whether a material has failed or undergone plastic yielding. Based on this criterion, it was found that CT/FEM can predict femoral bone strength and fracture site better compared with the von Mises yield criterion. Validation studies of simulating a sideways fall on the hip revealed that CT/FEM predicted fracture loads and fracture patterns with a high degree of accuracy.

There are two mechanisms of bone healing known as primary bone healing and secondary bone healing. Primary bone healing is characterized by a direct attempt of the cortex to re-establish itself after interruption without the formation of a callus, while the secondary bone-healing process includes four successive phases: inflammatory response, soft callus formation, hard callus development, and remodeling. The effect of the various mechanical and biological factors on bone healing has been examined using mechanobiological models. Nevertheless, the impact of the initial healing phase, which includes the inflammatory stage, the granulation tissue formation, and the initial callus formation during the first few days post-fracture, is generally neglected in such studies. In this direction, an FE model was presented in [28] considering different levels of diffusion coefficient for mesenchymal stem cell (MSC) migration, Young's modulus of granulation tissue, interfragmentary gap size and callus thickness aiming to provide new knowledge on the modulatory effects of the parameters which play a crucial role during the initial phase of bone healing. The results revealed how faster MSC migration, stiffer granulation tissue, thicker callus, and smaller interfragmentary gap enhanced healing to some extent. Nevertheless, after a certain threshold, a state of saturation was observed for MSC migration rate, callus thickness, and granulation tissue stiffness. The findings of [28] quantified the effects of the initial healing phase on healing outcomes to better understand the biological and mechanobiological mechanisms and their utilization in the design and optimization of treatment strategies. It was also shown that for the cases of fractures, in which bone segments are in close proximity, the callus development is not required. This finding is consistent with the concepts of primary and secondary bone healing.

In [29], a dynamic model regulated by both biphasic poroelastic finite element analysis and fuzzy logic control was established. Fuzzy logic control is applied in a comprehensive and easy manner to simulate the process of tissue differentiation. More specifically, it enhances the communication among medical experts and researchers as it is convenient for them to modify the fuzzy logic rules and improve the simulation of the tissue differentiation process. A three-dimensional fracture-healing model was presented with two different interfragmentary movements (case A: 0.25 mm and case B: 1.25 mm). It was found that with the progress of fracture healing, both simulated interfragmentary movements revealed a decrease. Also, it was shown that the starting time of the decrease for case B was later than that for case A. A good agreement was shown when both cases were compared with experimental data. Also, it was reported that the dynamic model of [29] could simulate the healing process under different mechanical environments and had the potential to extend to the multiscale healing model, which is essential for reducing the animal experiments and helping to characterize the complex dynamic interaction between tissue differentiations within the callus region.

In [30], three different kinds of fracture-healing models were presented: mechanoregulatory models, bioregulatory models, and coupled mechanobioregulatory models. From an overview of these models, we can conclude that multiscale models and coupled mechanobioregulatory models are needed in the future. More specifically, for bioregulatory healing models, a set of coupled partial differential equations is used to simulate the density and concentration of different cells and growth factors. The changes in cell concentration are modeled by cell functions (migration, proliferation, differentiation, and decay), which are regulated by growth factors. The following equations describe cell activity:

$$\frac{\partial \mathbf{c}_{\mathrm{m}}}{\partial t} = \nabla \cdot \left[ D_{\mathrm{cm}}\left(\mathbf{c}_{\mathrm{m}}\right)\nabla \mathbf{c}_{\mathrm{m}} - \mathbf{c}_{\mathrm{m}} \sum_{i=1}^{n} f_i\left(\mathbf{c}\right)\nabla \mathbf{c}_i \right] + \mathbf{f}_0\left(\mathbf{c}_{\mathrm{m}}, \mathbf{c}\right), \tag{8.32}$$

$$\frac{\partial \mathbf{c}}{\partial t} = D\Delta \mathbf{c} + \mathbf{g}\left(\mathbf{c}_{\mathrm{m}}, \mathbf{c}\right), \tag{8.33}$$

in which $\mathbf{c}_{\mathrm{m}}$ denotes the density of migration cell type (MSCs and fibroblasts) and $\mathbf{c}$ represents other $n$ cell concentrations, in which no directed migration is modeled. $\mathbf{c}_{\mathrm{m}}(t, \mathbf{x})$ and $\mathbf{c}(t, \mathbf{x})$ are functions of time $t$ and space $\mathbf{x}$. $D_{\mathrm{cm}}(\mathbf{c})$ and $D$ are diffusion coefficients, $f_i(\mathbf{c})$ represents the taxis coefficients for both chemotaxis and haptotaxis, and $\mathbf{f}_0(\mathbf{c}_{\mathrm{m}}, \mathbf{c})$ and $\mathbf{g}(\mathbf{c}_{\mathrm{m}}, \mathbf{c})$ are reaction terms for cell differentiation, proliferation, and decay, and matrix and growth factor production and decay. However, the bioregulatory healing models are subject to specific limitations as follows: (i) fracture healing is a complex process implying that as more biological processes are incorporated into the numerical models, the number of partial differential equations increases as well as the complexity of the numerical solution, (ii) the geometry of the callus has been simplified to a construction of rectangles, which does not represent the real boundary conditions of the callus region, (iii) only bioregulatory factors have been adopted to regulate the healing process, while mechanical stimuli, which play an important role during fracture healing, were not considered in the model. Also, the cause of the release of growth factors is not simulated in the mode, although, in the case of fracture healing, there is a strong possibility that mechanical stimuli are the cause of the release of these growth factors.

In our recent study [31], we presented the latest computational approaches for the ultrasonic assessment of fracture healing and osteoporosis. The axial transmission and the backscattering methods have been applied to investigate the variation of ultrasound (US) velocity, the propagation of guided waves, and the interaction of the complex scattering phenomena with healing bones. At the macrostructure level, computational models have been established mimicking healthy and pathologic bones, while the use of imaging modalities such as scanning acoustic microscopy (Figure 8.10) has opened new perspectives for the evaluation of the bone microstructure using quantitative US (QUS) [11, 32].

At the nanostructure level, a mechanobioregulatory model of bone solidification under the US presence was presented in [33] incorporating also the mechanical environment on the regeneration process, which is known to influence cellular processes (Figure 8.11). An iterative procedure was applied and FEM was used to derive the mechanical stimuli at the linear elastic phases of the poroelastic callus region. Also, a coupled system of partial differential equations was applied to

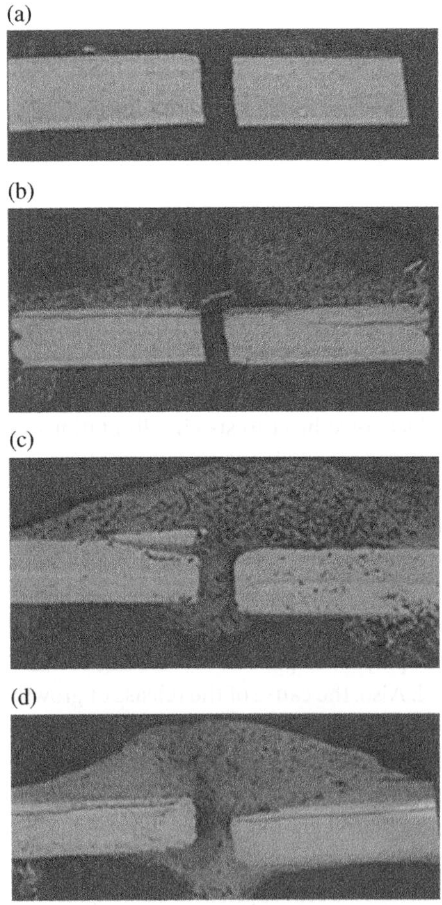

**Figure 8.10** Scanning acoustic microscopy images for: (a) week 2, (b) week 3, (c) week 6, and (d) week 9 after osteotomy.

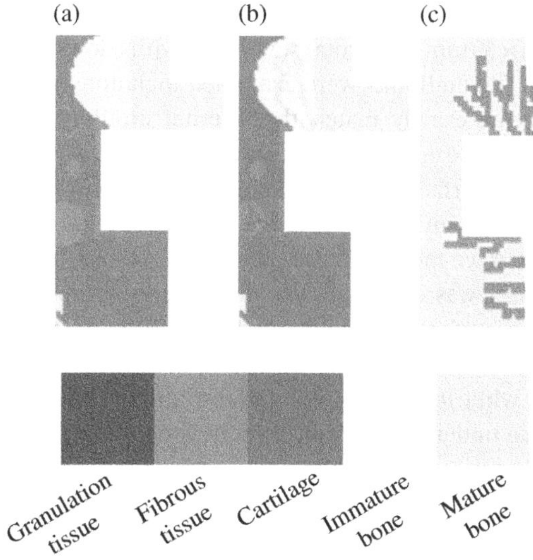


**Figure 8.11** Predicted healing patterns when cells are immediately distributed within the fracture after eight days: (a) the analysis was run without including US, (b) the US effect was taken into consideration, (c) the vascular network from the angiogenesis problem is given. The color palette indicates the different tissue phenotypes for (a) and (b).

simulate the enhancement of the cell angiogenesis process due to US and thus the oxygen concentration in the fractured area. Numerical simulations were presented with and without the impact of US which illustrate the effect of progenitor cells' origin in the healing pattern and the healing rate and simultaneously demonstrate the salutary role of US on bone repair. The results at different hierarchical levels have indicated the promising monitoring potential of QUS and its positive impact on the acceleration of the bone-healing process.

Computational modeling of bone implants has also gained significant interest especially for the cases of knee joint and hip joint arthroplasty in an orthopedic surgeon, as well as in the orthodontic sector. Successful development of implants for orthopedic surgical procedures depends on a comprehensive understanding of the interaction between implant biomaterials and the host tissues of the musculoskeletal system. Mechanical factors are a key part of this interaction [34]. Aseptic loosening, osteolysis, and infection are the most commonly reported causes of revision total knee arthroplasty. In [35], the role of implant design features and stems in resisting loosening, and the sensitivity of the implants to a loose surgical fit due to saw blade oscillation, were examined numerically. The FEM was used and computational models of the distal femur were established implanted with four different implant types: cruciate retaining, posterior stabilizing (PS), total stabilizing with a short stem (12 mm × 50 mm), and total stabilizing with a long stem (19 mm × 150 mm). Two different fit conditions were examined including a normal fit (the resections on the bone exactly match the internal profile of the implant) and a loose fit due to saw blade oscillation (removal of one millimeter of bone from the anterior and posterior surfaces of the distal femur). Frictional interfaces were considered at the bone-implant interfaces allowing relative motions to be recorded. It was found that interface motions increase with increasing flexion angle and loose fit. Additionally, it was observed that the features of implant design influence significantly the surface area under increased motion, while their effect on the values of peak motion is minor. Short uncemented stems behaved similarly to PS implants, while long canal-filling stems exhibited the least amount of motion at the interface under any fit condition. In conclusion, it was found that long-stemmed prostheses are less susceptible to surgical cut errors compared to short-stemmed and stemless implants.

According to [36], although experimental hip simulator testing under edge-loading conditions has produced increased, clinically relevant, wear rates for hard-on-hard bearings when compared to concentric conditions, such testing is time-consuming and costly. Thus, a computational edge-loading model (Python Edge Loading [PyEL] – quasistatic, rigid, and frictionless) was developed considering realistic bearing geometries. The aim of this study was to produce predictions of separation within the typical experimental measurement error of ~0.5 mm. The model was verified and validated against comparable FE models (including inertia and friction) and pre-existing experimental test data for 56 cases, covering a variety of simulated cup orientations, positions, tissue tensions, and loading environments. The PyEL model showed a good agreement with both the computational and experimental results. Compared to the FE models, it was found that the assumption of no inertia had a minor impact on the maximum separation prediction. With high contact force cases, the assumption of no friction had a larger effect (up to ~5% error). The PyEL model was able to predict the experimental maximum separations within ~0.3 mm. The results from this study suggest that a rapid, quasistatic mechanics-based model can provide a good estimate of the maximum separation and contact location generated under a total hip replacement ISO standard edge-loading protocol. It could therefore be used to optimize an experimental test plan and efficiently investigate a much wider range of scenarios and variables.

In another recent study [37], the aim was to predict the evolution of the resonance frequency of the bone–implant interface in a dental implant by means of FE simulation. To this end, a phenomenological interface model was developed which was able to simulate the mechanical effects of the osseointegration process at the bone–implant interface. The outcome of the numerical simulations was compared with experimental findings in rabbits. An early stage of slow bone ingrowth, followed by a faster osseointegration phase until final stability, was predicted by the simulations. The evolution of the resonance frequency of the implant and surrounding tissues along the simulation period was also derived, showing a threefold increase in the first principal frequency. The results were in agreement with the experimental measurements indicating that the model can be useful for the assessment of the effect of mechanical factors such as implant loading or implant geometry on the indirect evaluation of the process of implant osseointegration.

The application of multiscale modeling to scaffold design was presented in [38]. In tissue engineering design, it is, in general, hypothesized that the scaffold should provide a mechanical environment similar to the predegenerative one for initial function and have sufficient pore interconnectivity for cell migration and cell/gene delivery. In the case of bone tissue, the design of such scaffolds can be greatly improved by the knowledge of the bone adaptation process, thus facilitating the identification of scaffold microstructures compatible with the properties of real bone. In this direction, a multiscale model for bone adaptation was established in [38] that potentially can respond to some of the scaffold design requirements. The proposed model focused on the two top-scale levels of bone architecture. The macroscale (whole bone) is characterized by the bone's apparent density distribution and the microscale where the trabecular structure of bone in terms of its mechanical properties is characterized. At the global scale, bone was assumed as a continuum material characterized by equivalent mechanical properties. At the local scale, the bone trabecular anisotropy was approached considering a locally periodic porous material. The relevance of incorporating a micro design scale relies on the possibility of controlling morphometric parameters that not only characterize trabecular structure, and thus can help the design of bone substitutes but also allow a fine balance between bone tissue biological and mechanical functions.

The potential of computational modeling and simulation approaches at multiple length scales to elucidate the response of musculoskeletal tissues to orthopedic implants was investigated in [34] leading to improved treatment outcomes. As the most well-characterized musculoskeletal tissue is the osseous tissue, micro- and nanoscale modeling of bone and collagen were presented in [34]. More specifically, the elastoplastic constitutive response of the bone extracellular matrix was examined using FEM of bone nanoindentation combined with atomic force microscopy to map the inelastic deformation of the tissues. The use of serial milling and block face imaging techniques to determine soft tissue anatomy at multiple length scales for the definition of model geometry was also covered. The potential for further development of multiscale computational biomechanics through modeling cell and tissue mechanobiology was discussed, including more detailed mechanical characterization of the tissue–implant interface, modeling the strain fields experienced by cells within the extracellular matrix and within scaffolds and coupled modeling of other physical and biological phenomena within tissues and biomaterials.

### 8.5.2 Multiscale Modeling of Articular Cartilage

Across spatial scales, biological systems exhibit exquisite hierarchy in architecture and function, leading to complex, observable phenomena. In the articular cartilage of our joints, the organization of molecule- to tissue-level structures governs the interplay of macromolecules and determines the biological activity of embedded cells (chondrocytes), motivating the development of new computational models to provide insight and understanding [39]. While existing models provide insight into mechanical quantities like stress and strain, the leading edge of the field has yet to connect these data to known architectures of matrix molecules like collagen subtypes, proteoglycans, and glycoproteins, and to regional variations in water content. Multiscale models of cartilage could allow researchers to understand how complex interactions at the molecular scale (e.g. between the network of collagen and the densely packed proteoglycan) generate the remarkable macromechanics of cartilage.

Continuum mechanics-based models of the macro-scale, simulating response at the joint level, have generally focused on the macroscopic tissue-level mechanical response of the cartilage [40]. In these modeling approaches, the underlying physics of the tissue has been largely simplified. Cartilage was often modeled as an elastic material, although this assumption is only valid at certain loading and boundary conditions. More recent studies have integrated more realistic representations of cartilage constitutive behavior at the joint level. Modeling of cartilage at the cellular and microstructural levels provides the ability to predict chondrocyte deformations as a function of macroscopic tissue strains, to establish indicators of microstructural loading, and provides an understanding of the role of matrix composition on cellular and extracellular mechanics. Two modeling modalities are commonly used: (i) focusing on the micromechanics as dictated by the cartilage's composition and (ii) exploring the mechanical environment of chondrocytes, commonly relying on FEA and also informed by the former micromechanical modeling modality. Cartilage mechanical properties at the tissue level have been primarily determined by the combination of the mechanical contributions of proteoglycans and collagen. Proteoglycans, via their fixed negative charges, induce a Donnan osmotic swelling pressure which is counteracted by tensile stresses in the collagen network. The combination of the swelling and the restriction against swelling determines the mechanical properties of the cartilage tissue. Cartilage mechanical models have attempted to capture these effects of swelling with triphasic representations (solid, fluid, and ionic), fitting parameters to tissue-level consolidation experiments, or into computationally less expensive biphasic swelling models that assume ion flux is significantly faster than water flow. To better capture the nature of the tissue, models with fiber reinforcement have been developed. In fact, the contributions of swelling and fiber reinforcement are products of the primary biochemical constituents of the tissue (proteoglycans and collagens).

Next, we focus on cartilage injury and degeneration, and osteoarthritis (OA) models as described in [41]. The onset and progression of OA are related to the mechanical environment of the tissue. 3D models of the knee joint have potentially provided useful tools whereby the mechanics of cartilage degeneration can be better understood. In [42], the focal surface injury of articular cartilage was modeled using a patient-specific FE model loaded at 30° of knee flexion. The side effects of osteochondral defect on contact pressures were studied and a defect size of 10 mm was reported as a threshold for clinical considerations of focal articular surface injury repair. The effect of osteochondral defects on cartilage mechanical response was examined in [43]. A computational model was presented considering the depth-dependent properties of cartilage and fibers, and the calcified cartilage was assumed as linear elastic and isotropic. Four different cases were considered: localized bone damage, cartilage-bone interface damage, bone overgrowth, and absence of collagen fibers in the deep zone. Significant change in joint contact mechanics was reported specially in the case of bone damage combined with cartilage split (which resulted in the absence of deep collagen fibers). Moreover, the results for the cartilage-bone interface damage indicated an increased chance of OA onset and progression. The effect of cartilage defects on stress concentration was investigated in [44]. A hyperelastic transversely isotropic model was presented. The strain energy density function consisted of three parts: one represented the quasi-incompressibility of the tissue, one pertained to fibers in tension, and the third pertained to the matrix, which was assumed as Neo-Hookean. It was found that large cartilage defects produced high-stress concentrations as compared to small defects. In [45], healthy, osteoarthritic, and repaired cartilages were considered to establish a 2D knee joint model. Different material models were presented and compared for cartilage: isotropic poroelastic, transversely isotropic poroelastic, and fiber-reinforced poroviscoelastic (FRPVE). In the FRPVE, the fiber direction and fluid content fraction were depth-dependent, and a Neo-Hookean hyperelastic model was used for the nonfibrillar matrix. Their results demonstrated the important role of collagen fibers in controlling stress and strain distribution within cartilaginous tissues, which may be used in the design of artificial cartilage.

According to [40], few conventional models connect mechanical and biological/biochemical factors, let alone on varying spatial and temporal scales. Macroscale computational models may estimate stiffness in the bulk tissue matrix, and hierarchical models may refine our understanding of the local (substrate) stiffness experienced by individual cells. Nevertheless, stiffness is only one parameter driving cellular expression. The activity of ion channels, cell–cell connectivity and signaling, and intracellular cytoskeletal networks require more advanced models that predict distinct mechanotransduction and biochemical pathways. Emerging knowledge of chondrocyte biological activity in health and through the progression of disease may in part drive this need [40]. Alternatively, new computational formulations that incorporate growth and development may predict new biological activity. Predictive multiscale modeling of the mechanics and evolution of cartilage is challenging as much of the required knowledge (experiments, theories, and numerics) remain poorly understood. Additionally, no multiphase, multiscale models relevant to cartilage mechanics and mechanobiology currently exist – a paucity that offers enormous opportunity for future models that link patient-specific large-strain mechanics, biology, and biochemistry in 3D, and drive patient-oriented treatments and soft-tissue replacements in tissue engineering [40].

### 8.5.3 Multiscale Modeling of Tendons and Ligaments

#### 8.5.3.1 Advances in Multiscale Modeling of Tendons

Tendons and ligaments are soft tissues that support muscle and bone structures in the body. Tendons are responsible to transfer forces from muscles to bones and their role is essential for movement. Unfortunately, they are frequently injured, and their poor healing ability results in long-term loss of function. Medical interventions, including surgical and nonsurgical treatments, physical therapy, steroid injections, and anti-inflammatory medications have limited efficacy, and re-rupture is common [46]. These poor outcomes have motivated the research for alternative treatment strategies to prevent tendon injury, improve regenerative healing, develop tissue replacements from stem cells, and have paved the path for the evaluation of tendons using computational methods. The limited understanding of the factors that regulate tendon formation, injury, and healing indicates the need to further exploit the advantages of numerical approaches at multiscale levels.

Computational models of tendon pathology have been used for assessing the causes, onset, and progression of tendon damage at both the cell and bulk tissue levels. Computational models can provide useful information, such as the stress distributions in tendon, which cannot be easily obtained in an experimental environment. Tendon exhibits anisotropic, inhomogeneous, and viscoelastic mechanical properties that are determined by its complicated hierarchical structure and varying organization of different tissue constituents [47]. Although extensive research has been conducted to use modeling approaches to interpret tendon structure–function relationships in combination with experimental data, many issues remain unclear (i.e. the role of minor components such as decorin, aggrecan, and elastin), and the integration of mechanical analysis across different length scales has not been well applied to explore stress or strain transfer from macro- to microscale. Multiscale modeling approaches incorporating tendon features is considered an advantageous methodology to understand further the physiological mechanical response of tendon and corresponding adaptation of properties owing to unique *in vivo* loading environments. According to [47], a number of mathematical models of tendon mechanics at the tissue-, fiber- or fibril-level can be categorized into two general types: (i) the phenomenological approach, which uses a mathematical formulation that can closely match the mechanical response of tendon, but often with parameters that lack a clear physical interpretation, and (ii) the microstructural approach, which describes gross tendon mechanical behavior by combining or generalizing the mechanical behavior of different components in tendon (e.g. collagen fiber distribution). Many models, of both types, are developed based on a continuum theory, wherein the mathematical formulation used to represent tendon behavior is based on a continuum assumption on the tissue scale. In addition, several different modeling schemes have been developed to evaluate mechanical properties of individual tendon constituents, or their interactions, on the molecular level.

Next, we focus on computational and mathematical approaches for the modeling of tendons in the case of injury. To understand stress distributions associated with injury, 2D FE analysis has been used in [48, 49]. The FEM was used to model stress concentrations in partial-thickness defects in the rotator cuff, and highlighted the importance of limiting mechanical loading to prevent the worsening of partial tears. Another 2D FE model predicted locations of increased strain and isolated tendon fascicle damage in "jumpers' knee," a common patellar tendon injury with previously unknown etiology. Evaluation of the model using cadaveric patellar tendon-tibia samples showed that the predicted loading conditions with the highest local strain induced tendon fascicle disruption in three of the five samples, at the anatomical location of reported pain. More recent studies [50] deal with the evolution of the cellular, biochemical, and mechanical characteristics of tendons due to injury. These changes can be challenging to be expressed mathematically, but several tendon injury models are based on Hill equations, which are commonly used to model cellular responses, particularly secretion or degradation of molecules or ligands [51, 52]. A three-component Hill-type equation model was used to incorporate mechanical and strain-dependent proteolytic collagen fiber damage in a human Achilles soleus tendon unit [50]. The model predicted that proteolytic damage would result in collagen fiber shortening, while mechanical damage would result in overall fiber lengthening, thus showing that collagen fiber damage and resulting length after healing are modulated differently in overuse versus inflammation injuries. Predicting how collagen is altered by various damage and repair mechanisms could provide assistance to clinical decision-making and prevent reinjury during rehabilitation.

#### 8.5.3.2 Advances in Multiscale Modeling of Ligaments

The skeletal ligaments are short bands of tough fibrous connective tissue binding bones together across joints. Their mechanical function is to guide normal joint motion and restrict abnormal joint movement. Ligaments can be subjected to extreme stress while performing their role in restricting abnormal joint motions and can be damaged or completely disrupted when overloaded. Excessive stretching or disruption can result in gross joint instability, resulting in altered joint kinematics, altered load distribution, and increased vulnerability to injury of other ligaments and musculoskeletal tissues. Ligaments are highly anisotropic due to their fibrous structure. The anisotropy can vary significantly between different types of ligaments and the fiber orientation generally shows an adaptation to the mechanical environment. For instance, the collagen fibers in the cruciate ligaments of the knee are highly aligned with the long axis, while the organization of the inferior glenohumeral ligament of the shoulder varies considerably with location. Ligaments have time- and history-dependent viscoelastic properties that arise from the interaction of water with the ground substance matrix and the inherent viscoelasticity of the solid phase.

Strategies for FE modeling of ligament mechanics were described in [53], differentiating between whole-joint models and models of individual ligaments. Common approaches to obtaining 3D ligament geometry were presented with an emphasis on techniques that rely on volumetric medical image data. The acquisition of accurate geometry for the ligament(s) and possibly the bones is a fundamental requirement for the construction of 3D FE models of ligaments. Laser scanning and medical imaging are the primary techniques which have been used for this purpose. Both magnetic resonance imaging (MRI) and CT have been used to acquire ligament geometry. MRI can provide detailed images of soft tissue structure in diarthrodial joints. When compared to MRI, CT yields superior spatial resolution and a better signal-to-noise ratio, as well as images of high quality of the bones around the joint, which often must be included in FE models of ligaments to represent ligament wrapping and insertion site geometry. Soft tissue is visible in standard CT images, but there is little difference in the signal between soft tissues, and thus, it can be difficult to distinguish the boundaries of a specific ligament in an intact joint. Considerations for the 3D constitutive modeling of ligaments have been discussed in the context of ligament composition and structure. A novel approach to applying *in situ* strain to FE models of ligaments was described, and test problems were presented that demonstrate the efficacy of the approach. Approaches for the verification and validation of ligament FE models were outlined.

In addition, constitutive equations have been used to describe the mechanical behavior of ideal materials through specification of the dependence of stress on variables, such as the deformation gradient, rate of deformation, temperature, and pressure. The accurate description and prediction of the 3D mechanical behavior of ligaments by constitutive equations remain one of the challenges for computational modeling. The development and application of these constitutive models rely on the understanding of ligament structure and function, and knowledge of available experimental data. Several studies have based ligament viscoelasticity on nonlinear theories. In [54], a single integral finite strain model was presented to describe finite deformation of a nonlinearly viscoelastic material within the context of a 3D model. The specific form characterizing uniaxial extension was derived, and the concept of conversion from one material to another at the microscopic level was then applied to model the nonlinear behavior of ligaments and tendons. In [55], a framework was introduced based on elastic and viscous potentials. The resulting constitutive law is valid for large deformations and satisfies the principles of thermodynamics. A rheological network model was developed in [56] using an orthotropic hyperelastic constitutive model for fibrous tissue and a viscoelastic model for soft materials. Although a number of 3D theories for nonlinear ligament viscoelasticity have been developed, there is still a need for computational and experimental studies on ligament viscoelasticity that can provide the material coefficients that are necessary for anisotropic viscoelastic constitutive models.

In a more recent study [57], a multiscale modeling approach was presented to investigate the effects of the stochasticity of collagen-related structural properties on the biomechanical properties of ligaments and tendons. This approach enables the introduction of model parameters directly associated with tissue structural and biochemical features, opening to physically motivated parametric studies. Variance- and density-based global sensitivity analyses were presented, together with the quantification of output uncertainty due to stochastic variations of parameters. More specifically, the structural features of interest were the density of intermolecular cross-links, the fiber crimp amplitude, and the thickness and molecular persistence length. The main findings of this study can be summarized as follows: (i) the fiber crimp amplitude and intermolecular cross-link stiffness density are key parameters for stress obtained at a given stretch, (ii) tissue extensibility and stresses at the straightening of collagen fiber can be tuned in a one-by-one strategy by acting respectively on amplitude and thickness. In addition, a variation of nanoscale biochemical environment, that is molecular persistence length or intermolecular cross-links stiffness density, influences the tangent stiffness at transition stretch without affecting macroscale mechanical quantities,

(iii) in terms of tissue damage-risk, sensitivity analysis allows to highlight that a variation in fiber thickness affects fibril deformation modes at physiological loads, while an alteration of intermolecular cross-links stiffness density is a key parameter at over-physiological loads. On the other hand, uncertainty quantification indicated the reliability of the multiscale structural approach in reproducing experimental stress-strain data. As a proof of consistency, the outcomes from uncertainty quantification in terms of the identification of the dominant structural features corresponded to the ones obtained from sensitivity analysis. In addition, the uncertainty quantification provided information on the output quantities of interest whose stochasticity is amplified or reduced by input uncertainties.

### 8.5.4 Multiscale Modeling of the Skeletal Muscle

Only skeletal and smooth muscles (Section 8.5.5) are part of the musculoskeletal system and more specifically the skeletal muscle is the motor of the locomotion function of the human body. The skeletal muscle has a complex multiscale and hierarchical architecture (i.e. from fibers to myofibrils, sarcomeres, and contractile proteins (actin and myosin)) and function (e.g. voluntary contraction control) [58]. One of the milestone mathematical models describing skeletal muscle activation, contraction, and force has been presented based on the well-known experiment conducted by Hill in 1938 to elucidate the muscle work and contraction velocity phenomena. Based on this original finding, 1D lumped-parameter model of muscle contraction and force has been developed [59] and this model has been widely used in rigid-body musculoskeletal modeling [60]. In spite of the compact formulation with only five parameters and computational advantage, this 1D lumped-parameter model could not represent complex functional and structural relationships of the skeletal muscle accurately, especially in the case of diseases related to muscles such as dystrophy or spasticity.

The mathematical assessment of skeletal muscle behavior has attracted the interest of several research groups during the last century. Finite elastic theory has been used to establish 3D continuum models of skeletal muscles which take into account intrinsic activation mechanism. Viscoelasticity and hyperelasticity have been widely considered and incorporated into numerical models. It has been observed that elastic and viscous properties exhibit time-dependent strain when undergoing deformation. Also, single-scale and multiscale models have been presented. The application of multiscale muscle models enables the understanding of muscle properties at the macroscopic level (e.g. shape deformation) while considering mechanical and structural properties (e.g. fiber stretch or sarcomere length) at lower levels. The electrophysiological properties of the muscle contraction mechanism have been coupled with mechanical components to simulate the skeletal muscle features more realistically [58].

The proper integration of fibers and their activation function into realistic computational models is a challenging task from an experimental and numerical point of view. To improve the current 3D continuum models, recent studies focus on the exploitation of multimodal data derived from medical imaging, the integration of more biophysical phenomena, and model reproducibility. Different imaging techniques such as MRI-based ones (classical, dynamic, cine phase contrast, and elastography), CT, US, or optical microendoscopy should be examined to provide mechanical (e.g. viscosity and shear modulus), morphological (e.g. fiber and sarcomere length), and functional (e.g. contraction velocity) features of skeletal muscles to enhance model establishment and validation [58, 61–63]. Thus, a systematic multiscale characterization of the skeletal muscle should be conducted to provide a single coherent and consistent data set. More specifically, diffusion tensor imaging opens new horizons for reconstructing and tracking the fiber distribution in a more realistic manner [64]. Thus, robust data-processing protocols (e.g. model registration from multimodal and multiscale data or real-time tracking of muscle fiber distribution and contraction velocity) are required to comply with new multimodal data extraction. These multimodal and multiscale data can enhance robust model formulation, parameter identification, and validation. Nevertheless, a limited range of simulation case studies has been presented using current muscle models. Simple loading cases such as isometric activation and contraction or shortening and lengthening processes have been evaluated.

In a recent study [65], a novel multiscale modeling framework for skeletal muscles was presented based on analytical and numerical homogenization methods to evaluate the mechanical muscle response at finite strains under 3D loading conditions. An analytical microstructure-based constitutive model was developed and numerically implemented in an FE software. The analytical model considered explicitly the material properties, the volume fractions, and the spatial distribution of muscle's constituents by using homogenization techniques to bridge the different length scales of the muscle structure. Next, a numerical homogenization model was developed using periodic eroded Voronoi tessellation to virtually represent skeletal muscle microstructures. The eroded Voronoi unit cells were then resolved by FE simulations and were used to assess the analytical homogenization model. The material parameters of the analytical model were identified successfully based on available experimental data. It was shown that the proposed analytical model was in very good agreement with the numerical model for the full range of loadings, and a wide range of different volume fractions and heterogeneity contrasts between muscle's constituents. A qualitative application of the model on fusiform and pennate muscle structures revealed its efficiency to examine the effect of muscle fiber concentration variations in an organ-scale model simulation.

Most of the computational studies in this area mainly deal with the basic understanding of skeletal muscle properties in physiological conditions, while the application of skeletal muscle models in pathophysiological conditions still remains a challenging objective to be achieved [58]. To this end, in [66] a novel, multiscale muscle modeling framework was presented to elucidate the relationship between microstructural disease adaptations and modifications in both mechanical properties of muscle and strain in the cell membrane. An agent-based model was presented to randomly generate new muscle fiber geometries and mapped them into an FE model representing a cross section of a muscle fascicle. The variability in the shape and arrangement of fibers was examined, as well as the incorporation of disease-related changes. This method was applied to reveal the tradeoffs between mechanical properties and damage susceptibility in Duchenne muscular dystrophy (DMD). DMD is a fatal genetic disease caused by a lack of the transmembrane protein dystrophin, leading to muscle wasting and death due to cardiac or pulmonary complications. The most prevalent microstructural variations in DMD include lack of transmembrane proteins, fibrosis, fatty infiltration, and variation in fiber cross sectional area. A parameter analysis of these variations and a case study of DMD showed that the nature of fibrosis and density of transmembrane proteins strongly affected the stiffness of the muscle and susceptibility to membrane damage [66].

### 8.5.5 Multiscale Modeling of the Smooth Muscle

The use of computational modeling and multiscale approaches as a new means to evaluate the function of smooth muscle has been presented in the literature during the last few years [67–70].

In [67], a constitutive model was presented aiming to describe the main biochemomechanical features involved in the contractile response of smooth muscle cells, in which: (i) the biochemical response was modeled by extending the four-state Hai–Murphy model to isotonic contraction of the cells and (ii) the mechanical response was mainly modeled based on the phosphorylation-dependent hyperbolic relation between isotonic shortening strain rate and tension. The 1D version of the model was used to simulate shortening-induced deactivation with good agreement with selected experimental measurements. It was observed that the Hai–Murphy biochemical model ignores the strain rate effect on the kinetics of cross-bridge interactions with actin filaments in the isotonic contractions. The 2D and 3D versions of the model were established using the homogenization method under finite strain continuum mechanics framework. The 2D constitutive model was applied to simulate swine carotid media strips under electrical field stimulation and contraction of a hollow airway and a hollow arteriole buried in a soft matrix subjected to multiple calcium ion stimulations. It was shown that the transverse deformation may have a significant influence on the response of the swine carotid medium. In both cases, the orientation of the maximal value of attached myosin was aligned with the orientation of maximum principal stress.

Specific types of smooth muscles, such as the detrusor of the urinary bladder, reveal a variety of spikes which differ significantly in their amplitudes and time courses. Although the origin of this diversity has not been understood, it is often attributed to the syncytial nature of smooth muscle and its distributed innervation. In a recent study [68], a biophysically constrained computational model of the action potential of mouse urinary bladder smooth muscle was developed. Urinary incontinence is related with enhanced spontaneous phasic contractions of the detrusor smooth muscle (DSM). Although a complete understanding of the etiology of these spontaneous contractions has not been achieved yet, it has been suggested that the spontaneously evoked action potentials (sAPs) in DSM cells initiate and modulate the contractions. In order to provide new insights into the ionic mechanisms underlying sAP generation, a biophysically detailed computational model of a single DSM cell was presented in [68]. Mathematical models for nine ion channels found in DSM cells were established based on published experimental data. In addition, a 3D electrical model of syncytial smooth muscle was established in [69] using the compartmental modeling technique, with special reference to the bladder detrusor. Values of model parameters were sourced or derived from experimental data. The model was validated with experimental findings and the results were found to agree with both experimental observations and theoretical predictions. The outputs of the model also satisfied major parameters of electrical syncytia such as correlation between the spatial spread and temporal decay of electrotonic potentials as well as positively skewed amplitude frequency histogram for subthreshold potentials, and lead to interesting conclusions. This model was considered a unique and innovative platform for the interpretation of electrical activity in syncytial tissues.

Focusing on the multiscale and multiaxial mechanics of vascular smooth muscle, a multiscale mathematical model of arterial contractility was presented in [70] taking into consideration structural and functional constituents at molecular, cellular, and tissue levels. The artery was assumed to be a thick-walled incompressible cylinder described by an anisotropic model of the extracellular matrix and, to our knowledge, novel model of smooth muscle contractility. The latter included a 3D structural sensitivity to deformation, including microscale muscle filament overlap and filament lattice spacing. The overall model captured uniaxial and biaxial experimental contraction data, which was not possible when accounting for filament overlap alone. The model also enabled parameter sensitivity studies, which confirmed that uniaxial contraction tests are not as efficient as biaxial tests for identifying changes in vascular smooth muscle behavior.

## 8.6 Conclusion

For future improvements in computational approaches, the combination of computational models and corresponding experiments is important. At the same time, the validation of parameters in the models is also important. With the assistance of multidimensional computational models, the unknown mechanisms underlying pathologies of the musculoskeletal system can be tested and clinicians can conduct optimal treatment strategies.

## References


1 Jung, G.S. and Buehler, M.J. (2017). Multiscale modeling of muscular-skeletal systems. *Annu. Rev. Biomed. Eng.* 19 (1): 435–457. https://doi.org/10.1146/annurev-bioeng-071516-044555.

2 Potsika, V.T., Vavva, M.G., Protopappas, V.C. et al. (2013). Computational modeling of ultrasound wave propagation in bone. In: *Computational Medicine in Data Mining and Modeling* (ed. G. Rakocevic, T. Djukic, N. Filipovic and V. Milutinović), 349–376. New York, NY: Springer.

3 Marieb, E.N. and Hoehn, K.N. (2014). *Human Anatomy & Physiology*, 10e. Pearson.

4 Wang, X., Xu, S., Zhou, S. et al. (2016). Topological design and additive manufacturing of porous metals for bone scaffolds and orthopaedic implants: a review. *Biomaterials* 83: 127–141.

5 Sadat-Shojai, M., Khorasani, M.-T., Dinpanah-Khoshdargi, E., and Jamshidi, A. (2013). Synthesis methods for nanosized hydroxyapatite with diverse structures. *Acta Biomater.* 9 (8): 7591–7621.

6 Klika, V., Gaffney, E.A., Chen, Y.-C., and Brown, C.P. (2016). An overview of multiphase cartilage mechanical modelling and its role in understanding function and pathology. *J. Mech. Behav. Biomed. Mater.* 62: 139–157.

7 Maciel, A. (2002). Biomechanics of hip joint capsule. https://www.researchgate.net/publication/245587753_Biomechanics_of_Hip_Joint_Capsule/citation/download

8 Muscle tissue. In: *Anatomy and Physiology*. Chapter 10. Rice University is licensed under a Creative Commons Attribution 4.0 International License. https://openstax.org/books/anatomy-and-physiology-2e/pages/preface.

9 Gabella, G. (1994). Structure of smooth muscles. In: *Pharmacology of Smooth Muscle. Handbook of Experimental Pharmacology*, vol. 111 (ed. L. Szekeres and J.G. Papp). Berlin/Heidelberg: Springer https://doi.org/10.1007/978-3-642-78920-5_1.

10 Biga, L.M., Dawson, S., Harwell, A. et al. *Anatomy & Physiology*. Creative Commons Attribution-ShareAlike 4.0 International License. Oregon State University.

**11** Potsika, V. (2017). Computational modeling of long bones during the fracture healing process. PhD thesis.

**12** Knudson, D. (2007). Mechanics of the musculoskeletal system. In: *Fundamentals of Biomechanics*, 69–102. Springer.

**13** Gong, H., Wang, L., Zhang, M., and Fan, Y. (2014). Computational modeling of bone and bone remodeling. In: *Computational Modelling of Biomechanics and Biotribology in the Musculoskeletal System* (ed. Z. Jin, J. Li and Z. Chen), 244–267. https://doi.org/10.1533/9780857096739.2.244.

**14** Huiskes, R. and Chao, E.Y.S. (1983). A survey of finite element analysis in orthopaedic biomechanics: the first decade. *J. Biomech.* 16: 385–409.

**15** Parashar, S.K. and Sharma, J.K. (2016). A review on application of finite element modelling in bone biomechanics. *Perspect. Sci.* 8: 696–698. https://doi.org/10.1016/j.pisc.2016.06.062.

**16** Basu, P.K., Beall, A.G., Simmons, D.J., and Vannier, M. (1986). 3-D femoral stress analysis using CT scans and p-version FEM. *Biomater. Med. Devices Artif. Organs* 13: 163–186.

**17** Chand, R., Huge, E., and Rims, K. (1976). Stress in the human knee joints. *J. Biomech.* 9 (6): 417–422.

**18** Anderson, A.E., Ellis, B.J., Maas, S.A., and Weiss, J.A. (2010). A effects of idealized joint geometry on finite element predictions of cartilage contact stresses in the hip. *J. Biomech.* 43: 1351–1357.

**19** Bessho, M., Ohnishi, I., Matsuyama, J. et al. (2007). Prediction of strength and strain of the proximal femur by a CT-based finite element method. *J. Biomech.* 40: 1745–1753.

**20** Zysset, P.K., Dallara, E., Varga, P., and Pahr, D.H. (2013). Finite element analysis for prediction of bone strength. *Bonekey Rep.* 386: 1–8.

**21** Basafa, E., Armiger, R.S., Kutzer, M.D. et al. (2013). Patient-specific finite element modeling for femoral bone augmentation. *Med. Eng. Phys.* 35: 860–865.

**22** Brown, K.R., Tarsuslugil, S., Wijayathunga, V.N., and Wilcox, R.K. (2014). Comparative finite-element analysis: a single computational modeling method can estimate the mechanical properties of porcine and human vertebrae. *J. R. Soc. Interface* 11: 1–6.

**23** Dall'ara, E., Pahr, D., Varga, P. et al. (2012). QCT-based finite element models predict human vertebral strength in vitro significantly better than simulated DEXA. *Osteoporos. Int.* 23: 563–572.

**24** Scheiner, S., Pivonka, P., Smith, D.W. et al. (2014). Mathematical modeling of postmenopausal osteoporosis and its treatment by the anti-catabolic drug denosumab. *Int. J. Numer. Methods Biomed. Eng.* 30: 1–27.

**25** Gong, H., Fan, Y.B., and Zhang, M. Computational simulation for osteoporosis at the basic multicellular unit level. In: *7th Asian-Pacific Conference on Medical and Biological Engineering*, APCMBE 2008, Beijing, China (22–25 April 2008), 182–185. https://doi.org/10.1007/978-3-540-79039-6_47.

**26** Badilatti, S.D., Kuhn, G.A., Ferguson, S.J., and Müller, R. (2015). Computational modelling of bone augmentation in the spine. *J. Orthop. Transl.* 3 (4): 185–196. https://doi.org/10.1016/j.jot.2015.09.003.

**27** Imai, K. (2015). Computed tomography-based finite element analysis to assess fracture risk and osteoporosis treatment. *World J. Exp. Med.* 5 (3): 182. https://doi.org/10.5493/wjem.v5.i3.182.

**28** Ghiasi, M.S., Chen, J.E., Rodriguez, E.K. et al. (2019). Computational modeling of human bone fracture healing affected by different conditions of initial healing stage. *BMC Musculoskelet. Disord.* 20: 562. https://doi.org/10.1186/s12891-019-2854-z.

**29** Wang, M. and Yang, N. (2018). Three-dimensional computational model simulating the fracture healing process with both biphasic poroelastic finite element analysis and fuzzy logic control. *Sci. Rep.* 8 (1): https://doi.org/10.1038/s41598-018-25229-7.

**30** Wang, M., Yang, N., and Wang, X. (2017). A review of computational models of bone fracture healing. *Med. Biol. Eng. Comput.* 55 (11): 1895–1914. https://doi.org/10.1007/s11517-017-1701-3.

**31** Potsika, V.T., Protopappas, V.C., Grivas, K.N. et al. (2016). Ultrasonic evaluation of intact, healing and osteoporotic long bones. *J. Serb. Soc. Comput. Mech.* 10 (1): 168–189.

**32** Potsika, V.T., Protopappas, V.C., Grivas, K.N. et al. (2017). Numerical evaluation of ultrasonic backscattering in healing long bones. *J. Acous. Soc. Am.* 142 (2): 962–973.

**33** Grivas, K.N., Vavva, M.G., Polyzos, D. et al. (2019). Effect of ultrasound on bone fracture healing: a computational mechanobioregulatory model. *J. Acous. Soc. Am.* 145 (2): 1048–1059. https://doi.org/10.1121/1.5089221.

**34** Adam, C.J. (2017). Multiscale modelling and simulation of musculoskeletal tissues for orthopaedics, Chapter 1. In: *Biomaterials for Implants and Scaffolds, Springer Series in Biomaterials Science and Engineering*, vol. *8* (ed. Q. Li and Y.W. Mai), 1–44. Berlin, Heidelberg: Springer. https://doi.org/10.1007/978-3-662-53574-5_1.

**35** Conlisk, N., Howie, C.R., and Pankaj, P. (2017). Computational modelling of motion at the bone-implant interface after total knee arthroplasty: the role of implant design and surgical fit. *Knee* 24 (5): 994–1005. https://doi.org/10.1016/j.knee.2017.07.003.

**36** Etchels, L., Wang, L., Al-Hajjar, M. et al. (2019). Computationally efficient modelling of hip replacement separation due to small mismatches in component centres of rotation. *J. Biomech.* 95 (2019): 109296. https://doi.org/10.1016/j.jbiomech.2019.07.040.

**37** Pérez, M.A., Moreo, P., García-Aznar, J.M., and Doblaré, M. (2018). Computational simulation of dental implant osseointegration through resonance frequency analysis. *J. Biomech.* 41 (2): 316–325.

**38** Rodrigues, H.C., Coelho, P.G., and Fernandes, P.R. (2011). Multiscale modelling of bone tissue – remodelling and application to scaffold design, Chapter 2. In: *Advances on Modeling in Tissue Engineering*, Computational Methods in Applied Sciences, vol. 20 (ed. P. Fernandes and P. Bártolo), 15–34. Dordrecht: Springer. https://doi.org/10.1007/978-94-007-1254-6_2.

**39** Wang, X., Neu, C.P., and Pierce, D.M. (2019). Advances toward multiscale computational models of cartilage mechanics and mechanobiology. *Curr. Opin. Biomed. Eng.* 11: 51–57.

**40** Halloran, J.P., Sibole, S., Donkelaar, C.C. et al. (2012). Multiscale mechanics of articular cartilage: potentials and challenges of coupling musculoskeletal, joint, and microscale computational models. *Ann. Biomed. Eng.* 40 (11): 2456–2474.

**41** Kazemi, M., Dabiri, Y., and Li, L.P. (2013). Recent advances in computational mechanics of the human knee joint. *Comput Math. Methods Med.* 1–27. https://doi.org/10.1155/2013/718423.

**42** Papaioannou, G., Demetropoulos, C.K., and King, Y.H. (2010). Predicting the effects of knee focal articular surface injury with a patient-specific finite element model. *Knee* 17 (1): 61–68.

**43** Shirazi, R. and Shirazi-Adl, A. (2009). Computational biomechanics of articular cartilage of human knee joint: effect of osteochondral defects. *J. Biomech.* 42 (15): 2458–2465.

**44** Pena, E., Calvo, B., Martinez, M.A., and Doblare, M. (2007). Effect of the size and location of osteochondral defects in degenerative arthritis. A finite element simulation. *Comput. Biol. Med.* 37 (3): 376–387.

**45** Mononen, M.E., Julkunen, P., Toyras, J. et al. (2011). Alterations in structure and properties of collagen network of osteoarthritic and repaired cartilage modify knee joint stresses. *Biomech. Model. Mechanobiol.* 10 (3): 357–369.

**46** Theodossiou, S.K. and Schiele, N.R. (2019). Models of tendon development and injury. *BMC Biomed. Eng.* 1 (1): 1–32. https://doi.org/10.1186/s42490-019-0029-5.

**47** Fang, F. and Lake, S.P. (2016). Modelling approaches for evaluating multiscale tendon mechanics. *Interface focus* 6 (1): 20150044.

**48** Lavagnino, M., Arnoczky, S.P., Elvin, N., and Dodds, J. (2008). Patellar tendon strain is increased at the site of the Jumper's knee lesion during knee flexion and tendon loading. *Am. J. Sports Med.* 36 (11): 2110–2118.

**49** Sano, H., Wakabayashi, I., and Itoi, E. (2006). Stress distribution in the supraspinatus tendon with partial-thickness tears: an analysis using two-dimensional, finite element model. *J. Shoulder Elbow Surg.* 15 (1): 100–105.

**50** Young, S.R., Gardiner, B., Mehdizadeh, A. et al. (2016). Adaptive remodeling of Achilles tendon: a multi-scale computational model. *PLoS Comput. Biol.* 12 (9): e1005106.

**51** Ingalls, B.P. (2013). *Mathematical Modeling in Systems Biology: An Introduction*. Cambridge: The MIT Press.

52 Goutelle, S., Maurin, M., Rougier, F. et al. (2008). The Hill equation: a review of its capabilities in pharmacological modelling. *Fundam. Clin. Pharmacol.* 22 (6): 633–648.

53 Weiss, J.A., Gardiner, J.C., Ellis, B.J. et al. (2005). Three-dimensional finite element modeling of ligaments: technical aspects. *Med. Eng. Phys.* 27 (10): 845–861. https://doi.org/10.1016/j.medengphy.2005.05.006.

54 Johnson, G.A., Livesay, G.A., Woo, S.L., and Rajagopal, K.R. (1996). A single integral finite strain viscoelastic model of ligaments and tendons. *J. Biomech. Eng.* 118 (2): 221–226.

55 Pioletti, D.P., Rakotomanana, L.R., Benvenuti, J.F., and Leyvraz, P.F. (1998). Viscoelastic constitutive law in large deformations: application to human knee ligaments and tendons. *J. Biomech.* 31: 753–757.

56 Bischoff, J.E., Arruda, E.M., and Grosh, K. (2004). A rheological network model for the continuum anisotropic and viscoelastic behavior of soft tissue. *Biomech. Model. Mechanobiol.* 3: 56–65.

57 Hamdia, K.M., Marino, M., Zhuang, X. et al. (2019). Sensitivity analysis for the mechanics of tendons and ligaments: investigation on the effects of collagen structural properties via a multiscale modelling approach. *Int. J. Numer. Methods Biomed. Eng.* https://doi.org/10.1002/cnm.3209.

58 Dao, T.T. and M.-C. Ho Ba Tho. (2018). A systematic review of continuum modeling of skeletal muscles: current trends, limitations, and recommendations. *Appl. Bionics Biomech.* 1–17. https://doi.org/10.1155/2018/7631818.

59 Zajac, F.E. (1989). Muscle and tendon: properties, models, scaling, and application to biomechanics and motor control. *Crit. Rev. Biomed. Eng.* 17 (4): 359–411.

60 Dao, T.T. (2016). Rigid musculoskeletal models of the human body systems: a review. *J. Musculoskelet. Res.* 19 (3): 1630001.

61 Abe, T., Loenneke, J.P., and Thiebaud, R.S. (2015). Morphological and functional relationships with ultrasound measured muscle thickness of the lower extremity: a brief review. *Ultrasound* 23 (3): 166–173.

62 Bilston, L.E. and Tan, K. (2015). Measurement of passive skeletal muscle mechanical properties in vivo: recent progress, clinical applications, and remaining challenges. *Ann. Biomed. Eng.* 43 (2): 261–273.

63 Clarke, E.C., Martin, J.H., D'entremont, A.G. et al. (2015). A non-invasive, 3D, dynamic MRI method for measuring muscle moment arms in vivo: demonstration in the human ankle joint and Achilles tendon. *Med. Eng. Phys.* 37 (1): 93–99.

64 Brandão, S., Parente, M., Silva, E. et al. (2017). Pubovisceralis muscle fiber architecture determination: comparison between biomechanical modeling and diffusion tensor imaging. *Ann. Biomed. Eng.* 45 (5): 1255–1265.

**65** Spyrou, L.A., Brisard, S., and Danas, K. (2019). Multiscale modeling of skeletal muscle tissues based on analytical and numerical homogenization. *J. Mech. Behav. Biomed. Mater.* 92: 97–117.

**66** Virgilio, K.M., Martin, K.S., Peirce, S.M., and Blemker, S.S. (2015). Multiscale models of skeletal muscle reveal the complex effects of muscular dystrophy on tissue mechanics and damage susceptibility. *Interface Focus* 5 (2): 20140080.

**67** Liu, T. (2014). A constitutive model for cytoskeletal contractility of smooth muscle cells. *Proc. R. Soc. London, Ser. A* 470 (2164): 20130771–20130771. https://doi.org/10.1098/rspa.2013.0771.

**68** Mahapatra, C., Brain, K.L., Manchanda, R., and Guerrero-Hernandez, A. (2018). A biophysically constrained computational model of the action potential of mouse urinary bladder smooth muscle. *PLoS One* 13 (7): e0200712. https://doi.org/10.1371/journal.pone.0200712.

**69** Appukuttan, S., Brain, K.L., and Manchanda, R. (2015). A computational model of urinary bladder smooth muscle syncytium. *J. Comput. Neurosci.* 38 (1): 167–187. https://doi.org/10.1007/s10827-014-0532-6.

**70** Murtada, S.I., Humphrey, J.D., and Holzapfel, G.A. (2017). Multiscale and multiaxial mechanics of vascular smooth muscle. *Biophys. J.* 113 (3): 714–727.

# 9

# Multiscale Modeling of Cardiovascular System

## 9.1 Introduction

Cardiovascular disease (CVD) is the most common cause of death in western societies [1]. The main pathophysiologic phenotype is the atherosclerotic disease, which is characterized by the thickening of the arterial wall and the continuing lumen area narrowing which consequently will cause the blood flow restriction. In general, the material which is responsible for this mechanism is the atherosclerotic plaque. Atherosclerosis mechanism involves the accumulation of calcium, cholesterol and fats within the artery wall to form the atherosclerotic plaques, which reduce and may eventually block the blood flow through the artery, resulting in devastating consequences, such as heart attack. This atherosclerotic plaque is presented with high heterogeneity among different patients consisting of soft materials such as lipid or fibrotic tissue to more stiff material such as calcified esters. The process of atherosclerotic plaque growth and development is multifactorial with several risk factors to contribute to increased rates of progression. Such factors are not only hypertension, male gender, age, and diabetes but also the genetic phenotype or other external factors. Additionally, CVD is characterized as a systemic inflammatory disease, meaning that an inflammatory response is initiated and is responsible for the establishment and progression of the disease. On the other hand, the phenotype of the disease is characterized by local manifestations. The main cause of the local plaque development is the effect of hemodynamics on the vascular physiology. In particular, hemodynamics and especially the endothelial shear stress is found to be responsible for changes in the endothelial membrane, which affects the endothelial function, its permeability, as well as the local gene expression. An overall approach to understanding the pathophysiology of atherosclerotic disease requires

*Multiscale Modelling in Biomedical Engineering*, First Edition. Dimitrios I. Fotiadis, Antonis I. Sakellarios, and Vassiliki T. Potsika.

Published 2023 by John Wiley & Sons, Inc.

several levels of analysis starting from imaging analysis and 3D reconstruction of the arteries, to microscale modeling of biological functions using agent-based modeling and discrete particle dynamics approaches to macroscale modeling employing usually finite elements. Ideally, the interaction of these approaches is required for the most advanced simulation of all the biological phenomena taking place during atherosclerosis. Finally, the recent years' machine-learning approaches are being developed to utilize the computational results and the clinical data to provide an accurate prediction of disease progression, prognosis, and diagnosis.

## 9.2 Cardiovascular Mechanics

Vascular diseases with atherosclerosis being the one with the highest epidemiologic prevalence involve many physical phenomena from microscale and cell–cell interactions to macroscale mechanisms such as the blood fluid dynamics and its effect on the physiology of the artery. In this context, multiscale models have been developed where the effect of blood rheology is estimated and used to understand the response of the tissue and the cells to this. Accurate multiscale models incorporate 3D arterial geometries, consider patient-specific parameters and boundary conditions as well as implants such as the stent devices which further affect vascular physiology.

### 9.2.1 Visualization of the Cardiovascular System and 3D Arterial Reconstruction

To implement a comprehensive cardiovascular model, 3D arterial geometries are required. The 3D reconstruction of the arterial segments or trees requires the processing of medical imaging. Specifically, vessel visualization can be achieved using invasive or noninvasive imaging modalities. As presented in detail in Chapter 2, several imaging modalities can be used for the visualization of the arteries and using special imaging processing algorithms to generate arterial models. In particular, intravascular imaging (intravascular ultrasound [IVUS] and optical coherence tomography [OCT]) can be used in combination with X-ray angiography to generate accurate geometries of arteries. The most usual algorithm for this purpose is the segmentation of the lumen and outer wall borders using active contours. Moreover, two angiographic views are utilized to generate the arterial path in the 3D space, where the identified borders can be positioned perpendicularly. These imaging modalities are considered invasive and it can be assumed that it is extremely difficult to provide some real-time results. On the other hand, computed tomography coronary angiography (CTCA) and magnetic resonance angiography (MRA) can be used for the accurate and fast reconstruction of the coronary and peripheral arterial trees. Such methodologies are based usually on 3D-level set approaches. It is worth noting that in recent years, deep learning approaches have been implemented for the segmentation of all the available imaging modalities. As the final result, the reconstruction algorithms provide a macroscale representation of the arterial segments or even trees. These geometries consist of the lumen and outer wall domains and in some more advanced applications, the plaque components.

### 9.2.2 Blood Flow Modeling

Researchers are trying to perform accurate simulations of hemodynamics aiming to associate the effect of blood rheology and some flow parameters with the understanding of the disease evolution and subsequently the development of tools useful for the diagnosis, prognosis, and prediction of CVD. In order to understand the normal and pathologic behavior of the human vascular system, detailed knowledge of blood flow, and the response of blood vessels is required. The localization of the atherosclerotic lesion on the vessel walls is strongly correlated to hemodynamic disturbances [2]. Blood flow modelling has been a field of research of wide interest and for this purpose, numerous studies have been reported to investigate the aforementioned matter. One of the most important features of the arterial blood flow is its pulsatile nature. Normally, blood flow is considered as a laminar flow, but at the regions of stenosis, disturbances and circulations occur. Another parameter which is also considered, in the fluid-structure interaction problems, is the displacement of the vessel wall due to the heart motion and the fluid forces. Several animal and experimental studies used patented devices to calculate the blood flow velocity and pressure relationship [3, 4], and Doppler velocimetry is used to quantify the flow velocity *in vivo*. Ultrasound can also be used to measure blood flow velocity *in vivo*, especially at the peripheral arteries such as the carotids.

There are several risk factors for heart disease but hemodynamic factors are thought to be responsible for the local manifestations of the vascular disease and especially in areas of complex flow, e.g. bifurcations and curvature of the arteries in the coronary, carotid, abdominal, and femoral arteries. Investigators are concentrated on turbulence as a key to diagnostic characteristics. Although it still remains uncertain with regard to the exact mechanisms responsible for the initiation of these phenomena, it has been established that the development of atherosclerosis, even in the early stages of the disease, is strongly related to the characteristics of the blood flow in the arteries. Zarins et al. [5] noted that in carotid artery phantoms, atherosclerotic lesions localize along the outer wall and concluded that the thickening of the arterial wall and the formation of the atherosclerotic plaque is related to regions with low ESS. Areas of low ESS are generally prone to plaque development. Another similar study was presented by Ku et al. [6] who manage to correlate the location of the atherosclerotic plaque with the ESS. Positive correlations were noted between intimal thickening and the inverse of the maximum shear stress. Through Doppler laser optimization the blood velocity was calculated while in real carotid arteries taken from cadavers, the intimal thickness was measured. This study illustrated the *in vivo* complexities of flow patterns in the carotid bifurcation and it also demonstrated the accuracy and usefulness of rigid *in vitro* models as a technique for examining arterial flows in great detail. Glagov et al. [7] noted that human coronary arteries actually enlarge in cross section as atherosclerotic plaques increased in size. Friedman et al. [8] in a series of studies found that low shear stress location was involved in the earliest intimal thickness observed but high ESS was associated with the latter or more advanced intimal thickness.

#### 9.2.2.1 Steady and Pulsatile Flow of Blood

In most of the presented studies, the flow had been assumed steady. In Forrester and Young's study [9], the equations are developed to approximately describe the flow characteristics of an incompressible fluid through an axisymmetric converging-diverging tube. In the second part of their work [10], an experimental apparatus to study some of these flow characteristics is described, the theoretical and experimental results are compared, and some speculations are made on their implications in occlusive vascular disease. In their experiments, water and blood were used and the experimental results did not differ greatly and, in general, an agreement was found between the experimental and theoretical results. Pressure drop through arterial stenosis models in steady flow conditions was studied by Cavalcanti et al. [11]. Measures of pressure drop were made in two different plexiglass models of axial-symmetric arterial stenoses. The results of this study indicate that the upstream-middle pressure drop is independent of tapering degree but it is highly influenced by area reduction and it is much greater than the middle-downstream drop. Tamamidis and Assanis [12] computed steady flow in various curved tubes with a rectangular cross section.

As we mentioned earlier, one of the most important features of the arterial blood flow is its pulsatile nature. Experimental studies of pulsatile flow through rigid stenoses have been reported. Young and Tsai [13] experimentally developed an oscillating flow, with and without a steady component and the pressure drop was determined as a function of several parameters, through an equation of predicting the pressure drop across a stenosis. The effect of unsteadiness on the flow through stenosis and bifurcations was examined by Siouffi et al. [14] (Figure 9.1). Measurements were performed at 60 bifurcations which had permitted the reconstruction of 3D velocity profiles. The importance of the secondary flow in the

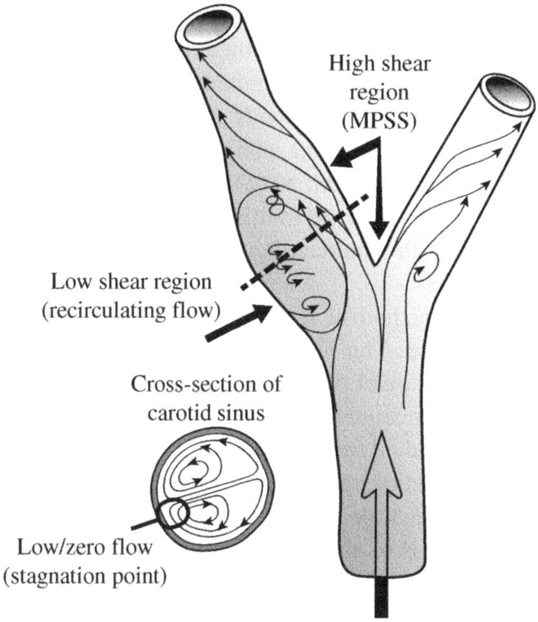


**Figure 9.1** Blood flow and shear stress at the carotid bifurcation. Change in ESS throughout the course of the cardiac cycle at two locations within the carotid bifurcation [14].

branching was analyzed for various values of the flow parameters. Results showed a strong influence of unsteadiness on the flow characteristics and the hemodynamic factors. So, for the microcirculation studies, it is necessary to take into account unsteadiness and the actual shape of the flow-time forcing factor.

An alternative decomposition using a frequency domain approach which combines both random and coherent stresses into a single apparent stress is presented by Lieber and Giddens [15]. This approach is preferable to the traditional ensemble averaging method when estimating fluctuating stresses in arterial flows.

#### 9.2.2.2 Computational Fluid Dynamics Modeling

Due to the complex nature of the vascular geometry, flow, and the compliant nature of the wall, closed-form mathematical solutions are not available and numerical solutions are the only practical way in order to simulate them in the last few years. Computational fluid dynamics (CFD) software platforms have been developed, which allow the modelling of realistic geometries with deformable walls. CFD modelling enables investigation of pressure and flow fields at a temporal and spatial resolution. Through these models, it is easy to overcome some difficulties such as the ESS measurements, a key factor in the development of atherosclerosis and in-stent restenosis (ISR). CFD models can be used for ESS calculation and map its spatial distribution establishing the link between hemodynamic disturbance and atherogenesis and has demonstrated the preferential deposition of atherosclerotic plaque at arterial bends and bifurcations. The latest computational dynamics techniques allow the comprehensive analysis of fluid and hemodynamic parameters involved, such as ESS, residence times, and particle path lines. Nowadays, much of the interest in this field is focused on the fluid-structure interactions occurring at the interface between the blood and the vessel wall.

The most common computational method in order to simulate the arterial blood flow is the finite element method. The FEM is efficient for simulating complex flow of the coronary artery because of the unstructured grids as opposed to structured grids used in finite difference method. This method makes use of spatial discretization and a weighted residual formulation to arrive at a system of matrix equations. Solution of these equations yields an approximate solution to the original boundary value problem. The Galerkin method is the most common weighted residual formation which leads to symmetric stiffness matrices. In a finite element framework, the upwinded terms can be developed in several different ways.

#### 9.2.2.3 Newtonian and Non-Newtonian Behavior of Blood

In order to perform computational blood flow simulations, *in vivo* or *in vitro* measurements derived from Doppler ultrasound or magnetic resonance imaging are required to be applied as boundary conditions. The unsteady entry blood flow in a 90° curved tube is numerically and experimentally investigated by comparing the Newtonian and non-Newtonian blood models [17]. The rheological properties of blood play an important role in the physiology of blood circulation. Blood can be considered a non-Newtonian fluid since its viscosity varies with shear rate. However, the Newtonian model is more suitable in mid-range to high shear rates. So, in rheological terms, in medium to large arteries, it can be assumed that the blood behaves as a Newtonian fluid having constant viscosity. The non-Newtonian model gives better results of ESS than the Newtonian model and mainly in small arteries, blood acts as a non-Newtonian fluid [18]. The blood behaves like a Newtonian fluid when shear rate is higher than a threshold value, while shear rate can be used as a function of the blood composition and it is primarily modulated by its hematocrit [19]. Due to the high shear rate in large arterial blood vessels (i.e. vessel diameter $\geq 1$ mm) [20], blood was modelled as a Newtonian fluid in many studies by accepting the viscosity of blood to be constant and equal to the blood viscosity (generally accepted 0.0035 Pa·s) corresponding to the high shear rate. The first computational studies modeled the blood as a Newtonian fluid and the flow as laminar and incompressible. The arteries were modelled as cylindrical pipes with non-deformable arterial walls. In general, blood flow modelling requires the solution of the 3D transient Navier–Stokes equations.

Taylor et al. [21] described a comprehensive finite element framework to enable computational vascular research. The developed software system provides an integrated set of tools to solve clinically relevant blood flow problems and test hypotheses regarding hemodynamic (blood flow) factors in vascular adaptation and disease. The validity of the computational method was established by comparing the numerical results to an analytic solution for pulsatile flow as well as to experimental flow data. A computational study of hemodynamics which was also validated experimentally was performed by Perktold et al. [22]. This research was based on the solution of the time-dependent Navier–Stokes equations employing the Galerkin method as well as a velocity pressure separation system. Time derivatives were discredited using Euler's differences backward, while the transport term was linearized using Picard iterations at each time step. The computational results were in good agreement with the experimental results in the expansion phase but not in the contraction phase. On the other hand, Santamarina et al. [23] used the finite volume method to solve the Navier–Stokes equations since they pursued studying the flow in a curved cylindrical model that represented a coronary artery. In the finite volume method, the flow domain is discretized into finite volumes instead of finite elements. The equations are first integrated and linearized over a finite volume and are then combined to produce the algebraic equations for the entire flow field. A well-documented study about image-based CFD modelling in realistic arterial geometries was presented by Steinman [24]. The medical imaging and image-processing techniques available to acquire the necessary geometric and functional boundary conditions were reviewed and the issues related to accuracy, precision, and modelling assumptions were also examined.

Banerjee et al. [25], using both experimental and computational simulations, investigated the hemodynamic diagnostics of coronary stenosis. Also, computational models were used to calculate the pressure drop at the same location and the results were compared with the experimental ones. The derived system of equations was solved using finite elements. The authors concluded that the calculation of the pressure drop could potentially be used in everyday clinical practice to assess the hemodynamic significance of coronary stenosis.

Formaggia et al. [26] suggested an alternative approach: the one-dimensional models for blood flow in arteries. These models could be used as an alternative approach against the complex three-dimensional models or in conjunction with them in a geometrical multiscale fashion. Their computational complexity was several orders of magnitude lower than that of multidimensional models based on the coupling of the Navier–Stokes equations for the flow field in the arterial lumen and a mechanical model for the vessel-wall displacement, at the price of providing just averaged information. They successfully investigated the effects on pulse waves of the geometrical and mechanical arterial modification but they lacked to provide flow-field details sufficient to permit a reliable calculation. In a multiscale approach, one-dimensional models may be coupled on the one hand with lumped-parameter models based on a system of ordinary differential equations or, on the other, to three-dimensional fluid-structure models [18]. However, these models had a problem in the calculation of bifurcations and possible discontinuous wall properties. To this purpose, a domain decomposition approach was adopted and the conditions which ensure the stability of the coupling were provided. This numerical method was based on a finite element Taylor–Galerkin scheme combined with operator-splitting techniques, to perform several test cases for the assessment of the proposed models.

F.N. van de Vosse et al. [27] presented a combined arbitrary Lagrange-Euler fictitious domain (ALE-FD) method for fluid-structure interaction problems in cardiovascular biomechanics. An arbitrary Lagrange-Euler (ALE) method originating from the modeling of metal-forming processes can be used to accommodate the deformation of the fluid domain surrounded by a moving/deforming solid and it has been used for fluid-structure interaction problems in vascular systems in the past. For both blood flow and solid mechanics of vascular tissue, the performance of tetrahedral and hexahedral Crouzeix–Raviart elements was evaluated. Comparable convergence results were found, although, for the test cases, the hexahedral elements were considered more accurate. Another study which uses the same technique was held by Figueroa et al. [28]. A new method was developed to simulate blood flow in 3D deformable models of arteries. The method couples the equations of the deformation of the wall with the equations governing the fluid motion using basic assumptions of a thin-walled structure. Furthermore, the coupling was done in such a way that the degrees of freedom of the wall and the fluid boundary are enforced to be identical. While the aforementioned method did not produce results that are equivalent to ALE formulation, since they utilized a fixed fluid mesh and linearized kinematics, the algorithmic simplicity enables the solution of larger sections of the vasculature as it is often required in surgical planning. Another study carried out by Politis et al. [29] examines the ESS distribution in idealized vessels with composite arterial coronary grafts. In this study, flow simulations were assumed to be transient and the finite element method was employed for the numerical solution. Several degrees of stenosis as well as the number and the configuration of the grafts were used to investigate the effect of the blood flow on the arterial stenosis. Kim et al. [30] developed a method that predicts coronary flow and pressure of 3D epicardial coronary arteries by representing the cardiovascular system using a hybrid numerical/analytic closed loop system comprising a 3D model of the aorta, lumped parameter coronary vascular models to represent the coronary vascular networks, three-element Windkessel models of the rest of the systemic circulation and the pulmonary circulation, and lumped parameter models for the left and right sides of the heart. This study simulated the coronary flow and pressure as well as aortic flow and pressure in subject-specific models by considering interactions between these models and the heart, impedance of the systemic arterial system and the pulmonary system, and impedance of coronary vascular beds. The computed coronary flow and pressure and the aortic flow and pressure waveforms were realistic.

A multiscale approach to model blood flow had been proposed by Moreno et al. [31]. This approach suggested a finite element discretization to represent the coarse scales (macro size) and it was coupled to smoothened dissipative particle dynamics that capture the fine-scale features (micro-scale). Blood was assumed to be incompressible, and flow was described through the Navier–Stokes equation. The proposed coupling was tested using two benchmark problems in fully coupled systems. The suggested algorithm can be used with any particle-based method able to solve the Navier–Stokes equations. Further referents of the model can be incorporated to explicitly include blood constituents and the non-Newtonian behavior of blood. A solver based on a confirming finite element method for 3D- and 1D-coupled incompressible flow problem was developed by Dobroserdova et al. [32]. In this study, a finite element method and a geometrical splitting numerical algorithm were studied, while coupling conditions were conserved for the energy of 1D-3D-coupled models of fluid flows and a simulation of a blood flow over an inferior vena cava filter was held.

In the first studies of blood flow modeling, the blood was treated as Newtonian with constant viscosity as described in the previous section. Experimental investigations over many years showed that blood flow exhibits non-Newtonian behavior such as shear thinning, thixotropy, viscoelasticity, and yield stress. Its rheology is influenced by many factors including plasma viscosity. It was suggested in several studies [33] that the blood at a low shear rate (0.1 seconds−1) exhibits yield stress and behaves like a Casson-model fluid. Theodorou and Bellet et al. [34] presented a numerical method for determining the velocity distribution, the pressure drop along the tube, and the wall stress. The authors accepted the "power-law" model by considering a steady or unsteady laminar, axisymmetric flow of a pseudoplastic fluid through a locally constricted tube. A finite element difference algorithm and an approximate solution were used.

Power-law model is a type of generalized Newtonian fluid (time-independent non-Newtonian fluid) for which the shear stress, $\tau$ is given by:

$$\tau = K\left(\frac{\partial u}{\partial y}\right)^{n}, \tag{9.1}$$

where $K$ is the is the flow consistency index (SI units Pa sn), $\left(\frac{\partial u}{\partial y}\right)$ is the shear rate or the velocity gradient perpendicular to the plane of shear (SI unit $s^{-1}$), and $n$ is the flow behavior index (dimensionless).

Generalized Newtonian fluids satisfy the following rheological equation:

$$\tau = \mu_{\text{eff}}(\dot{\gamma})\dot{\gamma}, \tag{9.2}$$

where $\tau$ is the shear stress and $\dot{\gamma}$ is the shear rate. The quantity:

$$\mu_{\text{eff}} = K\left(\frac{\partial u}{\partial y}\right)^{n-1}, \tag{9.3}$$

represents an apparent or effective viscosity as a function of the shear rate (SI unit Pa s).

Gijsen et al. [17] proceeded to a numerical and experimental investigation about the influence of the non-Newtonian properties of blood on the flow in large arteries with unsteady flow in a 90° curved tube. Due to the complexity of the problem, several studies applied computational methods to solve the equations of motion for the flow in curved tubes. In this investigation, the time-dependent flow rates for the Newtonian and the non-Newtonian blood analog were identical. For the numerical computation, a Carreau–Yasuda model was employed to accommodate the shear-thinning behavior of the Xanthan gum solution. The dimensionless viscosity ($\eta^*$) could be written as

$$\eta^* = \frac{\eta - \eta_\infty}{\eta_0 - \eta_\infty} = \left[1 + \left(\lambda_{\text{cy}}\dot{\gamma}\right)^{\alpha}\right]^{(n-1)/a}, \tag{9.4}$$

where ($\eta^*$) is viscosity and $\dot{\gamma}$ is the shear rate. This model describes the behavior of fluids with Newtonian plateaus at low ($\eta_0$), high shear rate ($\eta_\infty$), and a transition area governed by the parameters $n$, $a$, and $\lambda_{\text{cy}}$.

The experimental results showed that the non-Newtonian flow cannot be predicted with a Newtonian model with a viscosity of ($\eta_\infty$). Finally, a comparison between the non-Newtonian fluid model and a Newtonian fluid at a rescaled Reynolds number was presented. Comparison reveals that the character of flow of the non-Newtonian fluid was simulated quite well by using the appropriate Re number. The value of Re for the non-Newtonian flow, based on the characteristic viscosity, was about three times lower than the Re number for the Newtonian flow.

Leuprecht and Perktold in 2001 [35] had implemented computer simulation of non-Newtonian effects on blood flow in large arteries. The influence of the viscoelastic effects on blood flow was studied numerically. The present study focused on the influences of the shape of the flow domain on the viscoelastic velocity patterns in large arteries. The computer simulation was based on two different viscoelastic models to demonstrate non-Newtonian flow behavior of blood: an Ordroyd-B model and a modified Oldroyld-B model to include shear thinning effects. The advantage of the chosen model was that it permitted the separate investigation of the influence of shear thinning effects and elastic behavior. The governing equations were solved by means of an upwind stabilized Galerkin FEM. The results showed that the effect of the viscoelastic properties depends on the shape of the flow domain in large arteries and the indicated significant effect in the stenosed model. The flow through the curved tube showed a minor quantitative viscoelastic effect.

A comparative study of blood flow simulation in coronary arteries using Newtonian and non-Newtonian blood models was performed [36]. This study investigated the pulsatile blood flow through four different right coronary arteries, which have been reconstructed from biplane angiograms. A non-Newtonian blood model (the Generalized Power Law), as well as the usual Newtonian model of blood viscosity, was used to study the ESS in each of these arteries over the entire cardiac cycle. The results were presented in terms of the ESS distribution, non-Newtonian importance factor, and particle path simulation. It was found that the ESS was low at the proximal part and increasing toward the outlet due to the tapering effect of arteries. Based on the ESS distribution study it was found that Newtonian blood flow modeling was a good approximation.

Seven non-Newtonian molecular viscosity models, plus the Newtonian one, can be used. In particular, seven non-Newtonian models are applicable: The non-Newtonian power law, generalized power law, Carreau and Casson, modified cross-blood viscosity models (Carreau–Yasuda), and Walburn–Schneck Law. The capabilities and limitations of various molecular viscosity models, in the left coronary arterial tree, were previously analyzed [37] through molecular viscosity, local and global non-Newtonian importance factors, endothelial shear stress (ESS) and ESS gradient (ESSG). The results showed that the non-Newtonian power law, generalized power law, Carreau and Casson, and modified cross blood viscosity models gave comparable molecular viscosity, ESS, and ESSG values. The power law and Walburn–Schneck models overestimated the non-Newtonian global importance factor $I_G$ and underestimated the area-averaged ESS and ESSG values. The non-Newtonian power law and the generalized power law blood viscosity models were found to approximate the molecular viscosity and ESS calculations in a more satisfactory way.

The average global non-Newtonian importance factor $I_G$ was defined as:

$$I_G = \frac{1}{N} \cdot \frac{\left| \sum_{i=1}^{\text{totalgridnodes}} \left(\mu - \mu_\infty\right)^2 \right|^{1/2} * 100}{\mu_\infty}. \tag{9.5}$$

The $I_G$ expresses the relative difference from the Newtonian value in percentage terms of each value of molecular viscosity. Values of $I_G$ greater than 0.25 indicate non-Newtonian flow behavior.

During the last decade, with the simultaneous development in medical imaging blood flow simulation studies have been largely based on realistic geometric data using realistic patient-specific 3D models of coronary vasculature. The aforementioned 3D models were derived from data through coronary imaging modalities such as angiography (ICA), CTCA, magnetic MRA, IVUS, and OCT. These 3D imaging techniques have enabled the development of the quantification of blood flow in patient-specific anatomic and physiological models. Noninvasive *in vivo* methods include CTCA and MRI. Thus, they are important modalities for clinical and research investigations in cardiovascular fluid mechanics.

### 9.2.3 Plaque Growth Modeling

The model of Calvez et al. was refined later by Sakellarios et al. (2011) by the addition of the nitric oxide effect in the atherosclerotic process [38]. Nitric oxide (NO) is a signaling molecule, produced by the endothelial nitric oxide synthases (eNOS) that were found to control arterial contraction and dilation after enabling arterial SMC contraction and dilation, respectively. More specifically, Sakellarios et al. achieved improved plaque growth modeling by modeling LDL permeability as a multivariate function that also depends on nitric oxide concentration. Their reasoning lay in the assumption of a higher endothelial LDL flux in case of arterial dilation since endothelial cells' interspace increases during arterial dilation. Their effort also includes the implementation of their model in a 3D reconstructed artery and the validation of the predicted results with those from real follow-up arterial segments. A refined mathematical model by Calvez et al. was also proposed by Siogkas et al. (2011), presenting a more valid equation for cytokine production based on experimental data [39]. The refined model was implemented in a 3D reconstructed arterial segment and a validation strategy was proposed. According to that, areas of increased arterial growth were evaluated and compared between the realistic follow-up and the simulated segments. In another study, the effects of several biological factors on the arterial wall growth of a 3D arterial segment were examined [40]. Among the considered variables are tachycardia (implementation of a pulsatile blood flow as a boundary condition), hypertension, high LDL blood concentration, and high endothelial permeability. Their effort was based on the previously presented model of Siogkas et al., which was further refined regarding the function of endothelial permeability and the included parameters' values, using the latest experimental data. Cilla et al. (2013) proposed the inclusion of several biological factors in the atherosclerotic process, and using the model of Calvez et al., they resulted in a more complex mathematical model [41]. In the lumen region, the LDL and monocyte dynamics are considered using the convection-diffusion-reaction equations, changing the preserved assumption of a steady LDL and monocyte in blood. In the arterial wall, monocyte dynamics are now considered, releasing the model from the preserved assumption of monocyte instantaneous differentiation to macrophages. Other considered species are the contractile and synthetic SMCs and the collagen which greatly contribute to plaque growth. It was assumed that contractile SMCs differentiate into synthetic SMCs in case of cytokines' presence, in order to restore the areas of inflammation. In this process, synthetic SMCs secrete collagen, which acts as a tissue-bonding substance. In this instance, plaque is assumed to be consisted of foam cells, synthetic SMCs, and collagen. The arterial wall thickening is performed as in Calvez et al., while the model is implemented in a 3D ideal segment. Hao and Friedman developed a plaque growth model, considering the effects of several biological factors in the atherosclerotic process [42]. Their model considers the species dynamics of the arterial wall domain, neglecting the blood flow effects on endothelium, which can be justified in their case of a 3D-designed ideal segment that presents a uniform distribution of ESS. The species dynamics are modeled by the diffusion-reaction equations, while no solvent fluids are considered such as blood or plasma. In their model, LDL is also the culprit of atherosclerosis, which infiltrates and accumulates within the arterial wall. The infiltrated LDL is oxidized by the free radicals of the arterial wall, whose dynamics are considered, which is unique in literature. Another novelty of their work is the inclusion of the dynamics of the monocyte chemoattractant protein (MCP-1), which is secreted by endothelial cells and increases the endothelial monocyte flux. In their turn, monocytes are considered implicitly, as their instantaneous differentiation to macrophages is assumed in the arterial wall. Macrophages are attracted to the oxidized LDL, which they uptake and transform into foam cells. In their turn, foam cells secrete chemokines that increase macrophage fluxes in the arterial wall. The chemotactic forces of MCP-1 and platelet-derived growth factor (PDGF) along with the haptotaxis by extracellular matrix (ECM) move SMCs from media to intima. PDGF is secreted by macrophages, foam cells, and SMCs. ECM is produced due to the presence of matrix metalloproteinase (MMP), which is a product of SMCs, while it is inhibited by the tissue inhibitor of metalloproteinase (TIMP), which is a product of both macrophages and SMCs. Plaque growth is increased by the activation of T-cells, which occurs due to the presence of interleukin IL-12, which is a product of macrophages. Along with LDL, the available high-density lipoproteins (HDL) infiltrate within the arterial wall, where free radicals oxidize them. This reaction reduces the concentration of the free radicals, which are necessary for LDL oxidation, and therefore decreases the plaque growth rate. Using these biologic reactions in a system of equations and implementing them into a 3D-designed ideal segment, Hao et al. present illustrations of the arterial wall thickening and the lumen stenosis. Sakellarios et al. extended their previous plaque growth model, inserting the effects of HDL and macrophage migration inhibitory factor (MIF) [43]. Specifically, the oxidation rate of LDL was correlated with the arterial wall HDL concentration, and the MIF expression was correlated with the size of the lesion area. A unique mathematical model was developed by Yang et al., who coupled the fluid dynamics of blood and substances with the arterial wall deformation into a Lagrangian–Eulerian formulation [44]. Their mathematical model considers the convection-diffusion-reaction equations for the dynamics of monocytes, macrophages, and foam cells, and the coupled Navier–Stokes and stress–strain equations for the dynamics of the blood and the arterial wall. To simulate a more realistic deformation, they considered the arterial wall as a neo-Hookean material with parameters that depend on the substances' concentrations. Last but not least, the proposed model enables the simulation of the plaque evolution, while it considers a dynamic arterial wall elasticity. A specialized model for plaque localization was presented by Alimohammadi et al., which is based on a variable LDL permeability, which depends on several flow-related metrics [45]. Specifically, the LDL pathways into the arterial wall are described extensively, while an additional description of the derivation of the Kedem–Katchalsky equations is provided, which are used to model the endothelial flow rates of LDL. Using experimental data, they correlated the endothelial cells' shape with several flow-related metrics, and later to the correlation of leaky cells to specific shapes. They validated their results after the examination of real arterial segments. To be more precise, the flow-related metrics include the time-averaged wall shear stresses (TAWSS), the oscillatory shear index (OSI), and the shear stress index (SSI). The work of Yang et al. [44] was later extended by Kafi et al. who also presented transient lumen stenosis by plaque growth, considering although a non-Newtonian blood flow and different elastic properties of the arterial wall [46]. Specifically, the blood viscosity is described by the Carreau–Yasuda model, while the arterial wall was considered isotropic, except for the fibrous cap and the lipid pool, which are considered hyperelastic materials and modeled by the Mooney–Rivlin model. Finally, the differences between Newtonian and non-Newtonian blood flows and between a fixed and deformable arterial wall were presented after the implementation of their model to a 3D-designed axisymmetric segment that included a stenosed region. Anlamlert presented another model of plaque growth, specializing in fibrous cap growth [47]. In their model, the LDL is infiltrating the arterial wall with a steady endothelial flux. The considered substances of the atherosclerotic process include LDL, oxidized LDL, macrophages, foam cells, and SMCs. The evolution of the atherosclerotic process is based on the inflammatory signaling of two chemoattractant substances, a macrophage attractant, that derives from oxidized LDL and a smooth muscle cell attractant, that derives from macrophages. Finally, their model is implemented to a 2D-designed arterial segment, which includes a fibrous cap. Chalmers et al. introduced an analytical pathway of HDL in the atherosclerotic process [48]. The considered substances include the oxidized LDL, HDL, macrophages, foam cells, monocyte chemoattractants, and cytokines. Apart from these, the LDL and monocytes are also considered, but implicitly, as their instant differentiation to oxidized LDL and macrophages is considered respectively. Their novelty concerns the ability of the foam cells' differentiation back to macrophages after the reverse cholesterol transport by HDL. Therefore, using this approach, the decrease of plaque volume is enabled. Although their model was applied in one dimension, their mathematical model of the HDL effects is novel and can be easily applied in 3D segments. Guo et al. examined the effects of neovascularization and intraplaque hemorrhage on plaque growth [49]. Their rationale is that neovascularization and intraplaque hemorrhage can be simulated by separating the fluid dynamics of the arterial wall into intravascular, interstitial, and transvascular flow. In their model, the Poiseuille's law is implemented to define the intravascular flow, while the transvascular flow, which represents the fluid leakage into the arterial wall, is modelled by Starling's law. The interstitial flow is modeled by the Darcy's law that accounts for flow in porous domains, such as the arterial wall. The substances of the atherosclerotic process include endothelial cells, macrophages, SMCs, vascular endothelial growth factors, MMP, and ECM. Plasma dynamics are modeled by the convection-diffusion equation, while a special criterion was assumed that correlated the plasma concentration with the hemorrhage and neovasculature. Specifically, the high plasma concentration was correlated with an increased macrophage concentration due to the initiation of inflammation by intraplaque hemorrhage. Their work includes illustrations of the arterial wall's substance concentrations, after the implementation of their model in a 2D-designed cross section of a vessel. Watson et al. studied the fibrous cap growth and developed a mathematical model to simulate it [50]. In their approach, the arterial wall was modeled to consist of two discrete phases, a phase of normal concentrations of the substances and a phase of high concentration in SMCs that represents the fibrous cap. In both phases, the mass and momentum conservation equations are implemented that define the flux between the two phases. This process initiates by the PDGF-induced SMC proliferation that results in a large accumulation of SMCs in specific areas, transforming them into fibrous caps. Their effort included the coupling of the mass and momentum conservation equations with the PDGF convection–diffusion-reaction equations. To conclude, the results of this study include the SMC and PDGF volume fraction distributions in a case study of a one-dimensional domain.

The most complete multilevel continuum-based model of atherosclerotic plaque growth model has been presented by Pleouras et al. [22]. In particular, in the first level of the model, blood flow is simulated by adopting the Navier–Stokes equations to assess the ESS distribution on the endothelial membrane. The ESS values are used as input for the simulation of plaque growth and biological and biochemical processes as a second level. These include: (i) the LDL, HDL, and monocytes transport and penetration through the endothelial membrane in the arterial wall, (ii) the oxidation of LDL particles considering the atheroprotective effect of HDL, (iii) the inflammation triggering through the presence of macrophages, (iv) the formation of foam cells, and (v) the presence of collagen and proliferation of smooth muscle cells. These processes are simulated employing convection–diffusion-reaction equations. In the final level, the total plaque volume is estimated by the simulation of the smooth muscle cells, foam cells, and collagen and it is used to simulate the thickening of the arterial wall causing the lumen narrowing. A case example is presented in Figure 9.2.

### 9.2.4 Agent-Based Modeling

Another way of modeling the biological processes in the cardiovascular domain, especially at the microscale level is based on agent-based models (ABMs). This approach enables the simulation of interactions of autonomous agents evaluating their influence on the overall system. The use of different autonomous agents enables the modeling of complex systems and phenomena increasing the application of the ABM technique. The proper implementation of an ABM model requires the accurate definition of the agents' behaviors and interactions. In the cardiovascular domain, which by default is characterized by a complex space and time domain, ABM could be significantly useful to understand biological interactions. In particular, the processes taking part in the cardiovascular system can be in the range of meters (organism), mm (organs such as arteries, veins, or the heart) to nm (molecules and cells). Moreover, there are processes that can be performed in milliseconds, while others such as the evolution of a disease may take years. Though the definition of ABM characteristics is extremely difficult, some studies attempted to model the multiscale nature of CVD and atherosclerosis using ABM.

ABM is a powerful tool that is used to inform policy or decisions in many fields of practical importance. This approach enables the simulation of interactions of autonomous agents evaluating their influence on the overall system. The use of different autonomous agents enables the modeling of complex systems and phenomena increasing the application of the ABM technique. The proper implementation of an ABM model requires an accurate definition of the agents' behaviors and interactions. In the cardiovascular domain, which by default is characterized by a complex space and time domain, ABM could be significantly useful to understand biological interactions. In particular, the processes taking part in the cardiovascular system can be in the range of meters (organism), mm

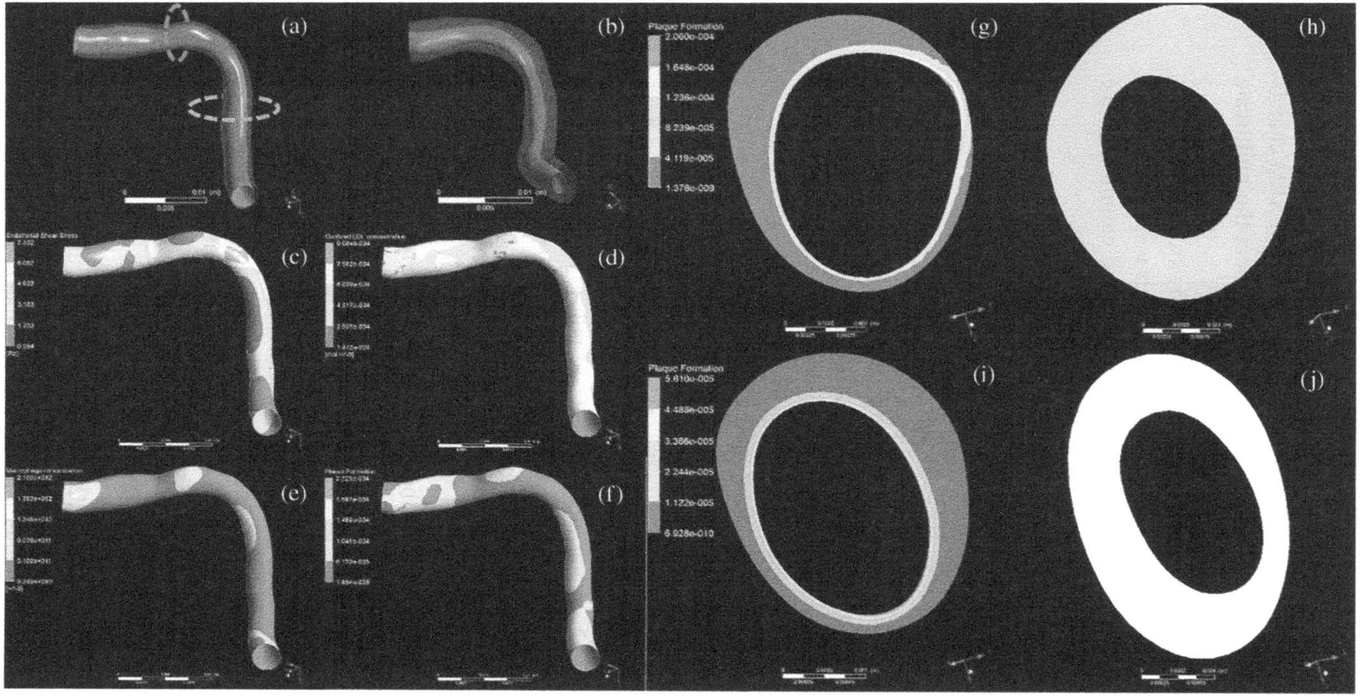


**Figure 9.2** Case example of plaque growth modeling [22]. (a and b) show the reconstructed arteries (red: lumen; transparent blue: arterial wall) for baseline and follow-up time point, respectively. The lumen stenosis is presented clearly at the follow-up reconstruction (b). (c–f): Distribution of endothelial shear stress, oxidized LDL concentration accumulation in the arterial wall, macrophages concentration, and plaque formation. Regions of low ESS present higher accumulation of oxidized LDL and inflammatory molecules. (g, h): Cross section with plaque formation variable at the baseline (g) and the corresponding follow-up cross section (h). This cross section corresponds at the red dotted line of panel (a). (i, j): Cross section with plaque formation variable at the baseline (i) and the corresponding follow-up cross section (j), where this cross section is located at the green dotted line of pane (a). (h, j): Cross sections of increased plaque area and lumen area reduction as found realistically at the reconstructed arteries.

(organs such as arteries, veins, or the heart) to nm (molecules and cells). Moreover, there are processes that can be performed in milliseconds, while others such as the evolution of a disease may take years. Though the definition of ABM characteristics is extremely difficult, some studies attempted to model the multiscale nature of CVD and atherosclerosis using ABM.

Hence, ABM has become a major modeling trend in many areas, such as epidemics, public health, biology, supply chains, consumer goods markets, financial markets, land use/cover and agricultural policy, ecosystem and natural resource management, environmental planning and policy, and many more.

#### 9.2.4.1 Key Components of Agent-Based Modelling

The fundamental components of any ABM are:

i) ***Agents***
Agents are autonomous entities which interact with their environment and other agents. They are characterized by attributes and behaviors. Their attributes are external or external states that can be dynamically changed by their behavior (mutable or immutable over time). There are two types of agents: (i) mobile or (ii) stationary. The first one can move within the simulation space but the second has no moving capability. Furthermore, according to cognitive level of agents, they can be classified into four types – reflective or myopic agents, utility-based agents, goal-based agents, and adaptive agents [51].

ii) ***Rules***
The interaction between agents and the agents' responses to changes in their environment is defined by a set of behavioral rules. Each rule describes an agent's reaction to set a specific environmental condition (agent–environmental interactions), its neighbor's state (agent–agent interactions), and/or its own internal state (agent–self interactions). The rules may navigate not only individual decisions but also changes in the environment. These rules are generally derived from the literature or experimentally [51].

iii) ***Environment***
The simulation space represents the agents' external environment, i.e. the space in which agents exist and it contains nonagent parameters for the experimental question being addressed. Usually, most simulations have more than one environment. Environments can be either static or dynamic [51].

iv) ***Time-scale***
Simulations go through several discrete time steps which are called ticks. A tick can have different durations, e.g. milliseconds, seconds, days, and years. At each time step, a core set of processes is repeated: repetitively agents make decisions and exhibit behaviors. The choice of how each time step is scaled is an important decision and drives the type of questions an individual model can answer [51].

#### 9.2.4.2 Agent-Based Modelling and Simulation Approach

Mathematical models and computational simulations are powerful tools that can facilitate the improvement of preventive solutions and the development of new strategies. Systems science methodologies such as system dynamics, discrete-event simulation, network analysis, and agent-based modelling have the potential to inform decision-makers on how chronic health conditions develop and their consequences. Particularly, by simulating the behavior of multiple autonomous entities (agents or cellular components), as well as their interactions with each other and with their local environment, they can predict higher-level "emergent" patterns [52].

The key features of ABM and their simulation approach are the following [51]:

1) **Bounded rationality:** to take into account real-world complexity, characterized by nonstationary systems and endogenously changing environments, it is possible to consider only bounded rationality assumption, that is defined by Simon [53] as a choice that takes into account the cognitive limitations of the decision-maker – limitations of both knowledge and computational capacity.
2) **Exhibition of emergence:** the emergence concept is central to complex systems and refers to how the system behavior arises from the interaction among its parts and cannot be predicted from knowledge about the parts in isolation [54].
3) **Heterogeneity:** by explicitly modelling every single agent, their specific characteristics and behaviors, ABM allows a high degree of heterogeneity, which characterized real-world complex systems [55].
4) **Bottom-up perspective:** the complex macrosystems emerge from repeated interactions among their constituents' units.
5) **Comprehensiveness:** by modelling each agent and its behaviors, it is possible to analyze the evolution over time of any agents in the model.
6) **Networked interactions:** agents interact directly because their current decisions depend on the previous choices made by other agents in the population. These interactions may contain structures, such as subgroups of agents or local networks [56].
7) **Uncertainty:** decision making in complex real-world systems showed to be cumbersome because there are various sources of uncertainty, which are difficult to be considered [56].

#### 9.2.4.3 Problem Definition

Arterial walls can develop plaques (atheroma) comprising derived foal cells, lipids, fatty substances, cholesterol, cellular waste products, elastin, collagen, fibrin, calcium, and other constituents. At the beginning, the plaque progression induces artery outward remodeling to accommodate the volume of the plaque without lumen area reduction. After the initial progression, remodelling happens toward inside the lumen causing the lumen narrowing which is called stenosis.

Initial damages to the endothelium cause an increase in permeability of the layer. Lipoprotein LDL particles are absorbed into the intima by passing the endothelium lining. Some of these LDL particles become oxidized and this attracts monocytes which differentiate into macrophages that uptake the particles. Cytokines are released, which, in turn, attract more monocytes. The sticky cytokines on the artery wall capture platelets that accumulate at the site where monocytes infiltrated into the artery. Macrophages and foam cells, platelets and endothelium release growth factors stimulate the migration and proliferation of smooth cells and fibroblasts.

The problem stated here describes numerous components contributing to plaque creation and progression, each one with proper characteristics, behavior, and rulesets. For this reason, we selected ABMS as the proper approach to reproduce the evolution of plaque progression and artery reshaping (dynamical system) by simulating the behavior of autonomous cellular components (agents).

Traditionally, systems biology has focused on the study of gene and protein interactions at the single-cell level. However, biological systems are made up of many spatial and temporal scales, each rich and complex. The global behaviors of multicellular biological systems result from individual cell behavior and the interactions among the cells. Cells are live creatures that move, interact, proliferate, and die, and their behaviors determine the dynamics of multicellular biological systems. Particularly, interactions at lower scales of biological organization induce new, emergent properties that are observable at higher scales [57]. Hence, modelling biological systems requires incorporating these complex interactions among the individual cells and environmental factors. In this context, several research groups used ABMs. Agent-based modelling is a computational modelling approach useful not only for understanding complex biological systems but also for facilitating *in silico* experiments in a cost- and time-efficient manner.

#### 9.2.4.4 ABM Applications in Cardiovascular Systems

When the response of complex biological systems strictly depends on cellular behaviors, which, in turn, are affected by the changes in the microenvironmental factors, a multiscale modeling approach is mandatory. Therefore, multiscale approaches utilizing discrete methods such as cellular automata (CA) models and ABMs have been used to analyze different aspects of cardiovascular tissues/systems regulations [58].

One example is the two-dimensional ABM that has been presented by Pappalardo et al. [59]. A model to describe the very early stage of atherosclerosis was developed. This model reproduced clinical and laboratory parameters associated with atherogenesis, considering all the crucial entities, and their interactions, that play an important role in the immune system processes that control atherogenesis. Furthermore, they considered small molecules like interleukins or signaling molecules and more complex molecules.

In the cardiovascular domain, ABM was used mainly for the simulation of cell interaction or movement. Bailey et al. presented an ABM to simulate the trafficking of inflammatory cells at the microvasculature [60]. To this purpose, a network of the microvasculature was constructed using the extracted architecture of the mouse muscle vasculature. Blood flow dynamics were performed whose results were integrated into the ABM to simulate the trafficking of the monocytes. Monocyte's transport was based on the flow parameters as well as on the local concentrations of the cytokines. A two-way interaction was simulated between the monocytes and the local mechanical effects. The model predicts the monocyte adhesion, rolling, and extravasation and its results were validated using molecular knockout models with good agreement between them. A more detailed ABM model was presented by Papalardo et al. [2] considering four different clinical scenarios for a set of 200 virtual patients: (i) normal LDL concentration level, (ii) high level of LDL concentration, (iii) high level of LDL concentration treated with lipid-lowering medication, (iv) increased oxidative stress due to alcohol, smoke, etc. The developed SimAthero model considers cellular and molecular entities and their interactions. The development is based on literature-found parameters as well as on fine-tuning of the model regarding the unknown set parameters. Though the main atherosclerotic mechanisms are simulated using SimAthero, few things are available about the assumptions of the model. The results of SimAthero demonstrate the differences between the four clinical scenarios, without however being validated using real clinical data. In the same year, Deo et al. proposed an ABM for the simulation of the inflammatory processes in atherosclerosis, particularly focusing on plaque formation and rupture.

A 3D ABM was developed by Olivares et al. [61] to simulate relevant cellular and molecular phenomena involved in the early formation of foam cells in the intima layer of arteries. The vascular wall often maintains mechanical homeostasis through short-term changes in vascular tone and long-term changes in ECM microstructure and cell phenotype. Thorne et al. [62] developed a 2D-ABM at the cell level suitable for coupling with a continuum-level model to describe cell-mediated changes in the geometry, composition, and properties of an adapting vascular wall.

A hybrid biological modeling framework that combined a two-dimensional lattice-free ABM and finite element model was proposed by Zahedmanesh et al. [63]. They studied the collagen matrix turnover following stent-induced arterial injury and the subsequent development of ISR. Specifically, they evaluated the influence of stent deployment diameter and stent strut thickness on the level of ISR.

Another hybrid computational framework that integrates both ABM and partial differential equations (PDEs) was implemented by Garbey et al. [64]. On one hand, the fixed grid ABM describes well the discrete elements that comprise the tissue at a cellular level. The model tracks cell dynamics, focusing on cell proliferation, cell apoptosis, and ECM production/degeneration. On the other hand, PDE accurately describes continuum mechanics, particularly computes the hemodynamic forces and stress-strain relationships that define the environment. A more realistic model has been presented by Garbey et al. [65] which includes several assumptions, particularly related to cells' motion. This motion, in fact, should be computed based on a continuum space rather than on a discrete grid.

A recent example of the application of ABM to plaque progression is provided by Bhui and Hayenga [66]. They presented a multiscale model of leukocyte TEM and plaque evolution in the left anterior descending (LAD) coronary artery. The approach integrates cellular behaviors via ABM and hemodynamic effects via CFD. This multiscale and multiphysics approach appropriately captured and integrated the spatiotemporal events occurring at the cellular level to predict leukocyte transmigration and plaque evolution. Finally, a multiscale modelling framework of atherosclerosis that replicates the hemodynamic-driven arterial wall remodelling and plaque formation is developed by Corti et al. [67]. The CFD simulation computed the hemodynamic in a 3D artery model, while 2D ABMs simulated cell, ECM, and lipid dynamics in multiple vessel cross sections. The ABM outputs were mostly driven by cell and ECM dynamics, largely affecting the lumen area. A subset of parameters was found to affect the final lipid core size, without influencing cell/ECM or lumen area trends.

### 9.2.5 Discrete Particle Dynamics

A general goal of biological systems is to acquire a detailed understanding of the dynamics of living systems by relating functional properties of the whole systems with the interactions of their constituents. This goal is tackled through mathematical and algorithmic approaches often based on differential equations. Within this area, models that define the cell as the fundamental unit of model development, as well as discrete representations of different model entities are important to plan *in vitro* experiments and generate new testable hypotheses.

To date, approaches for computational modelling of multicellular, multiscale biological systems can be grouped into two categories: continuum and discrete approaches [68]. The former are mathematical models since they express the interaction between systems properties in the form of mathematical equations and continuum models of time and space [69] as the approaches presented in the first half of this chapter. These models have been used to describe the behavior of macroscopic systems in which a great number of molecules are assumed to exist to enable their description with equations and average changes [70]. There are many advantages of this approach, but it may not capture the discrete nature of biological systems or solve complex processes which involve many variables [71].

In contrast to this strategy, the discrete approach does not attempt to represent the cells as a single continuous variable but rather determines complex system behavior, which is determined by the specification of discrete states instead of continuous variables. In particular, the discrete models can indicate how and why the system evolves, while continuous models describe the numerical changes of the variables defining the system [72]. The two most widely used discrete methodologies as it refers above for modeling multiscale biological systems are the CA models and the ABMs.

### 9.2.6 Multiscale Model of Drug Delivery/Restenosis

Drug-eluting stents (DES) are vascular devices used for the treatment of coronary artery disease (CAD). The global coronary stent devices' market size is estimated at approximately eight billion euros (2019). Currently, while the cardiovascular biomedical industry invests in developing new DES, several time- and the cost-consuming obstacles occur; three to seven years are required on average for bringing these devices to the market, while there are associated high costs estimated between 31 and 94 million euros. This is due to those essential but costly elements of the development and commercialization lifecycle for DES, which establish their safety and efficacy prior to the product being released to the market. Provided that billions of stent deployment procedures are being carried out worldwide, even rates of complications in low percentage, represent a large number of patients. The motivation for developing new DES has been driven by the need for satisfactory acute/short- and mid-/long-term results and reduction of the ISR rate.

Coronary stent implantation has significantly improved percutaneous coronary intervention (PCI) and enabled the management of early complications of plain balloon angioplasty. However, several complications have been observed, such as ISR arising from neointimal hyperplasia. The DES were specifically designed to reduce hyperproliferation within stents as a mechanism to overcome the high rate of ISR that is estimated to vary from 3% to 20%. DES have attracted great interest and have been utilized in several clinical studies evaluating their efficacy and safety [73, 74]. The design and development of effective and safe DES require the identification and selection of appropriate drugs and their delivery under suitable release kinetics so as to achieve the retention of the drug at a sufficient concentration [44]. In addition, stent design and coating polymers also play an important role in the drug release mechanisms.

Computational models of drug delivery are an effective tool for providing additional information related to drug-release kinetics for enhancing the performance of novel implants and developing strategies to improve drug therapy. These models can capture tissue barriers and drug uptake despite the limitations of the thus-far-developed continuum framework of pharmacokinetics and pharmacodynamics [75]. Early studies, originally focused on linear binding relations, modeled the drug delivery in two cases in order to have a small computational domain and avoid the complexity of 3D [76]. The objective of such studies was to compare different mechanisms of drug transport within tissue and determine the dominant process on clinical indicators based on drug and model specifications. Later these approaches were extended to two-dimensional studies in idealized geometries [77] focusing on the effect of three-dimensional hemodynamics, relative positioning, and drug distribution in the arterial tissue [78].

Other studies incorporate the coating and model the drug release from the coating and the subsequent drug uptake in the arterial wall. In those models, several factors are investigated such as the effect of drug diffusivity in the coating and the arterial wall and the coating thickness. There also exist some studies that focus on the creation of a tissue-growth model, integrating the ISR with the FE approach [79]. For instance, another study [63] performed a two-dimensional (2D) computational analysis coupled with an ABM of smooth muscle cell proliferation and ECM generation. A research team [80] integrated a 3D FEM stent expansion model into a lattice-based cell growth model, while another team [16] developed a computational three-dimensional (3D) multiscale model of ISR by accounting for smooth muscle cell proliferation and re-endothelialization fully coupled to blood flow simulations.

#### 9.2.6.1 Benefits of Multiscale Model of Drug Delivery/Restenosis

The development of a computational drug-delivery/restenosis model simulating the medium-/long-term DES performance related to the interaction and the effect of the stent and the drug on the arterial wall physiology can open the door to a whole new concept of personalized treatment approach integrating multidisciplinary efforts; computational modelling, materials science and engineering, and systems biology. It is expected to revolutionize the DES design by incorporating physiological and biological knowledge and capturing the features of individual patients, while in parallel, introducing the concept of the patient-specific model. This approach will overcome the existing limitations of the *in vivo* and animal experiments and provide a much deeper, safer, and cost-effective understanding.

The creation of this multiscale computational model can combine the drug delivery model with a plaque growth model and simulate the effect of drug on arterial physiology in order to predict the ISR progression. In particular, in this advanced *in silico* model of drug delivery/restenosis model for arteries in which

DES have been implanted, it will also be deployed in the arteries and the drug release will be investigated as the drug's interaction with the biological tissue plays an important role. The development of a model of ISR, which is affected by the stent drug, simulates the elimination of the inflammation in the arterial wall. The FE method and/or the finite volumes can be employed to create the advanced multiscale computational model. More specifically, convection–diffusion-reaction equations can be employed to simulate the drug delivery, as well as the drug's effect within the arterial wall and specifically on the pathophysiologic mechanisms of ISR.

In simple words, this model enables the interaction of the stent, the drug and the physiologic mechanisms taking place in the arterial wall. Diffusion–reaction equations will be used for modeling the mass balance between biological species in the arterial wall. It can be therefore formulated in terms of densities/concentrations of a number of important species (SMCs, ECs, matrix-degrading metalloproteinases (MMPs), growth factors (GFs), and ECM). The main parameters used in the model to define the biological interaction between the different species will be calibrated from experimental data.

## 9.3 Conclusions

It has been clear that in the last years, a lot of effort is given to the computational modeling of the CVD. In particular, the complexity of the disease requires a multidisciplinary approach to understand the pathophysiology and the mechanisms that underlie the disease evolution. It has been initially demonstrated that hemodynamics affect the progression of the disease and its establishment and for this reason, the first studies focus on the accurate simulation of the blood flow and the estimation of the wall or endothelial shear stress. However, this hypothesis regards only the initiation stage of the disease and for this reason, more advanced computational approaches are needed such as mass transport modeling or plaque growth simulations. All these approaches in the current literature are based on continuum mechanics and mainly finite elements. Their purpose is to investigate macroscopically the disease progression. In parallel, however, many studies were presented that focus on the microscopic environment of the arterial walls and the cells' behavior. Discrete dynamics and agent-based modeling approaches are the most efficient for this simulation. The future, nevertheless, requires the combination of these two scaling approaches into one in order to become possible to simulate most of the pathophysiology and biology of the arteries. The results will be beneficial for the development of advanced state-of-the-art predictive models for CVD as these will be described in the next chapter.

## References


1 J. Mackay and G. Mensah. "Atlas of heart disease and stroke." http://www.who.int/cardiovascular_diseases/en/cvd_atlas_16_death_from_stroke.pdf 2004

2 Pappalardo, F., Cincotti, A., Motta, A., and Pennisi, M. (2009). Agent based modeling of atherosclerosis: a concrete help in personalized treatments. In: *Emerging Intelligent Computing Technology and Applications. With Aspects of Artificial Intelligence*, 386–396. Berlin, Heidelberg: https://doi.org/10.1007/978-3-642-04020-7_41.

3 Khouri, E.M., Gregg, D.E., and Lowensohn, H.S. (1968). Flow in the major branches of the left coronary artery during experimental coronary insufficiency in the unanesthetized dog. *Circ. Res.* 23 (1): 99–109.

4 Fam, W.M. and McGregor, M. (1969). Pressure-flow relationships in the coronary circulation. *Circ. Res.* 25 (3): 293–301.

5 Zarins, C.K., Giddens, D.P., Bharadvaj, B.K. et al. (1983). Carotid bifurcation atherosclerosis. Quantitative correlation of plaque localization with flow velocity profiles and wall shear stress. *Circ. Res.* 53 (4): 502–514.

6 Ku, D.N., Giddens, D.P., Zarins, C.K., and Glagov, S. (1985). Pulsatile flow and atherosclerosis in the human carotid bifurcation. Positive correlation between plaque location and low oscillating shear stress. *Arteriosclerosis* 5 (3): 293–302.

7 Glagov, S., Weisenberg, E., Zarins, C.K. et al. (1987). Compensatory enlargement of human atherosclerotic coronary arteries. *N. Engl. J. Med.* 316 (22): 1371–1375. https://doi.org/10.1056/NEJM198705283162204.

8 Friedman, M.H., Hutchins, G.M., Bargeron, C.B. et al. (1981). Correlation between intimal thickness and fluid shear in human arteries. *Atherosclerosis* 39 (3): 425–436. https://doi.org/10.1016/0021-9150(81)90027-7.

9 J. H. Forrester and D. F. Young, "Flow through a converging-diverging tube and its implications in occlusive vascular disease. I. Theoretical development," *J. Biomech.*, vol. 3, no. 3, pp. 297–305, May 1970, doi: https://doi.org/10.1016/0021-9290(70)90031-x.

10 J. H. Forrester and D. F. Young, "Flow through a converging-diverging tube and its implications in occlusive vascular disease — II: theoretical and experimental results and their implications," *J. Biomech.*, vol. 3, no. 3, pp. 307–316, May 1970, doi: https://doi.org/10.1016/0021-9290(70)90032-1.

11 S. Cavalcanti, P. Bolelli, and E. Belardinelli, "Pressure drops through arterial stenosis models in steady flow condition," *J. Biomech. Eng.*, vol. 114, no. 3, pp. 416–418, Aug. 1992, doi: https://doi.org/10.1115/1.2891404.

12 Tamamidis, P. and Assanis, D.N. (1993). Three-dimensional incompressible flow calculations with alternative discretization schemes. *Numer. Heat Transfer, Part B* 24 (1): https://www.tandfonline.com/doi/abs/10.1080/10407799308955882 (accessed 29 November 2022).

**13** D. F. Young and F. Y. Tsai, "Flow characteristics in models of arterial stenoses – II. Unsteady flow," *J. Biomech.*, vol. 6, no. 5, pp. 547–559, Sep. 1973, doi: https://doi.org/10.1016/0021-9290(73)90012-2.

**14** M. Siouffi, R. Pelissier, D. Farahifar, and R. Rieu, "The effect of unsteadiness on the flow through stenoses and bifurcations," *J. Biomech.*, vol. 17, no. 5, pp. 299–315, Jan. 1984, doi: https://doi.org/10.1016/0021-9290(84)90025-3.

**15** B. B. Lieber and D. P. Giddens, "Apparent stresses in disturbed pulsatile flows," *J. Biomech.*, vol. 21, no. 4, pp. 287–298, Jan. 1988, doi: https://doi.org/10.1016/0021-9290(88)90259-X.

**16** P. S. Zun, T. Anikina, A. Svitenkov, and A. G. Hoekstra, "A comparison of fully-coupled 3D in-stent restenosis simulations to in-vivo data," *Front. Physiol.*, vol. 8, p. 284, 2017, doi: https://doi.org/10.3389/fphys.2017.00284.

**17** F. J. H. Gijsen, E. Allanic, F. N. van de Vosse, and J. D. Janssen, "The influence of the non-Newtonian properties of blood on the flow in large arteries: unsteady flow in a 90° curved tube," *J. Biomech.*, vol. 32, no. 7, pp. 705–713, Jul. 1999, doi: https://doi.org/10.1016/S0021-9290(99)00014-7.

**18** B. Thomas and K. S. Sumam, "Blood flow in human arterial system-a review," *Proc. Technol.*, vol. 24, pp. 339–346, Jan. 2016, doi: https://doi.org/10.1016/j.protcy.2016.05.045.

**19** T. A. Crowley and V. Pizziconi, "Isolation of plasma from whole blood using planar microfilters for lab-on-a-chip applications," *Lab Chip*, vol. 5, no. 9, pp. 922–929, Aug. 2005, doi: https://doi.org/10.1039/B502930A.

**20** Cho, Y.I. and Kensey, K.R. (1991). Effects of the non-Newtonian viscosity of blood on flows in a diseased arterial vessel.1. Steady flows. *Biorheology* 28 (3, 4): 241–262.

**21** C. A. Taylor, T. J. R. Hughes, and C. K. Zarins, "Finite element modeling of blood flow in arteries," *Comput. Meth. Appl. Mech. Eng.*, vol. 158, no. 1, 2, pp. 155–196, May 1998, doi: https://doi.org/10.1016/S0045-7825(98)80008-X.

**22** D. S. Pleouras, Antonis I. Sakellarios, Panagiota Tsompou, et al., "Simulation of atherosclerotic plaque growth using computational biomechanics and patient-specific data," *Sci. Rep.*, vol. 10, no. 1, Art. no. 1, Oct. 2020, doi: https://doi.org/10.1038/s41598-020-74583-y

**23** A. Santamarina, E. Weydahl, J. M. Siegel, and J. E. Moore, "Computational analysis of flow in a curved tube model of the coronary arteries: Effects of time-varying curvature," *Ann. Biomed. Eng.*, vol. 26, no. 6, pp. 944–954, Nov. 1998, doi: https://doi.org/10.1114/1.113.

**24** D. A. Steinman, J. B. Thomas, H. M. Ladak, J. S. Milner, B. K. Rutt, and J. D. Spence, "Reconstruction of carotid bifurcation hemodynamics and wall thickness using computational fluid dynamics and MRI," *Magn. Reson. Med.*, vol. 47, no. 1, pp. 149–59, Jan. 2002, doi: https://doi.org/10.1002/mrm.10025.

**25** Banerjee, R.K., Back, L.H., Back, M.R., and Cho, Y.I. (2003). Physiological flow analysis in significant human coronary artery stenoses. *Biorheology* 40 (4): 451–476.

**26** Formaggia, L., Lamponi, D., and Quarteroni, A. (2003). One-dimensional models for blood flow in arteries. *Journal of Engineering Mathematics* 47: 251–276. https://doi.org/10.1023/B:ENGI.0000007980.01347.29.

**27** F. N. van de Vosse *et al.*, "Finite-element-based computational methods for cardiovascular fluid-structure interaction," *J. Eng. Math.*, vol. 47, no. 3, pp. 335–368, Dec. 2003, doi: https://doi.org/10.1023/B:ENGI.0000007985.17625.43.

**28** C. A. Figueroa, I. E. Vignon-Clementel, K. E. Jansen, T. J. R. Hughes, and C. A. Taylor, "A coupled momentum method for modeling blood flow in three-dimensional deformable arteries," *Comput. Meth. Appl. Mech. Eng.*, vol. 195, no. 41, pp. 5685–5706, Aug. 2006, doi: https://doi.org/10.1016/j.cma.2005.11.011.

**29** A. K. Politis, G. P. Stavropoulos, M. N. Christolis, F. G. Panagopoulos, N. S. Vlachos, and N. C. Markatos, "Numerical modeling of simulated blood flow in idealized composite arterial coronary grafts: steady state simulations," *J. Biomech.*, vol. 40, no. 5, pp. 1125–1136, 2007, doi: https://doi.org/10.1016/j.jbiomech.2006.05.008.

**30** H. J. Kim, I. E. Vignon-Clementel, J. S. Coogan, C. A. Figueroa, K. E. Jansen, and C. A. Taylor, "Patient-specific modeling of blood flow and pressure in human coronary arteries," *Ann. Biomed. Eng.*, vol. 38, no. 10, pp. 3195–3209, Oct. 2010, doi: https://doi.org/10.1007/s10439-010-0083-6.

**31** N. Moreno, P. Vignal, J. Li, and V. M. Calo, "Multiscale modeling of blood flow: coupling finite elements with smoothed dissipative particle dynamics," *Proc. Comput. Sci.*, vol. 18, pp. 2565–2574, Jan. 2013, doi: https://doi.org/10.1016/j.procs.2013.05.442.

**32** T. K. Dobroserdova and M. A. Olshanskii, "A finite element solver and energy stable coupling for 3D and 1D fluid models," *Comput. Meth. Appl. Mech. Eng.*, vol. 259, pp. 166–176, Jun. 2013, doi: https://doi.org/10.1016/j.cma.2013.03.018.

**33** S. Charm and G. S. Kurland, "Tube flow behavior and shear stress-shear rate characteristics of canine blood," *Am. J. Physiol.*, vol. 203, no. 3, pp. 417–421, Sep. 1962, doi: https://doi.org/10.1152/ajplegacy.1962.203.3.417.

**34** G. Theodorou and D. Bellet, "Laminar flows of a non-newtonian fluid in mild stenosis," *Comput. Meth. Appl. Mech. Eng.*, vol. 54, no. 1, pp. 111–123, Jan. 1986, doi: https://doi.org/10.1016/0045-7825(86)90038-1.

**35** A. Leuprecht and K. Perktold, "Computer simulation of non-Newtonian effects on blood flow in large arteries," *Comput. Meth. Biomech. Biomed. Eng.*, vol. 4, no. 2, pp. 149–163, Feb. 2001, doi: https://doi.org/10.1080/10255840008908002.

**36** B. M. Johnston, P. R. Johnston, S. Corney, and D. Kilpatrick, "Non-Newtonian blood flow in human right coronary arteries: transient simulations," *J. Biomech.*, vol. 39, no. 6, pp. 1116–1128, 2006, doi: https://doi.org/10.1016/j.jbiomech.2005.01.034.

**37** J. V. Soulis, G. D. Giannoglou, Y. S. Chatzizisis, K. V. Seralidou, G. E. Parcharidis, and G. E. Louridas, "Non-Newtonian models for molecular viscosity and wall shear stress in a 3D reconstructed human left coronary artery," *Med. Eng. Phys.*, vol. 30, no. 1, pp. 9–19, Jan. 2008, doi: https://doi.org/10.1016/j.medengphy.2007.02.001.

**38** Sakellarios, A.I., Siogkas, P., Exarchos, T. et al. (2011). Journal of the Serbian Society for. *Computational Mechanics* 5 (2): 90–100.

**39** P. Siogkas *et al.*, "Multiscale - patient-specific artery and atherogenesis models," *IEEE Trans. Biomed. Eng.*, vol. 58, no. 12, pp. 3464–3468, Dec. 2011, doi: https://doi.org/10.1109/TBME.2011.2164919.

**40** A. I. Sakellarios *et al.*, "Patient-specific computational modeling of subendothelial LDL accumulation in a stenosed right coronary artery: effect of hemodynamic and biological factors," *Am. J. Physiol. Heart Circ. Physiol.*, vol. 304, no. 11, pp. H1455–1470, Jun. 2013, doi: https://doi.org/10.1152/ajpheart.00539.2012.

**41** M. Cilla, E. Pena, and M. A. Martinez, "Mathematical modelling of atheroma plaque formation and development in coronary arteries," *J. R. Soc. Interface*, vol. 11, no. 90, pp. 20130866, Nov. 2013, doi: https://doi.org/10.1098/rsif.2013.0866.

**42** W. Hao and A. Friedman, "The LDL-HDL profile determines the risk of atherosclerosis: a mathematical model," *PLoS One*, vol. 9, no. 3, p. e90497, Mar. 2014, doi: https://doi.org/10.1371/journal.pone.0090497.

**43** A. I. Sakellarios *et al.*, "Prediction of atherosclerotic plaque development in an in vivo coronary arterial segment based on a multilevel modeling approach," *IEEE Trans. Biomed. Eng.*, vol. 64, no. 8, pp. 1721–1730, Aug. 2017, doi: https://doi.org/10.1109/TBME.2016.2619489.

**44** Y. Yang, W. Jäger, M. Neuss-Radu, and T. Richter, "Mathematical modeling and simulation of the evolution of plaques in blood vessels," *J. Math. Biol.*, vol. 72, no. 4, pp. 973–996, Mar. 2016, doi: https://doi.org/10.1007/s00285-015-0934-8.

**45** M. Alimohammadi, C. Pichardo-Almarza, O. Agu, and V. Díaz-Zuccarini, "A multiscale modelling approach to understand atherosclerosis formation: a patient-specific case study in the aortic bifurcation," *Proc. Inst. Mech. Eng. H*, vol. 231, no. 5, pp. 378–390, May 2017, doi: https://doi.org/10.1177/0954411917697356.

**46** O. Kafi, N. El Khatib, J. Tiago, A. Sequeira, "Numerical simulations of a 3D fluid-structure interaction model for blood flow in an atherosclerotic artery," *Math. Biosci. Eng.*, vol. 14, no. 1, pp. 179–193, 2017, doi: https://doi.org/10.3934/mbe.2017012.

**47** Anlamlert, W. (2017). Modeling fibrous cap formation in atherosclerotic plaque development: stability and oscillatory behavior. *Adv. Differ. Equ.* 2017: 195.

**48** A. D. Chalmers, C. A. Bursill, and M. R. Myerscough, "Nonlinear dynamics of early atherosclerotic plaque formation may determine the efficacy of high density lipoproteins (HDL) in plaque regression," *PLoS One*, vol. 12, no. 11, p. e0187674, Nov. 2017, doi: https://doi.org/10.1371/journal.pone.0187674.

**49** M. Guo, Y. Cai, X. Yao, and Z. Li, "Mathematical modeling of atherosclerotic plaque destabilization: role of neovascularization and intraplaque hemorrhage," *J. Theor. Biol.*, vol. 450, pp. 53–65, Aug. 2018, doi: https://doi.org/10.1016/j.jtbi.2018.04.031.

**50** Watson, M.G., Byrne, H.M., Macaskill, C., and Myerscough, M.R. (2018). A two-phase model of early fibrous cap formation in atherosclerosis. *arXiv:1805.07058 [q-bio]* http://arxiv.org/abs/1805.07058 (accessed 29 November 2022).

**51** Sabzian, H., Shafia, M.A., Maleki, A. et al. (2019). Theories and practice of agent based modeling: some practical implications for economic planners. *arXiv:1901.08932 [econ]* http://arxiv.org/abs/1901.08932 (accessed 29 November 2022).

**52** Epstein, J.M. (2006). *Generative Social Science: Studies in Agent-Based Computational Modeling*. Princeton University Press.

**53** Taylor, S. (2014). *Agent-Based Modeling and Simulation*. Springer.

**54** Bouarfa, S., Blom, H.A., Curran, R., and Everdij, M.H. (2013). Agent-based modeling and simulation of emergent behavior in air transportation. *Complex Adapt. Syst. Model* 1: 15.

**55** Hammond, R.A. (2015). *Considerations and Best Practices in Agent-Based Modeling to Inform Policy*. National Academies Press (US) https://www.ncbi.nlm.nih.gov/books/NBK305917/ (accessed 29 November 2022).

**56** B. Vermeulen and A. Pyka, "Agent-based modeling for decision making in economics under uncertainty," *Economics*, vol. 10, no. 1, Dec. 2016, doi: https://doi.org/10.5018/economics-ejournal.ja.2016-6.

**57** Youssef, B.B. (2018). Simulating cell–cell interactions using a multicellular three-dimensional computational model of tissue growth. In: *Computational Science and Its Applications – ICCSA 2018*, Cham, 215–228. https://doi.org/10.1007/978-3-319-95165-2_15.

**58** Naik, G.R. (2012). *Applied Biological Engineering: Principles and Practice*. BoD – Books on Demand.

**59** F. Pappalardo, S. Musumeci, and S. Motta, "Modeling immune system control of atherogenesis," *Bioinformatics*, vol. 24, no. 15, pp. 1715–1721, Aug. 2008, doi: https://doi.org/10.1093/bioinformatics/btn306.

**60** A. M. Bailey, B. C. Thorne, and S. M. Peirce, "Multi-cell agent-based simulation of the microvasculature to study the dynamics of circulating inflammatory cell trafficking," *Ann. Biomed. Eng.*, vol. 35, no. 6, pp. 916–936, Jun. 2007, doi: https://doi.org/10.1007/s10439-007-9266-1.

**61** A. L. Olivares, M. A. González Ballester, and J. Noailly, "Virtual exploration of early stage atherosclerosis," *Bioinformatics*, vol. 32, no. 24, pp. 3798–3806, Dec. 2016, doi: https://doi.org/10.1093/bioinformatics/btw551.

62 B. C. Thorne, H. N. Hayenga, J. D. Humphrey, and S. M. Peirce, "Toward a multi-scale computational model of arterial adaptation in hypertension: verification of a multi-cell agent based model," *Front. Physiol.*, vol. 0, 2011, doi: https://doi.org/10.3389/fphys.2011.00020.

63 H. Zahedmanesh, H. V. Oosterwyck, and C. Lally, "A multi-scale mechanobiological model of in-stent restenosis: deciphering the role of matrix metalloproteinase and extracellular matrix changes," *Comput. Meth. Biomech. Biomed. Eng.*, vol. 17, no. 8, pp. 813–828, Jun. 2014, doi: https://doi.org/10.1080/10255842.2012.716830.

64 M. Garbey, M. Rahman, and S. Berceli, "A multiscale computational framework to understand vascular adaptation," *J. Comput. Sci.*, vol. 8, pp. 32–47, May 2015, doi: https://doi.org/10.1016/j.jocs.2015.02.002.

65 M. Garbey, S. Casarin, and S. A. Berceli, "A versatile hybrid agent-based, particle and partial differential equations method to analyze vascular adaptation," *Biomech. Model. Mechanobiol.*, vol. 18, no. 1, pp. 29–44, Feb. 2019, doi: https://doi.org/10.1007/s10237-018-1065-0.

66 R. Bhui and H. N. Hayenga, "An agent-based model of leukocyte transendothelial migration during atherogenesis," *PLoS Comput. Biol.*, vol. 13, no. 5, p. e1005523, May 2017, doi: https://doi.org/10.1371/journal.pcbi.1005523.

67 A. Corti, C. Chiastra, M. Colombo, M. Garbey, F. Migliavacca, and S. Casarin, "A fully coupled computational fluid dynamics – agent-based model of atherosclerotic plaque development: multiscale modeling framework and parameter sensitivity analysis," *Comput. Biol. Med.*, vol. 118, p. 103623, Mar. 2020, doi: https://doi.org/10.1016/j.compbiomed.2020.103623.

68 H. Byrne and D. Drasdo, "Individual-based and continuum models of growing cell populations: a comparison," *J. Math. Biol.*, vol. 58, no. 4, 5, pp. 657–687, Apr. 2009, doi: https://doi.org/10.1007/s00285-008-0212-0.

69 F. Azuaje, "Computational discrete models of tissue growth and regeneration," *Briefings Bioinf.*, vol. 12, no. 1, pp. 64–77, Jan. 2011, doi: https://doi.org/10.1093/bib/bbq017.

70 F. Milde, M. Bergdorf, and P. Koumoutsakos, "A hybrid model for three-dimensional simulations of sprouting angiogenesis," *Biophys. J.*, vol. 95, no. 7, pp. 3146–3160, Oct. 2008, doi: https://doi.org/10.1529/biophysj.107.124511.

71 M. Hwang, M. Garbey, S. A. Berceli, and R. Tran-Son-Tay, "Rule-based simulation of multi-cellular biological systems – a review of modeling techniques," *Cell. Mol. Bioeng.*, vol. 2, no. 3, pp. 285–294, Sep. 2009, doi: https://doi.org/10.1007/s12195-009-0078-2.

72 C. Priami, "Algorithmic systems biology," *Commun. ACM*, vol. 52, no. 5, pp. 80–88, May 2009, doi: https://doi.org/10.1145/1506409.1506427.

**73** S. S. Goel *et al.*, "Management of drug eluting stent in-stent restenosis: a systematic review and meta-analysis," *Catheter. Cardiovasc. Interv.*, vol. 87, no. 6, pp. 1080–1091, 2016, doi: https://doi.org/10.1002/ccd.26151.

**74** A. D. Singh *et al.*, "Recurrent drug-eluting stent in-stent restenosis: a state-of-the-art review of pathophysiology, diagnosis, and management," *Cardiovasc. Revasc. Med.*, vol. 21, no. 9, pp. 1157–1163, Sep. 2020, doi: https://doi.org/10.1016/j.carrev.2020.01.005.

**75** V. G. Ng, C. Mena, C. Pietras, and A. J. Lansky, "Local delivery of paclitaxel in the treatment of peripheral arterial disease," *Eur. J. Clin. Invest.*, vol. 45, no. 3, pp. 333–345, 2015, doi: https://doi.org/10.1111/eci.12407.

**76** C. M. McKittrick *et al.*, "Combining mathematical modelling with in vitro experiments to predict in vivo drug-eluting stent performance," *J. Controlled Release*, vol. 303, pp. 151–161, Jun. 2019, doi: https://doi.org/10.1016/j.jconrel.2019.03.012.

**77** A. Cornelissen and F. J. Vogt, "The effects of stenting on coronary endothelium from a molecular biological view: time for improvement?," *J. Cell. Mol. Med.*, vol. 23, no. 1, pp. 39–46, Jan. 2019, doi: https://doi.org/10.1111/jcmm.13936.

**78** P. P. Karjalainen, W. Nammas, and J. K. Airaksinen, "Optimal stent design: past, present and future," *Int. Cardiol.*, vol. 6, no. 1, pp. 29–44, Feb. 2014, doi: https://doi.org/10.2217/ica.13.84.

**79** R. He, L. Zhao, V. V. Silberschmidt, and Y. Liu, "Mechanistic evaluation of long-term in-stent restenosis based on models of tissue damage and growth," *Biomech. Model. Mechanobiol.*, vol. 19, no. 5, pp. 1425–1446, 2020, doi: https://doi.org/10.1007/s10237-019-01279-2.

**80** C. J. Boyle, A. B. Lennon, and P. J. Prendergast, "Application of a mechanobiological simulation technique to stents used clinically," *J. Biomech.*, vol. 46, no. 5, pp. 918–924, Mar. 2013, doi: https://doi.org/10.1016/j.jbiomech.2012.12.014.

**81** M. K. Fitts, D. B. Pike, K. Anderson, and Y.-T. Shiu, "Hemodynamic shear stress and endothelial dysfunction in hemodialysis access," *Open Urol. Nephrol. J*, vol. 7, no. Suppl 1 M5, pp. 33–44, 2014, doi: https://doi.org/10.2174/1874303X01407010033.

# 10

# Risk Prediction

## 10.1 Introduction

In the last decade, we are witnessing two scientific explosions related to the healthcare field: the first one is related to the acquisition of large amounts of data collected through large-scale clinical trials sometimes in multicentric studies. The purpose of collecting the data is related to the understanding of the pathophysiology of a disease, but, most importantly, with the development of new predictive models either for risk assessment or diagnosis, prognosis, and treatment. Another evolution which originates from the computational power made available is the implementation of novel machine-learning algorithms for efficient data handling and analysis. This is based on the cloud platforms and the frameworks created for privacy of data (GDPR, Health Insurance Portability and Accountability Act [HIPPA]) which allow data sharing and analysis [1, 2].

The development of predictive models requires many preprocessing steps before employing selected algorithms for prediction. Most of the efforts in the preprocessing step are related to data curation, discretization, and feature selection. However, even before these steps, further effort is needed such as for data sharing which is necessary before contacting any medical data-based analysis. Moreover, to benefit from the availability of cohorts coming from different sites, harmonization of data is also required as a preprocessing step [3]. Predictive modeling based on data analytics can then be applied when the data quality is guaranteed.

*Multiscale Modelling in Biomedical Engineering*, First Edition. Dimitrios I. Fotiadis, Antonis I. Sakellarios, and Vassiliki T. Potsika.

Published 2023 by John Wiley & Sons, Inc.

## 10.2 Medical Data Preprocessing

### 10.2.1 Data Sharing

The development of data analytics approaches requires the collection of data which are usually provided by various clinical sites. This requires considering the protection of the privacy of patient data. It is essential for data sharing that the data are collected following the regulations, ethics, and the satisfaction of rules for the data quality in terms of metrics, structure, and information management [4]. It is the responsibility of the clinical site to request agreements of data use with the data processors. The rules governing data sharing are described in the Health Insurance Portability and Accountability (HIPA) guidelines in the United States and in the General Data Protection Regulation (GDPR) guidelines in Europe [5].

As soon as the ethical and legal concerns for data sharing are satisfied, a process known as quality assessment follows [6]. The purpose of this step is the enhancement of the data quality. Quality is related to the accuracy, consistency, completeness, etc., of the available data. Data quality can be improved by data curation or cleansing which attempts to identify and treat missing values, outliers, duplicates, and other common errors in the acquired data [7]. It is also highly important that the whole process is fully automated since the data curation process may be extremely time-consuming when large amounts of data are processed.

Data curation will transform the acquired data into a new set of data which satisfies several quality criteria: (i) accuracy, (ii) completeness, (iii) consistency, (iv) interpretability, (v) relevancy, and (vi) ease of manipulation [8]. The mechanisms for curating collected data include [9]: (i) the detection and elimination of duplicate fields (i.e. deduplication), (ii) the characterization of data according to their context (i.e. data annotation), (iii) the identification of duplicate fields with highly similar distributions (i.e. similarity detection), (iv) the transformation of data into standardized formats (i.e. standardization), (v) dealing with missing values (i.e. data imputation), and (vi) outlier detection for detecting values that deviate from the standard data range.

After the data curation step, another preprocessing step might be required especially when data come from different sites and harmonization is needed. This step is related to the data standardization which aims to transform further the data based on a standard model. This model consists of a template specific to the field under discussion and it is defined according to international standards including parameters, their ranges, and the types of variables. Standardization enables the semantic representation of the model based on ontologies in order to describe the domain knowledge [3]. An example of a standard semantic model consists of the main class which is the "patient" and many subclasses usually related to the collected types of data, e.g. "laboratory exams," "demographics," "clinical," "omics," etc. The subclasses may include subsequent subclasses or the specific subclass variables, e.g. "age," "gender," "cholesterol," etc. The process of standardization includes both the identification of the same variables and the normalization of the detected matches using standard or known ranges of each variable. Several approaches to matching exist, such as lexical matching, which uses string similarity measures [10].

### 10.2.2 Data Harmonization

The structural heterogeneities (Heterogeneities: variabilities in the data structure – e.g. parameters with different value ranges – due to the different data collection protocols) of the collected clinical data from different sites is a common issue when analysis of multiple sources of data is required. Medical data harmonization provides a tool to enhance the analysis of heterogeneous data sources. This heterogeneity causes substantial difficulties in the analysis and the conclusion-making due to the limited statistical power [3, 11]. Data harmonization focuses on the generation of homogenized versions of the original datasets which are characterized by their heterogeneity. Moreover, the concept of data harmonization is related to the analysis of data in the same or similar medical domain. The rationale behind data harmonization is to utilize computational approaches and data-driven techniques such as semantics to remodify the heterogeneous data in a format which consists of the same parameters and the same range values. This increases the statistical power of the analysis enabling the integrative analysis of different datasets. In this context, acquisition and analysis of the available datasets may answer unmet needs of significant diseases where data are available in significant sizes.

The emerging need for integrative analysis of the available retrospective medical data (which is a category of data needed to be exploited) with heterogeneity has led to the development of many data harmonization frameworks.

Biobank Standardisation and Harmonisation for Research Excellence in the European Union (BioSHaRE-EU) is a EU-funded project [12] where harmonization of eight cohorts was attempted. During this project, the DataSHaPER (DataSchema and Harmonization Platform for Epidemiological Research) framework [13] was developed. A reference model called DataSchema or Generic DataSchema includes about 180 variables from 45 domains and 13 themes. The theme is in concept similar to the class (e.g. clinical risk factors), while the domains are considered subclasses (e.g. cardiovascular risk factors). This framework was developed to be applied to both retrospective and prospective cohorts under the specific condition that the reference model has been set in a DataSchema format. The whole process consists of three steps: (i) definition of pairing rules for the variables, (ii) the application of the rules to classify the variables according to the generation or not of the reference model variables, and (iii) the employment of an algorithm for the generation of the DataSchema variables using the positively identified variables. The Opal software and servers are used for the harmonization process [14]. This is a secured process based on the DataShield cloud infrastructure [15]. During the harmonization analysis of the eight cohorts, 70% harmonization accuracy was achieved in terms of variable identification with shared terminology.

The same project also published the BiobankConnect software which can be used for the integration of heterogeneous biobanks. Lexical matching is used for this purpose in combination with semantics to achieve a semiautomated harmonization. Using the BiobankConnect, the potential of the harmonized terms is quantified using the elements defined by the user and by searching for any lexical matching. The manual annotation by the user is performed through the bioportal [16], which is a repository with ontologies in the biomedical domain. In the next step, matches between the raw data and the selected ones are performed yielding a matching score in a form of percentages. The software was evaluated in the harmonization of six biobanks with about 0.75 precision rate.

BioSHaRE project also developed the System for Ontology-based Re-coding and Technical Annotation (SORTA) applied to biomedical phenotype data [17]. The heterogeneity of the data is handled using the SORTA tool in a high level of analysis using international coding systems. In particular, the specified international codes are compared to the data values seeking potential matches. Ontologies or even .csv formatted documents can be used as the reference international coding system. Matching algorithms are employed to test SORTA's efficacy to do the matching between the data; for example, its testing to a dataset of 5210 entries of LifeLines cohort compared to the Dutch Uniform Multicenter Registration system reaches 97% and 58% for genetic disorders and malformation syndromes, respectively.

Another framework for semantic matching is provided by the S-Match tool [18]. Also, this tool is open source. It is based on taxonomies or, more specifically, lightweight ontologies, where tree-like structures are used and the nodes represent specific means. Semantic-matching algorithms are used to identify and estimate the relation between the nodes of the trees of two taxonomies and classify them into four categories: (i) equivalent, (ii) less general, (iii) more general, and (iv) disjointness. Its performance was evaluated using the TaxMe2 cohort [19] with 46% precision of the accurate detection of semantic relations.

Finally, another tool which was developed for the classification of gene functions of microorganisms is the functional ontology assignments for metagenomes (FOAM) [20]. This tool is based on the hidden Markov models (HMMs). The tool was trained using a database of sequences [21]. The process is performed semiautomatically, with the minimum user interaction of defining the target space.

As has already been described, several data harmonization frameworks have been proposed to co-analyze heterogeneous medical data, including the DataSHaPER [13] framework, the SORTA tool [17], the BiobankConnect software tool [22], the S-match semantic matching framework [18], and finally the FOAM [20].

A typical harmonization workflow consists of six stages, including the: (i) metadata extraction, (ii) construction of the reference semantic data models (i.e. reference ontologies), (iii) extraction of the terminologies from the reference ontologies, (iv) construction of a general-purpose medical dictionary, (v) lexical matching between the extracted terminologies and the reference ontology, and (vi) semantic matching between the extracted terminologies and the entities of the ontology. The outcome of the data harmonization workflow is a harmonized dataset which includes the lexically and/or semantically matched variables which are normalized according to the predefined range values in the dictionary. To avoid information loss by clinical datasets which have more information than the average clinical dataset, in a particular clinical domain, a second reference template is also developed including the terminologies that represent the maximal domain knowledge for each clinical domain of interest.

## 10.3 Machine Learning and Data Mining

Before applying a machine learning algorithm, as a final preprocessing step and especially when many features are collected, feature extraction is performed to identify a set of the most significant features as those having the highest contribution to the outcome (target feature). Discriminating a set of features can be achieved using feature ranking (gain ratio [23], information gain [24], and Gini index [25]) or feature selection methods, such as the fast correlation-based filter (FCBF) method [26] and exhaustive feature selection [27]. The former ranks the features in terms of their significance to predict the outcome whereas the latter identifies the most important features by incrementally adding features until predefined criteria are met.

Feature selection aims to reduce the complexity and dimensionality of a large amount of input medical data and at the same time minimize overfitting effects during the training of the ML algorithms. Moreover, feature selection is also necessary for the medical domain because it can be used for the identification of features which are highly correlated with the target and less correlated among them, commonly referred to as biomarkers. The biomarkers can improve our understanding regarding the pathophysiology of the disease of interest or can directly be used in clinical practice as indicators for risk stratification, diagnosis, prognosis, and treatment.

Traditionally, in the medical domain, statistical models are used for the prediction of a disease outcome or event through the detection of significant differences among a set of independent variables (e.g. features) given a target feature (i.e. a dependent variable). These statistical models are usually based on regression analysis (e.g. logistic regression) for the detection of significant associations among the dependent variable and the set independent variables (i.e. the potential biomarkers). However, the recent years, it has been demonstrated that machine learning can improve the predictive accuracy of the statistical models since the ML models can shed light on the complexity of the biological pathways which usually represent a disease condition. For this purpose, machine learning can handle different types of data from conventional data types, such as the clinical characteristics of a disease to complex ones, such as genetic data or other types of omics or imaging data.

Supervised learning algorithms are used for the prediction of an outcome either for diagnostic or prognostic purposes based on training steps using the collected data and then provided by the clinical experts' annotations. There are many supervised algorithms for this purpose, such as regression [28], support vector machines (SVMs) [29], decision trees [30], Naïve Bayes [31], and artificial neural networks (ANNs) [32]. All these can be used for the prediction of the patient's risk to present or develop a disease outcome such as a fatal event. For example, in the cardiovascular disease (CVD), common outcome targets are the death by CVD or the disease progression or the disease severity.

Another major category of machine learning algorithms is the unsupervised learning models in which the major difference with the supervised ones is that they do not need the prior knowledge of the target outcome. The unsupervised models cluster data structures with similar patterns [33]. Such algorithms are the hierarchical clustering and the k-means [34], the spectral clustering [35] and the hierarchical clustering [36]. Another difference of the two categories is that the development of the former one is based on the training on a set of data assuming one feature as the target outcome and evaluating the models' performance using mainly k-fold cross-validation or train/test split ratio approaches.

Besides these two major categories, the recent years and due to the large amount of data collection especially when omics data are concerned, the mining requires advanced approaches based on deep learning algorithms [37] such as the multi-layer neural networks [38] of the convolution neural networks [39]. These approaches are being used to identify patterns and motifs which are invisible in the big data and can be especially used for image segmentation or the development of predictive models using omics data.

### 10.3.1 Supervised Learning Algorithms

#### 10.3.1.1 Regression Analysis

The linear dependence between one (univariate or simple) or more (multivariate or multiple) independent variables and a dependent variable can be found using a linear regression model. This is the most widely used type of analysis toward the identification of the dependence of independent variables with the dependent variable (target) in the medical domain. If we assume that we have the dependent variable $y$ and a set of independent variables $\{x_1, x_2, \ldots, x_n\}$, then the linear association model is defined as in:

$$y = w_0 1 + w_1 x_1 + w_2 x_2 + \cdots + w_n x_n + \varepsilon = x^T w + \varepsilon, \tag{10.1}$$

where $w_1, w_2, \ldots, w_n$ are the coefficients of each independent variable, $w_0$ is the interception parameter and ε is the disturbance parameter of the model $x^T$ and $w$ is the inner product of the weight vector $w$. As it is clear by the equation, a regression model can be used for the prediction of the dependent variable when the coefficients are known. The estimation of the weight vector can be achieved using several approaches such as using the ordinary least squares (OLS) [40], the generalized least squares (GLS) [41], the weighted least squares (WLS) [42], and the recursive least squares (RLS) [43], among others.

A special subcategory of the regression analysis is the logistic regression. In that case, the dependent variable is a categorical one and the model provides class probabilities. Such an example is when we want to predict the existence of a disease or to predict an event. Logistic regression estimates the log-odds as a linear combination of the input independent variables. In logistic regression, the weight vector is estimated based on the maximization of the log-likelihood using, for example, the Bernoulli distribution [44].

#### 10.3.1.2 Support Vector Machines

The support vector machine (SVM) algorithm aims to define an hyperplane which can be used for the separation of the data points based on given points aiming to solve high-dimensional problems either for regression or classification [29]. The hyperplanes work as boundaries used for the decision making. Data points located close to the hyperplane are named support vectors and the rationale is to identify vectors by which the margin of the hyperplane is maximized. Such a hyperplane can be written as:

$$x^T w - b = 0, \tag{10.2}$$

where $b$ is a scalar that determines the offset of the hyperplane in the 2D space. The rationale is to find the region between the two dashed lines (upper and lower hyperplane) and consequently the hyperplane can be defined at the halfway between the upper and lower ones. Based on these assumptions, a data point is classified as 1 when $x^T w - b \geq 1$ and –1 when $x^T w - b \leq -1$. Thus, a weight vector **w** can be found minimizing the function:

$$\tilde{\mathbf{x}} = \text{sign}\left(\mathbf{x}^{\mathbf{T}}\mathbf{w} - b\right), \tag{10.3}$$

where sign() is a function that returns 1 in the case of a positive value and 1 otherwise.

This concept applies for linearly separable data points, where, in this case, the SVM is referred to as hard margin classifier. When the data points are not linearly distinguishable, the SVM is called soft margin classifier [45] and the hinge loss function is introduced:

$$\max\left(0, 1 - y_i\left(\mathbf{x}_{\mathbf{i}}^{\mathbf{T}}\mathbf{w} - b\right)\right) = 0. \tag{10.4}$$

In the nonlinear soft margin, data points are allowed to be located between the two margins of the hyperplanes in contrast to the hard margin where the data points must be located always on the marginal hyperplanes. The advantage of the nonlinear classification is obvious in the case where the data points are not linearly separable. In this case, the hyperplane can be defined as in:

$$\mathbf{k}\left(\mathbf{x}, \mathbf{x}^{\mathbf{T}}\right)\mathbf{w} - b = 0, \tag{10.5}$$

where now the kernel function $\mathbf{k}(\mathbf{x}, \mathbf{x}^{\mathbf{T}})$ is introduced. Several types of kernels can be used, such as the linear kernel, the polynomial kernel [46], and the Gaussian radial basis function (RBF) [47], among others.

#### 10.3.1.3 Naïve Bayes

Another supervised learning approach is the Naïve Bayes algorithm [31] which aims to identify the data points that belong to a class based on the estimation of a conditional probability. More specifically, given a set of $n$-features, say $\{x_1, x_2, \ldots, x_n\}$, and a target outcome (feature), say $y$, the Bayes theorem [48] is defined as in:

$$P\left(y \mid x_1, x_2, \ldots, x_n\right) = \frac{P(y)P\left(x_1, x_2, \ldots, x_n \mid y\right)}{P\left(x_1, x_2, \ldots, x_n\right)}, \tag{10.6}$$

where $P(y|x_1, x_2, \ldots, x_n)$ is the posterior probability which is the conditional probability of $y$ given the set of input features, $P(x_1, x_2, \ldots, x_n|y)$ is the likelihood, $P(x_1, x_2, \ldots, x_n)$ is the joint probability of the input features, and $P(y)$ is the class probability. An estimation can be given by solving the following objective function:

$$\tilde{y} = \text{argmax}_y\left(P(y)\prod_{i=1}^{n}P\left(x_i \mid y\right)\right). \tag{10.7}$$

The maximum a posteriori (MAP) rule [49] can be applied to maximize the product of the likelihood and the prior probabilities. There are different types of Naïve Bayes classifiers based on the precise definition of the conditional probability $P(x_i | y)$. Such variations include the multinomial Naïve Bayes classifier and the Bernoulli classifier.

#### 10.3.1.4 Decision Trees

Another supervised learning algorithm which is widely used for classification or regression tasks are the decision trees [30]. A decision tree consists of nodes (i.e. features) which are recursively separated into more nodes based on a splitting function to yield computationally efficient models with favorable prediction results. Moreover, the decision trees-based models are highly interpretable because they can be easily visualized using dendrograms. The node splitting is usually achieved using an impurity function, e.g. information entropy and the Gini impurity index. These methods can be handled both numerical and categorical variables with integrated preprocessing steps, such as, data normalization or discretization.

One of the first introduced decision trees algorithms is the CART algorithm (classification and regression trees). Given a set of $n$-features $\{x_1, x_2, \ldots, x_n\}$, and a target outcome (feature), say $y$, using the CART algorithm, a large subset (i.e. the parent node, assume $C_{\mathrm{P}}$) of the training set is partitioned into two smaller subsets (i.e. the child nodes, assume $C_{\mathrm{L}}$ and $C_{\mathrm{R}}$), according to a splitting criterion $S_{\mathrm{c}}$. The criterion is the value that the information entropy of the gained information is being maximized given the target outcome. The rationale is that the parent note splits into child nodes (left and right) according to each criterion as a decision rule; for example, when a value is equal or more than the criterion follow the right child, otherwise follow the left. The CART algorithm can be used both with categorical and numerical features. Moreover, it can be used for regression problems, where the aim is to find the best criterion by which the error loss function is minimized as well as for classification problems in which the Gini index can be used for the evaluation of the decision.

Apart from the ID3, the C4.5, and its newer version the C5.0 [50] implementations can be used for the determination of the information entropy and information gain measures using both numerical and categorical data types. These algorithms are using the information entropy for splitting the nodes based on the uncertainty in each subset and always according to the target outcome/feature. Furthermore, the C4.5 algorithm integrates imputation and pruning steps to prevent possible overfitting [51]. In each case, the optimal criterion value is estimated based on the rule that in this, the entropy is minimized or, on the other hand, the information gain is maximized.

The process which is the same in CART algorithm as well, follows a top-down sequence, where the tree construction process starts from the root node and goes down to the leaf nodes. The process initiates using the input of a training set of $n$ features. First, a parent node $C$ is created and if the class labels are the same, an empty node is returned as a leaf node, otherwise the algorithm returns the node as a leaf node with the class label being equal to the majority class. In the next step, one of the aforementioned algorithms (Gini, ID3, etc.) is applied to identify the best splitting criterion. When the criterion is found, the training samples are partitioned after iteration on all possible values. The algorithm returns the node as a leaf node with the class label being equal to the majority class. The process is repeated until a stop criterion is met.

#### 10.3.1.5 Ensemble Classifiers

In the case of complex data, the application of decision trees may introduce bias during the decision making process since they cannot capture complex data patterns. Ensemble classifiers can overcome this limitation by combining multiple decision trees into a single robust classifier. The methods that can be used for the development of ensemble classifiers can be categorized into bagging (or bootstrap aggregation) and boosting methods.

Regarding the bagging method, specific subsets of the original data are selected for the training of a decision tree [52]. A representative example of such algorithms is the random forest (RF) algorithm. Using RF, $N$ decision trees are built not only using random occurrences of the training set but also using random feature subsets. Majority voting or averaging is then used to combine the prediction results of each tree. RF performs perfectly when high-dimensional data are analyzed and decreases the adjustment effect during the decision-making process based on the combination of the different patterns which are captured by the single trees.

The boosting algorithms follow a sequential strategy in contrast to the parallel one of the bagging ones. The rationale of this approach is that a set of weak learners is trained and based on their errors, there I s a next model with improved results. In the bagging approach, $N$ denotes the number of decision trees, but now refers to the number of boosting rounds. In each round and using the misclassification results, the algorithm reweights the features on each boosting round, where features with the highest misclassification weight are the focus of the next boosting round until the number of the boosting rounds is met. The most well-known boosting algorithms are: (i) the AdaBoost [53] where the final result is a weighted sum of the weak classifiers, (ii) the gradient boosting trees [54], where the boosting process is used for the optimization of the error loss function, and (iii) the extreme gradient boosting trees [55] which reduce the overfitting adopting a regularized approach enhancing in this way the classifier's performance.

#### 10.3.1.6 Artificial Neural Networks

ANNs consist of three layers: the input layer, the hidden layer, and the output layer. Each layer includes a set of nodes. The input layer consists of four input nodes which regard input feature observations. Five nodes compose the hidden layer and each one corresponds to each input node using specific weights. In the same concept, the output node is related through specific weights to the hidden nodes. If we assume the hidden node $h$, each one can be expressed as:

$$h_i = w_{i,1}x_1 + w_{i,2}x_2 + w_{i,3}x_3 + w_{i,4}x_4, \tag{10.8}$$

where $i = 1,2,\ldots,5$, $w_{i,j}$ are the weight values for each input node. In the same concept, the linear combination of the hidden layer with the output layer can be achieved. The weighted sums are estimated using specific functions referred to as activation functions. The neuronal activation can be simulated by the weighted sum of the input nodes with a step function which is applied to the weighted sum. ANNs can be used for multiclass problems by using more than one output layer. Also, ANNs can be categorized into feedforward neural networks (FNNs) [56] where no loops are allowed between the input and hidden layers and the output and hidden layers. Unsupervised learning algorithms do not require annotated data to make predictions; rather they cluster the input features based on their similarity into a predefined number of clusters. Some of the most widely applicable unsupervised learning algorithms are K-means, spectral clustering, and hierarchical clustering.

#### 10.3.1.7 K-Means

The K-means algorithm [34] minimizes the variance within the clusters to group the input features into a set of $k$-predefined clusters. A clustering centroid is searched by k-means for each of the defined clusters. The centroid is the one with the minimum distance between an observation and the rest of them in the cluster. The distance of the clustering centroids from the observations is computed and the algorithm clusters each observation to specific clusters based on the criterion that its centroid and the observation have the minimum distance, i.e. the Euclidean distance is usually used.

#### 10.3.1.8 Spectral Clustering

This algorithm is based on Graph Theory and especially the minimization of the similarity between clusters and maximization of the similarity within the clusters to achieve the clustering of the input samples [35]. For this purpose, the input data are first projected into a different dimensional space to improve the clustering formation procedure. An affinity matrix is first calculated which is then transformed into a Laplacian format. The dimensional space is reduced using eigenvalue decomposition (EVD). The Euclidean distance can also be used in this case to measure the similarity for the formation of the affinity matrix. The aim is to identify similar points meaning that they have the lowest distance between them. As soon as the Laplacian matrix is constructed and the eigenvectors are estimated, the k-means algorithm is applied to the $k$-largest eigenvectors to assign them to clusters. Using the inverse process to go back to the original space, each observation is assigned to specific clusters. Apart from the original spectral clustering, other variants include the N-cut algorithm [57] and spatially constrained spectral clustering [58].

#### 10.3.1.9 Hierarchical Clustering

A dendrogram, or else a hierarchical data clustering schema, is constructed using the hierarchical clustering algorithm implementing a nested approach. The root of the dendrogram is usually a cluster with all the observations and the leaves consist of observations which correspond to unique clusters. Toward this direction, both top-down and bottom-up approaches can be used for hierarchical clustering. In the former, all input observations are assigned to the root clusters which are split into pairs of additional clusters, until the leaves' clusters are identified which include only one observation. In the second approach, the inverse strategy is followed: each observation is assigned to one cluster and then pairs of similar clusters are constructed until a root cluster which includes all the observations is defined. In both cases, the number of clusters is predefined.

## 10.4 Explainable Machine Learning

In the current years, there is increasing use of machine learning algorithms and the development of predictive models. Moreover, as the collected data are characterized by significant complexity especially when omics data are concerned, the ML algorithms are also presented with increased complexity which subsequently affects the interpretation of the final results [59]. The implementation of novel approaches for predictive modeling in healthcare faces the challenge of interpretation and answering questions such as how the developed model works, what affects the decisions or, more importantly, can we trust the final model? All these questions regard the explainability of a predictive model. Traditionally, when the predictive models were based on statistical tests, the interpretation was mainly based on the probabilities or correlation coefficients. Using ML algorithms, however, explainability is not always a simple step. In particular, specific concerns which come from any developed predictive model, especially in biomedical domain, are the (i) correctness, (ii) robustness, (iii) bias, (iv) improvement, (v) transferability, and (vi) comprehensibility. Perhaps, the most important task is to explain the model's processing engine to a medical expert or physician.

### 10.4.1 Transparency

The first thing that we have to deal with when we attempt to explain an ML model is transparency, which regards the understanding of the way that a model is working and produces the prediction. In particular, transparency has to do with: (i) simulatability, which is the ability of the model to be simulated by a human, (ii) decomposability, which is the ability to decompose the model into parts, e.g. input, computation, output, and explain these parts, and (iii) the transparency of the applied algorithm which is the understanding of the process that the algorithm is following to produce results. For example, algorithms such as K-means which are explained here are characterized by increased transparency because the process of producing the clusters is clear. On the other hand, when neural networks are employed, constructing a loss function is difficult to be transparent. In this way, some algorithms are much more transparent (decision trees or regression models) than others which may look like black boxes (random forests, deep learning, etc.).

### 10.4.2 Evaluation and Types of Explanation

It is a fact that another important aspect of the explainable AI deals with the evaluation performance of the predictive models. Evaluation criteria are mainly related to comprehensibility which means how close to the reality are the findings and make sense to human physiology; fidelity which is the extent to which the results correctly describe the models which have been developed; the accuracy as a metric either in the form of specificity and sensitivity; the scalability and the generality which regard how the developed model is generalized and works well in other populations.

When a predictive model is developed and an explanation is attempted, one way to present the explanation is using natural language text or symbols. Another way of explanation is through visual analytics which concerns the generation of graphs or visual schemes, which not only help explain the results but also interpret the potential interactions between the variables. Another type of explanation is the degree of generalization. In particular, some models may work in a particular area and not globally. This makes sense especially, in the biomedical domain, where many factors may affect the model's performance. Another way to explain a model is by using examples or specific clinical scenarios. When complex models are developed, an explanation can be achieved using simplifications constructing simpler models. However, this may be tricky and the accuracy of each model has to be used for comparing the performance of the simple model against the complex one. Finally, another useful way of explaining a model is through interpretation and quantification of the effect that each input variable may have on the model's accuracy. This is related to the feature ranking approaches explained earlier.

There are several approaches and frameworks which summarize the types of machine learning approaches and their explainability. A detailed review is presented in [59]. Figure 10.1 [59] presents a way of summarizing the different algorithms based on a taxonomy approach. This approach starts with the classification

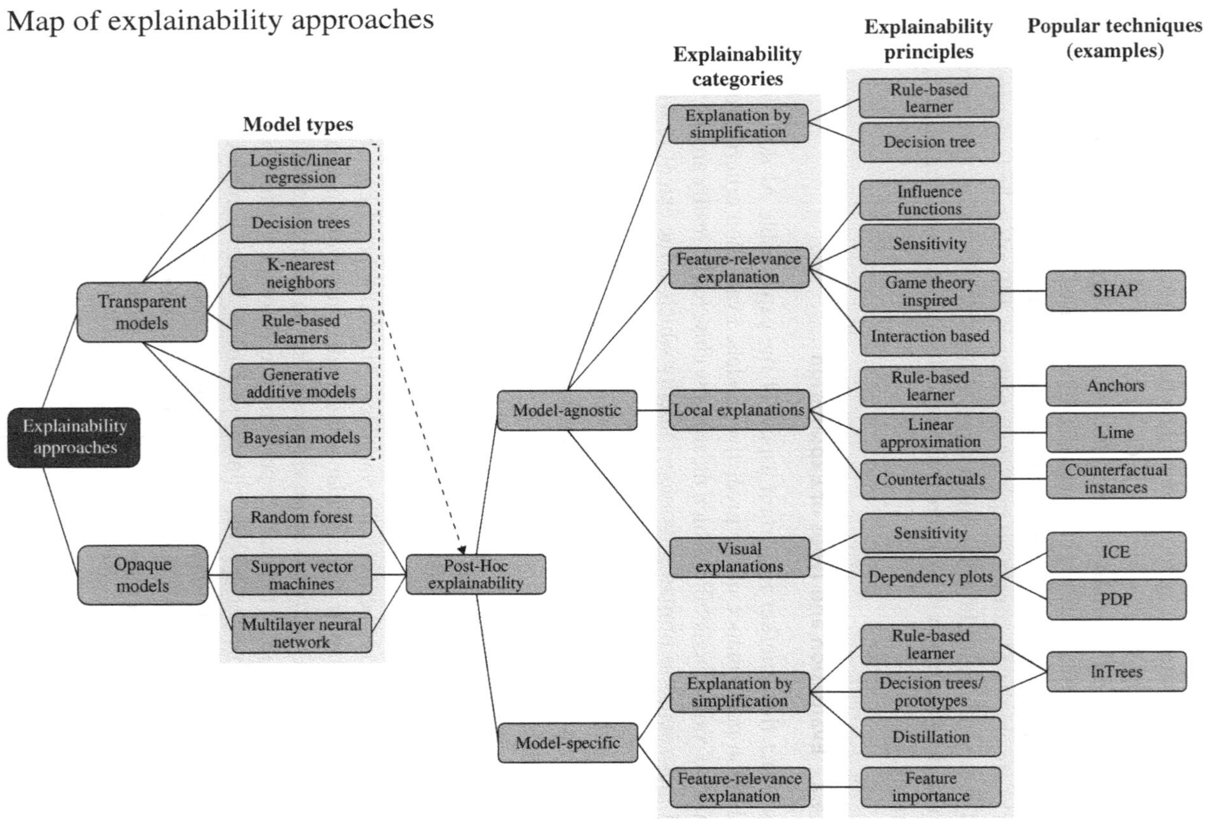


**Figure 10.1** A taxonomic view on XAI. *Source:* Belle and Papantonis [59]/Frontiers Media. S. A/CC BY.

between transparent and opaque models and the presentation of the respective ML algorithms. Further classification is achieved on the specific types of explanation and evaluation. The transparent models are the ones whose architecture enables the interpretation by satisfying at least one transparency criterion of the previous section. Such models are the linear/logistic regression, the decision trees, the K-nearest neighbors, the rule-based learning, the generalized additive models, and the Bayesian networks. On the other hand, the opaque models are quite difficult to be explained, but when present in some cases, increased predictive value. These models are the RFs, the SVMs, and the multilayer neural networks.

A major category of machine learning, which presents a significant use in recent years, is deep learning such as using multilayer neural networks. These models, on the one hand, are computationally costly, but, on the other hand, they are capable of presenting high performance outperforming other algorithms. Nevertheless, they also present minimum to no transparency. So, their explainability is tricky. To this purpose, specific strategies have been developed, i.e. NN-specific XAI methods, which attempt to explain the outcomes either using explanation by simplification or by feature-relevance explanation [60–65].

## 10.5 Example of Predictive Models in Cardiovascular Disease

Statistical modeling is the traditional way to assess CAD using mainly logistic regression models. Though they present good discrimination, in the final analysis, they fail to have high accuracy mainly because they are based on very few markers [66]. On the other hand, ML approaches seem to be promising to identify the high-risk patients successfully enabling the precision medicine and the use of the huge availability of data such as electronic health records (EHRs), personal health records (PHRs), and omics big. In brief, the main advantages of ML-driven PTP models rely on: (i) the adaptability of ML to the analysis of heterogeneous (multi-view) data, (ii) the strength of nonlinear ML-based analysis of multivariate data over linear regression approaches, (iii) the adequacy of today's ML techniques to deliver improved performance as data sets get larger, and (iv) the rapid diffusion of open-source ML software. Such attempts have been published in the current literature performing classification and clustering of the patients or even regression [67–74]. It is evident nowadays that big data and omics data can increase the accuracy of the developed models [75].

Usually, the detection of obstructive CAD regards the identification of patients with stenosis over at least 50%. We should clarify that the negative class is defined: either as (i) the complement of the positive class, or as (ii) individuals with no presence of CAD. On this basis, parametric (e.g. neural networks) or nonparametric (e.g. kernel methods) classification algorithms have been combined with feature evaluation techniques, aiming at specifying an accurate decision hyperplane and, in parallel, identifying the most informative features with respect to CAD severity. However, the reliance of these methods on both imaging and nonimaging features precludes their applicability as real PTP models.

Kurt et al. demonstrated that irrespective of the classification algorithm, typical CVD risk factors yield a high number of false positive (FP) predictions [76]. In particular, a logistic regression-based model, a CART, an FFNN, and an RBF network have all resulted in a relatively high sensitivity (ranging from 89.5 to 92.3), but their specificity was lower than 50%.

An FFNN was used to integrate data such as smoking, HDL, and diabetes with a stress test and achieved 88.4% accuracy [77]. Alizadehsani et al. by exploring the predictive potential of kernel-based methods, and particularly of SVM, both in feature selection and classification tasks [78] using 54 features including not only clinical but also electrocardiographic and biochemical data, demonstrated 93.4% accuracy. Knowledge-based feature selection can increase further the sensitivity of predictions [41]. Ensemble learning of the specified dataset, using a rotation forest (RTF) with FFNN as the base classifier, reached 91.2% accuracy, 95.6% sensitivity, and 86.7% specificity [48].

ML naturally arises as a favorable solution to CAD severity diagnosis when omics big data need to be exploited. The Corus® CAD algorithm, which constitutes an exemplary case, was developed previously by conventional statistical methodology, using gene expression data [79]. The Corus® CAD method is decomposed into the following phases: (i) feature selection based on an unsupervised cluster analysis, which yielded the identification of meta-genes (defined as cluster centroids), (ii) a ridge linear regression model of the age, sex, and the selected genes. In [79], a 0.75 (0.70–0.80) ROC, AUC is reported with respect to the identification of obstructive CAD patients. Moreover, by applying three arbitrary thresholds to Corus® CAD algorithm score (gene expression score), stratification of patients, according to the probability of CAD by maximal stenosis percentage, has also been proposed and validated in [80].

A major category of predictive models in CVD is when baseline characteristics are combined with computational continuum mechanics results to predict the disease progression. Though the entire vasculature system is exposed to atherogenic effects such as hypertension, diabetes mellitus, hyperlipidemia or smoking, and atherosclerotic lesions form at specific regions of the arterial tree, such as at the bifurcation and the inner wall of curvatures, where disturbed flow occurs. Hemodynamic forces play a major role in the localization of atherosclerosis because they create a pathologic environment. These local forces are modulators of vascular endothelial cell activity. The ESS is one of these forces, which is the tangential stress derived from the friction of blood on the wall and it plays the most fundamental role in atherosclerosis [81].

However, in the last decade, there has been increasing interest in biomechanical aspects of arterial disease: the role of ESS and fluid dynamics in the localization, growth, and prediction of atherosclerosis. Wilson et al. [82] studied the prediction of coronary heart disease using different categories of factors in which blood pressure was also included. The study included 2856 women and 2489 men from 30 to 74 years old at baseline and 12 years of follow-up. During the follow-up procedure, 227 women and 383 men developed CHD and one of the associated factors was the blood pressure combined, however, with other biological factors. The evaluation of this approach was found to be comparable to CHD when the continuous variables themselves were used. Through this investigation, a simple coronary disease prediction algorithm was developed which used variables such as blood pressure, LDL, and HDL to allow physicians to predict CHD in patients without CHD.

In the last few years, with the rapid growth in medical imaging, CFD codes have been developed, which allow the modeling of blood flow dynamics using realistic geometric data. CFD models provide the ESS distribution and its spatial distribution was used to link the hemodynamics with atherosclerosis. Feldman et al. [83] proposed a methodology to predict the progression of coronary atherosclerosis through the determination of the *in vivo* velocity and ESS in human coronary arteries. Areas of minor flow reversals and significant variations of the shear stress and velocity within the artery even in significant luminal obstruction were able to be identified using this methodology. The predictive power gained from this method was also complementary to other novel diagnostic modalities.

Stone et al. [84–86] in a series of studies investigate the role of ESS in the progression and prediction of CAD. In those studies, using the first pilot study as a proof of concept that the measurements of ESS *in vivo* could predict subsequent plaque progression, they calculated ESS and identified the areas with similar ESS to make it possible to evaluate and treat early, high-risk portions of artery. "*The PREDICTION Study*" [86] determined the role of local hemodynamics and vascular characteristics in plaque progression and the relationship with the plaque changes to clinical events and it is unique because the largest and most comprehensive serial anatomic natural history of atherosclerosis was presented and it was considered one of the most innovative methodologies to investigate the pathological mechanisms of CAD progression. For the validation of this study, 506 patients with acute coronary syndrome (ACS) undergoing percutaneous coronary intervention (PCI) were enrolled at the baseline. Vascular profiling was used to reconstruct each artery using intravascular ultrasound (IVUS) and coronary angiography and for the calculation of the ESS and plaque characteristics. Each reconstructed artery was divided into segments of 3 mm starting from the ostium. The same coronary arteries were used for the evaluation at the time of 6–10 months of follow-up in order to determine the changes in vascular characteristics. Clinical events were infrequent during the 1-year follow-up: 1 cardiac death, 4 patients with ACS in nonstented arteries, 15 hospitalizations of patients with stable angina, 3 patients with both native and stented segments treated with revascularization, and 13 patients who occurred an ACS. Thus, the number of patients with symptomatic clinical events was small to make significant statistical inferences between baseline and atherosclerotic events. The increase in plaque area was predicted by the large plaque burden which was observed at baseline, while the low ESS could predict the decrease in lumen area. These two factors (low ESS and large plaque burden) enabled to predict the exploratory end points of increased plaque burden and worsening of clinical luminal obstructions treated with a PCI at follow-up. The combination of these two independent predictors at baseline had a 41% positive and 92% negative predictive value for the prediction of the progression of obstructive artery disease treated with PCI.

As we mentioned earlier, the development in medical imaging offered the opportunity for more advanced CFD models and, consequently, improved realistic geometric models. Rybicki et al. [87] performed a study with the purpose to describe the 320-detector row CT, as a new tool to predict coronary artery plaque progression and potential rupture. To compute the flow patterns, the Lattice-Boltzmann method had been adopted. Although this approach overcame the limitations of other CFD methods for the proper evaluation of multiple coronary arteries including bifurcations noninvasively, ESS maps from CT vascular profile had not been demonstrated to be reproducible.

Rikhtegar et al. [88] presented a similar patient-specific computational study with the difference that this study was focusing on left coronary arteries (LCAs). For 96 total plaques, ASS had a higher sensitivity for the prediction of the plaque than the other factors and RRT had a higher PPV. Parameter performance was comparable to that in RCA except for one part at the start of the LCA. The main limitations of these two studies which used the optimal wall shear parameter for the prediction of plaque location were: (i) the generic inflow profiles and outlet boundary conditions which accept the assumption that there were not any artery motion and rigid vessel walls, and (ii) the reconstruction of the respective artery's healthy state anatomy from CTCA data of the diseased state.

In 2010, Vlachopoulos et al. [89] presented a systematic review study aiming to calculate robust quantitative estimates on the predictive values of blood pressure and derived hemodynamics markers for cardiovascular outcomes and all-cause mortality. For this purpose, 11 previous relevant studies were meta-analyzed and used 5648 subjects for a mean of 45 months' follow-up in which one or more indexes of hemodynamics like systolic blood pressure (SBP), pulse pressure (PP), and augmentation index (Alx) appeared and looked for a combined cardiovascular (CV) outcome or CV mortality or total mortality was evaluated. This analysis indicated that the central pressures and indices confessed a significant predictive ability of CV events and all-cause mortality (only Alx) in a range of population. Hemodynamic indexes proved to be independent predictors of future CV events and all-cause modality, while PP had no significantly better predictive value in comparison with branchial PP.

Another study in which the aim was the association between coronary artery ESS and the progression or transformation of atherosclerotic plaque and arterial remodeling in patients with CAD was held by Samady et al. [90]. Twenty patients underwent baseline and six-month follow-up VH-IVUS and biplane angiography. Also, baseline CFD models for ESS were developed. An increase in plaque burden was observed in low ESS areas than the high and intermediate ESS. Furthermore, a greater reduction of vessel and lumen area was noticed at low ESS segments while at high ESS segments, an increase in vessel and lumen area was found. This complex interplay between ESS categories and plaque progression enables the understanding of the role of the biomechanical factors to contribute to human coronary atherosclerosis progression and plaque transformation.

Stone et al. [91] utilized multimodality intravascular imaging aiming to detect high-risk patients using clinical and lesion-related factors. In this prospective study, 697 patients with ACS underwent coronary angiography followed by both intravascular ultrasonography of the LCA. The follow-up period was three years. In this three-year rate, major adverse cardiovascular events were found to be culprit lesions (CLs) (12.9%) and nonculprit lesions (NCLs) (11.6%), which were responsible for follow-up events and were angiographically mild at baseline ($32.3\% \pm 20.6\%$). On multivariate analysis, these lesions were associated with recurrent events and were characterized by a plaque burden of ≥70% or were classified according to the radiofrequency intravascular ultrasonography as thin-cap fibroatheroma. This led to the conclusion that in patients who underwent coronary intervention, major cardiovascular events which made their appearance during the follow-up were attributable to recurrence at the sites of culprit and NCLs with similar probability. Through this prospective study, the hypothesis that ACS may arise from atheroma with certain histopathological characteristics is established, even if these characteristics are not dependent on the degree of angiographic stenosis. Many more studies were presented in the same concept applying blood flow modeling and statistical modeling to predict disease progression [92–94].

A more advanced approach was presented by the Unit of Medical Technology and Intelligent Information Systems, Ioannina, Greece. In particular, for the first time, a complex approach to disease prediction was proposed and achieved by combining multilevel computational modeling of plaque growth with predictive models either based on statistical models or machine learning. Initially, the computational models included information on ESS and LDL concentration as calculated using finite elements. These results at the arterial level were combined with system-based molecular markers to predict disease progression using data from 186 patients (480 vessels) [95]. The binary logistic model depicts over 80% accuracy to predict disease progression when imaging, modeling, and lipidomics markers are combined into one model. When only computational results are used, the accuracy is about 60–65% [96, 97]. When advanced modeling of plaque growth is applied as presented in Chapter 9 of this book, including the major mechanisms of atherosclerosis, the accuracy to predict the disease progression is around 85% [98]. These models have been integrated into a unique platform which can be used by cardiologists and clinicians for the risk stratification, diagnosis, and prognosis of CVD [99–101]. It is expected that by applying machine learning or deep learning approaches to such datasets, the predictive accuracy of disease progression will further increase [102].

## 10.6 Conclusion

In conclusion, the final level of multiscale modeling could be the development of predictive models for the prediction of disease outcomes such as events, or the classification of patients into high- vs. low-risk patients. Besides the fact that traditionally statistical modeling in the form of linear or logistic regression was performed, these models had many limitations. First, it was difficult to include many parameters and second due to low accuracy, the researchers applied thresholding in order to optimize the models in terms of sensitivity or specificity. On the other hand, in the last years, ML techniques are utilized for this purpose enabling the classification of patients with increased accuracy. Moreover, deep learning algorithms allow the use of big data such as omics data going beyond the state of the art in the prediction of events or the identification of risk. Finally, the lack of explainability of the ML models was overcome through the concept of explainable AI. For this purpose, many algorithms and approaches have been developed, which enables among others the explanation of "which" and "how" each independent variable affects the target under identification.

## References


1 Nosek, B.A. et al. (2015). SCIENTIFIC STANDARDS. Promoting an open research culture. *Science* 348 (6242): 1422–1425. https://doi.org/10.1126/science.aab2374.

2 Iqbal, S.A., Wallach, J.D., Khoury, M.J. et al. (2016). Reproducible research practices and transparency across the biomedical literature. *PLoS Biol.* 14 (1): e1002333. https://doi.org/10.1371/journal.pbio.1002333.

3 Kourou, K.D. et al. (2019). Cohort harmonization and integrative analysis from a biomedical engineering perspective. *IEEE Rev. Biomed. Eng.* 12: 303–318. https://doi.org/10.1109/RBME.2018.2855055.

4 Khatri, V. and Brown, C.V. (2010). Designing data governance. *Commun. ACM* 53 (1): 148–152. https://doi.org/10.1145/1629175.1629210.

5 Krishnan, K. (2013). Introduction. In: *Data Warehousing in the Age of Big Data* (ed. K. Krishnan), xix–xxiii. Boston, MA: Morgan Kaufmann https://doi.org/10.1016/B978-0-12-405891-0.00016-7.

6 J. L. Kulikowski, "Data quality assessment," *Encyclopedia of Database Technologies and Applications*, 2005. https://www.igi-global.com/chapter/data-quality-assessment/20722 (accessed 29 November 2022).

7 M. Stonebraker Bruckner, D., Ilyas, I.F., Beskales, G., Cherniack, M., Zdonik, S.B., Pagan, A., & Xu, S., "Data Curation at Scale: The Data Tamer System. Conference on Innovative Data Systems Research."

8 Pipino, L.L., Lee, Y.W., and Wang, R.Y. (2002). Data quality assessment. *Commun. ACM* 45 (4): 211–218. https://doi.org/10.1145/505248.506010.

9 Pezoulas, V.C., Exarchos, T.P., and Fotiadis, D.I. (2020). *Medical Data Sharing, Harmonization and Analytics*, 1e. Academic Press https://www.elsevier.com/books/medical-data-sharing-harmonization-and-analytics/pezoulas/978-0-12-816507-2 (accessed 29 November2022).

10 Cohen, W., Ravikumar, P., and Fienberg, S. (2003). A comparison of string metrics for matching names and records. In: *KDD Workshop on Data Cleaning and Object Consolidation.*

11 Fortier, I., Doiron, D., Burton, P., and Raina, P. (2011). Invited commentary: consolidating data harmonization – how to obtain quality and applicability? *Am. J. Epidemiol.* 174 (3): 261–264; author reply 265–266. https://doi.org/10.1093/aje/kwr194.

12 Doiron, D. et al. (2013). Data harmonization and federated analysis of population-based studies: the BioSHaRE project. *Emerg. Themes Epidemiol.* 10 (1): 12. https://doi.org/10.1186/1742-7622-10-12.

13 Fortier, I. et al. (2010). Quality, quantity and harmony: the DataSHaPER approach to integrating data across bioclinical studies. *Int. J. Epidemiol.* 39 (5): 1383–1393. https://doi.org/10.1093/ije/dyq139.

14 Doiron, D., Marcon, Y., Fortier, I. et al. (2017). Software application profile: Opal and Mica: open-source software solutions for epidemiological data management, harmonization and dissemination. *Int. J. Epidemiol.* 46 (5): 1372–1378. https://doi.org/10.1093/ije/dyx180.

15 Doiron, D., Raina, P., Ferretti, V. et al. (2012). Facilitating collaborative research: implementing a platform supporting data harmonization and pooling. *Nor. Epidemiol.* 21: 221–224. https://doi.org/10.5324/nje.v21i2.1497.

**16** Welcome to the NCBO BioPortal. NCBO BioPortal. https://bioportal.bioontology.org/ (accessed 29 November 2022).

**17** Pang, C. et al. (2015). SORTA: a system for ontology-based re-coding and technical annotation of biomedical phenotype data. *Database J. Biol. Databases Curation* 2015: https://doi.org/10.1093/database/bav089.

**18** S-Match: An Open Source Framework for Matching Lightweight Ontologies. www.semantic-web-journal.net. http://www.semantic-web-journal.net/content/s-match-open-source-framework-matching-lightweight-ontologies (accessed 29 November 2022).

**19** Giunchiglia, F., Yatskevich, M., Avesani, P., and Shivaiko, P. (2009). A large dataset for the evaluation of ontology matching. *Knowl. Eng. Rev.* 24: 137–157. https://doi.org/10.1017/S026988890900023X.

**20** Ehrig, M. and Sure-Vetter, Y. (2005). FOAM - Framework for Ontology Alignment and Mapping - Results of the Ontology Alignment Evaluation Initiative. vol. 156.

**21** The Gene Ontology Consortium (2019). The Gene Ontology Resource: 20 years and still GOing strong. *Nucleic Acids Res.* 47 (D1): D330–D338. https://doi.org/10.1093/nar/gky1055.

**22** Pang, C. et al. (2014). BiobankConnect: software to rapidly connect data elements for pooled analysis across biobanks using ontological and lexical indexing. *J. Am. Med. Inform. Assoc.* 22: https://doi.org/10.1136/amiajnl-2013-002577.

**23** Karegowda, A., Manjunath, A., and Jayaram, M. (2010) Comparative study of attribute selection using gain ratio and correlation based feature selection. 2010. https://www.semanticscholar.org/paper/COMPARATIVE-STUDY-OF-ATTRIBUTE-SELECTION-USING-GAIN-Karegowda-Manjunath/4c21153c28d29c2d190d18b7666a95e4ef3ecdda (accessed 29 November 2022).

**24** A. Jović, K. Brkić, and N. Bogunović, "A review of feature selection methods with applications," in *2015 38th International Convention on Information and Communication Technology, Electronics and Microelectronics (MIPRO)*, May 2015, pp. 1200–1205. doi: https://doi.org/10.1109/MIPRO.2015.7160458.

**25** Liu, H., Zhou, M., Lu, X.S., and Yao, C. Weighted Gini index feature selection method for imbalanced data. In: *2018 IEEE 15th International Conference on Networking, Sensing and Control (ICNSC)*, March 2018, 1–6. https://doi.org/10.1109/ICNSC.2018.8361371.

**26** Yu, L. and Liu, H. (2003). Feature selection for high-dimensional data: a fast correlation-based filter solution. In: *Proceedings of the Twentieth International Conference on International Conference on Machine Learning (ICML'03)*, 856–863. AAAI Press.

**27** Peng, H., Long, F., and Ding, C. (2005). Feature selection based on mutual information criteria of max-dependency, max-relevance, and min-redundancy. *IEEE Trans. Pattern Anal. Mach. Intell.* 27 (8): 1226–1238. https://doi.org/10.1109/TPAMI.2005.159.

**28** Gunst, R.F. and Mason, R.L. (1980). *Regression Analysis and its Application: A Data-Oriented Approach. Routledge & CRC Press* https://www.routledge.com/Regression-Analysis-and-its-Application-A-Data-Oriented-Approach/Gunst-Mason/p/book/9780367403430 (accessed 29 November 2022).

**29** Huang, S., Cai, N., Pacheco, P.P. et al. (2018). Applications of Support Vector Machine (SVM) learning in cancer genomics. *Cancer Genom. Proteom.* 15 (1): 41–51. https://doi.org/10.21873/cgp.20063.

**30** Gunst, R.F. and Mason, R.L. (1980). *Classification and Regression Trees. Routledge & CRC Press* https://www.routledge.com/Classification-and-Regression-Trees/Breiman-Friedman-Stone-Olshen/p/book/9780412048418 (accessed 29 November 2022).

**31** Kubat, M. (2015). *An Introduction to Machine Learning.* Springer International Publishing https://doi.org/10.1007/978-3-319-20010-1.

**32** Hertz, J., Krogh, A., and Palmer, R.G. (2019). *Introduction to the Theory of Neural Computation.* CRC Press https://doi.org/10.1201/9780429499661.

**33** Simeone, O. (2017)"*A Brief Introduction to Machine Learning for Engineers.*" ArXiv, abs/1709.02840.

**34** Jain, A.K. (2010). Data clustering: 50 years beyond K-means. *Pattern Recognit. Lett.* 31 (8): 651–666. https://doi.org/10.1016/j.patrec.2009.09.011.

**35** von Luxburg, U. (2007). A tutorial on spectral clustering. *ArXiv07110189 Cs* http://arxiv.org/abs/0711.0189 (accessed 29 November 2022).

**36** Bouguettaya, A., Yu, Q., Liu, X. et al. (2015). Efficient agglomerative hierarchical clustering. *Expert Syst. Appl.* 42 (5): 2785–2797. https://doi.org/10.1016/j.eswa.2014.09.054.

**37** Miotto, R., Wang, F., Wang, S. et al. (2018). Deep learning for healthcare: review, opportunities and challenges. *Briefings Bioinf.* 19 (6): 1236–1246. https://doi.org/10.1093/bib/bbx044.

**38** Z. Che, S. Purushotham, R. Khemani, and Y. Liu (2015), "Distilling knowledge from deep networks with applications to healthcare domain," ArXiv, abs/1512.03542.

**39** Ronneberger, O., Fischer, P., and Brox, T. (2015). U-net: convolutional networks for biomedical image segmentation. *ArXiv150504597 Cs* http://arxiv.org/abs/1505.04597 (accessed 29 November 2022).

**40** Bun, M.J.G. and Harrison, T.D. (2019). OLS and IV estimation of regression models including endogenous interaction terms. *Econom. Rev.* 38 (7): 814–827. https://doi.org/10.1080/07474938.2018.1427486.

**41** Kida, Y. (2020). Generalized linear models. *Medium* https://towardsdatascience.com/generalized-linear-models-9cbf848bb8ab (accessed 29 November 2022).

**42** Carroll, R.J. and Ruppert, D. (2019). *Transformation and Weighting in Regression. Routledge & CRC Press* https://www.routledge.com/Transformation-and-Weighting-in-Regression/Carroll-Ruppert/p/book/9780367403379 (accessed 29 November 2022).

**43** Liu, Z. and Li, C. (2017). Recursive least squares for censored regression. *IEEE Trans. Signal Process.* 65 (6): 1565–1579. https://doi.org/10.1109/TSP.2016.2646660.

**44** Jing Li & Sunil K. Dhar (2013) Modeling with Bivariate Geometric Distributions, *Communications in Statistics - Theory and Methods*, 42:2, 252–266

**45** Chen, Y., Georgiou, T.T., and Tannenbaum, A.R. (2020). Probabilistic Kernel support vector machines. *ArXiv190406762 Cs Eess Stat* http://arxiv.org/abs/1904.06762 (accessed 29 November 2022).

**46** Gu, B., Quan, X., Gu, Y. et al. (2018). Chunk incremental learning for cost-sensitive hinge loss support vector machine. *Pattern Recognit.* 83: 196–208. https://doi.org/10.1016/j.patcog.2018.05.023.

**47** D. Meyer, "*Support Vector Machines * The Interface to Libsvm in Package*," 2001. https://www.semanticscholar.org/paper/Support-Vector-Machines-%E2%88%97-The-Interface-to-libsvm-Meyer/59422b2de46f45b34a9811b4773d4725199682f5 (accessed 29 November 2022).

**48** Karabulut, E.M. and İbrikçi, T. (2012). Effective diagnosis of coronary artery disease using the rotation forest ensemble method. *J. Med. Syst.* 36 (5): 3011–3018. https://doi.org/10.1007/s10916-011-9778-y.

**49** Bassett, R. and Deride, J. (2019). Maximum a posteriori estimators as a limit of Bayes estimators. *Math. Program.* 174: 129–144.

**50** Song, Y. and Lu, Y. (2015). Decision tree methods: applications for classification and prediction. *Shanghai Arch. Psychiatry* 27 (2): 130–135. https://doi.org/10.11919/j.issn.1002-0829.215044.

**51** Lee, S.-J., Xu, Z., Li, T., and Yang, Y. (2018). A novel bagging C4.5 algorithm based on wrapper feature selection for supporting wise clinical decision making. *J. Biomed. Inform.* 78: 144–155. https://doi.org/10.1016/j.jbi.2017.11.005.

**52** Baskin, I.I., Marcou, G., Horvath, D., and Varnek, A. (2017). Bagging and boosting of classification models. In: *Tutorials in Chemoinformatics*, 241–247. Wiley https://doi.org/10.1002/9781119161110.ch15.

**53** Schapire, R.E. (2013). Explaining AdaBoost. In: *Empirical Inference* (ed. B. Schölkopf, Z. Luo and V. Vovk), 37–52. Berlin, Heidelberg: Springer Berlin Heidelberg https://doi.org/10.1007/978-3-642-41136-6_5.

**54** Friedman, J.H. (2001). Greedy function approximation: a gradient boosting machine. *Ann. Stat.* 29 (5): 1189–1232.

**55** Chen, T. and Guestrin, C. (2016). XGBoost: a scalable tree boosting system. In: *Proceedings of the 22nd ACM SIGKDD International Conference on Knowledge Discovery and Data Mining (KDD '16)*, 785–794. New York, NY, USA: Association for Computing Machinery https://doi.org/10.1145/2939672.2939785.

**56** Lipton, Z.C., Berkowitz, J., and Elkan, C. (2015). A critical review of recurrent neural networks for sequence learning. *ArXiv150600019 Cs* http://arxiv.org/abs/1506.00019 (accessed 29 November 2022).

**57** Ding, C.H.Q., He, X., Zha, H. et al. (2001). A min-max cut algorithm for graph partitioning and data clustering. In: *Proceedings 2001 IEEE International Conference on Data Mining*, 107–114. https://doi.org/10.1109/ICDM.2001.989507.

**58** Pezoulas, V.C., Michalopoulos, K., Klados, M.A. et al. (2019). Functional connectivity analysis of cerebellum using spatially constrained spectral clustering. *IEEE J. Biomed. Health Inform.* 23 (4): 1710–1719. https://doi.org/10.1109/JBHI.2018.2868918.

**59** Belle, V. and Papantonis, I. (2021). Principles and practice of explainable machine learning. *Front. Big Data* 4: 688969. https://doi.org/10.3389/fdata.2021.688969.

**60** Özbakır, L., Baykasoğlu, A., and Kulluk, S. (2010). A soft computing-based approach for integrated training and rule extraction from artificial neural networks: DIFACONN-miner. *Appl. Soft Comput.* 10 (1): 304–317. https://doi.org/10.1016/j.asoc.2009.08.008.

**61** Sato, M. and Tsukimoto, H. (2001). Rule extraction from neural networks via decision tree induction. In: *IJCNN'01. International Joint Conference on Neural Networks. Proceedings (Cat. No.01CH37222)*, vol. 3, 1870–1875. https://doi.org/10.1109/IJCNN.2001.938448.

**62** Zilke, J.R., Loza Mencía, E., and Janssen, F. (2016). DeepRED – rule extraction from deep neural networks. In: *Discovery Science*, Lecture Notes in Computer Science, vol. 9956 (ed. T. Calders, M. Ceci and D. Malerba), 457–473. Cham: Springer https://doi.org/10.1007/978-3-319-46307-0_29.

**63** Micaelli, P. and Storkey, A. (2019). Zero-shot knowledge transfer via adversarial belief matching. *ArXiv190509768 Cs Stat* http://arxiv.org/abs/1905.09768 (accessed 29 November 2022).

**64** Montavon, G., Bach, S., Binder, A. et al. (2017). Explaining nonlinear classification decisions with deep taylor decomposition. *Pattern Recognit.* 65: 211–222. https://doi.org/10.1016/j.patcog.2016.11.008.

**65** Sundararajan, M., Taly, A., and Yan, Q. (2017). Axiomatic attribution for deep networks. *ArXiv170301365 Cs* http://arxiv.org/abs/1703.01365 (accessed 29 November 2022).

**66** He, T. et al. (2017). Diagnostic models of the pre-test probability of stable coronary artery disease: a systematic review. *Clin. Sao Paulo Braz.* 72 (3): 188–196. https://doi.org/10.6061/clinics/2017(03)10.

**67** Georga, E.I. et al. (2019). Artificial intelligence and data mining methods for cardiovascular risk prediction. In: *Cardiovascular Computing—Methodologies and Clinical Applications* (ed. S. Golemati and K.S. Nikita), 279–301. Singapore: Springer https://doi.org/10.1007/978-981-10-5092-3_14.

**68** Völzke, H. et al. (2013). Personalized cardiovascular medicine: concepts and methodological considerations. *Nat. Rev. Cardiol.* 10 (6): 308–316. https://doi.org/10.1038/nrcardio.2013.35.

**69** Ng, K., Steinhubl, S.R., deFilippi, C. et al. (2016). Early detection of heart failure using electronic health records: practical implications for time before diagnosis, data diversity, data quantity, and data density. *Circ. Cardiovasc. Qual. Outcomes* 9 (6): 649–658. https://doi.org/10.1161/CIRCOUTCOMES.116.002797.

**70** Karaolis, M.A., Moutiris, J.A., Hadjipanayi, D., and Pattichis, C.S. (2010). Assessment of the risk factors of coronary heart events based on data mining with decision trees. *IEEE Trans. Inf. Technol. Biomed. Publ. IEEE Eng. Med. Biol. Soc.* 14 (3): 559–566. https://doi.org/10.1109/TITB.2009.2038906.

**71** Austin, P.C., Tu, J.V., Ho, J.E. et al. (2013). Using methods from the data-mining and machine-learning literature for disease classification and prediction: a case study examining classification of heart failure subtypes. *J. Clin. Epidemiol.* 66 (4): 398–407. https://doi.org/10.1016/j.jclinepi.2012.11.008.

**72** Hassan, N., Sayed, O.R., Khalil, A.M., and Ghany, M.A. (2017). Fuzzy soft expert system in prediction of coronary artery disease. *Int. J. Fuzzy Syst.* 19 (5): 1546–1559. https://doi.org/10.1007/s40815-016-0255-0.

**73** Motwani, M. et al. (2017). Machine learning for prediction of all-cause mortality in patients with suspected coronary artery disease: a 5-year multicentre prospective registry analysis. *Eur. Heart J.* 38 (7): 500–507. https://doi.org/10.1093/eurheartj/ehw188.

**74** Weng, S.F., Reps, J., Kai, J. et al. (2017). Can machine-learning improve cardiovascular risk prediction using routine clinical data? *PLoS One* 12 (4): e0174944. https://doi.org/10.1371/journal.pone.0174944.

**75** Rumsfeld, J.S., Joynt, K.E., and Maddox, T.M. (2016). Big data analytics to improve cardiovascular care: promise and challenges. *Nat. Rev. Cardiol.* 13 (6): 350–359. https://doi.org/10.1038/nrcardio.2016.42.

**76** Kurt, I., Ture, M., and Kurum, A.T. (2008). Comparing performances of logistic regression, classification and regression tree, and neural networks for predicting coronary artery disease. *Expert Syst. Appl.* 34 (1): 366–374. https://doi.org/10.1016/j.eswa.2006.09.004.

**77** Verma, L., Srivastava, S., and Negi, P.C. (2016). A hybrid data mining model to predict coronary artery disease cases using non-invasive clinical data. *J. Med. Syst.* 40 (7): 178. https://doi.org/10.1007/s10916-016-0536-z.

**78** Alizadehsani, R. et al. (2013). A data mining approach for diagnosis of coronary artery disease. *Comput. Meth. Programs Biomed.* 111 (1): 52–61. https://doi.org/10.1016/j.cmpb.2013.03.004.

**79** Elashoff, M.R. et al. (2011). Development of a blood-based gene expression algorithm for assessment of obstructive coronary artery disease in non-diabetic patients. *BMC Med. Genet.* 4: 26. https://doi.org/10.1186/1755-8794-4-26.

**80** Thomas, G.S. et al. (2013). A blood-based gene expression test for obstructive coronary artery disease tested in symptomatic nondiabetic patients referred for myocardial perfusion imaging the COMPASS study. *Circ. Cardiovasc. Genet.* 6 (2): 154–162. https://doi.org/10.1161/CIRCGENETICS.112.964015.

**81** Chatzizisis, Y.S., Coskun, A.U., Jonas, M. et al. (2007). Role of endothelial shear stress in the natural history of coronary atherosclerosis and vascular remodeling – molecular, cellular, and vascular behavior. *J. Am. Coll. Cardiol.* 49 (25): 2379–2393. https://doi.org/10.1016/j.jacc.2007.02.059.

**82** Wilson, P.W., D'Agostino, R.B., Levy, D. et al. (1998). Prediction of coronary heart disease using risk factor categories. *Circulation* 97 (18): 1837–1847. https://doi.org/10.1161/01.cir.97.18.1837.

**83** Feldman, C.L., Ilegbusi, O.J., Hu, Z. et al. (2002). Determination of in vivo velocity and endothelial shear stress patterns with phasic flow in human coronary arteries: a methodology to predict progression of coronary atherosclerosis. *Am. Heart J.* 143 (6): 931–939. https://doi.org/10.1067/mhj.2002.123118.

**84** Stone, P.H. et al. (2007). Regions of low endothelial shear stress are the sites where coronary plaque progresses and vascular remodelling occurs in humans: an in vivo serial study. *Eur. Heart J.* 28 (6): 705–710. https://doi.org/10.1093/eurheartj/ehl575.

**85** Stone, P.H. et al. (2003). Effect of endothelial shear stress on the progression of coronary artery disease, vascular remodeling, and in-stent restenosis in humans - in vivo 6-month follow-up study. *Circulation* 108 (4): 438–444. https://doi.org/10.1161/01.Cir.0000080882.35274.Ad.

**86** Stone, P.H. et al. (2012). Prediction of progression of coronary artery disease and clinical outcomes using vascular profiling of endothelial shear stress and arterial plaque characteristics the PREDICTION study. *Circulation* 126 (2): 172–181. https://doi.org/10.1161/Circulationaha.112.096438.

**87** Rybicki, F.J. et al. (2009). Prediction of coronary artery plaque progression and potential rupture from 320-detector row prospectively ECG-gated single heart beat CT angiography: lattice Boltzmann evaluation of endothelial shear stress. *Int. J. Cardiovasc. Imaging* 25 (2): 289–299. https://doi.org/10.1007/s10554-008-9418-x.

**88** Rikhtegar, F. et al. (2012). Choosing the optimal wall shear parameter for the prediction of plaque location-A patient-specific computational study in human left coronary arteries. *Atherosclerosis* 221 (2): 432–437. https://doi.org/10.1016/j.atherosclerosis.2012.01.018.

**89** Vlachopoulos, C., Aznaouridis, K., O'Rourke, M.F. et al. (2010). Prediction of cardiovascular events and all-cause mortality with central haemodynamics: a systematic review and meta-analysis. *Eur. Heart J.* 31 (15): 1865–1871. https://doi.org/10.1093/eurheartj/ehq024.

**90** Samady, H. et al. (2011). Coronary artery wall shear stress is associated with progression and transformation of atherosclerotic plaque and arterial remodeling in patients with coronary artery disease. *Circulation* 124 (7): 779–788. https://doi.org/10.1161/Circulationaha.111.021824.

**91** Stone, G.W. et al. (2011). A prospective natural-history study of coronary atherosclerosis. *N. Engl. J. Med.* 364 (3): 226–235. https://doi.org/10.1056/NEJMoa1002358.

**92** Bourantas, C.V. et al. (2019). Implications of the local haemodynamic forces on the phenotype of coronary plaques. *Heart* 105 (14): 1078–1086. https://doi.org/10.1136/heartjnl-2018-314086.

**93** Bourantas, C.V. et al. (2016). Noninvasive prediction of atherosclerotic progression: the PROSPECT-MSCT study. *JACC Cardiovasc. Imaging* 9 (8): 1009–1011. https://doi.org/10.1016/j.jcmg.2015.07.005.

**94** Bourantas, C.V. et al. (2019). Utility of multimodality intravascular imaging and the local hemodynamic forces to predict atherosclerotic disease progression. *JACC Cardiovasc. Imaging* https://doi.org/10.1016/j.jcmg.2019.02.026.

**95** Sakellarios, A.I. et al. (2021). Non-invasive prediction of site-specific coronary atherosclerotic plaque progression using lipidomics, blood flow, and LDL transport modeling. *Appl. Sci.* 11 (5), Art. no. 5: https://doi.org/10.3390/app11051976.

**96** Sakellarios, A. et al. (2017). The effect of coronary bifurcation and haemodynamics in prediction of atherosclerotic plaque development: a serial computed tomographic coronary angiographic study. *EuroIntervention* 13 (9): e1084–e1091. https://doi.org/10.4244/EIJ-D-16-00929.

**97** Sakellarios, A. et al. (2016). Prediction of atherosclerotic disease progression using LDL transport modelling: a serial computed tomographic coronary angiographic study. *Eur. Heart J. Cardiovasc. Imaging* https://doi.org/10.1093/ehjci/jew035.

**98** Pleouras, D.S. et al. (2020). Simulation of atherosclerotic plaque growth using computational biomechanics and patient-specific data. *Sci. Rep.* 10 (1), Art. no. 1: https://doi.org/10.1038/s41598-020-74583-y.

**99** Sakellarios, A. et al. (2018). A clinical decision support platform for the risk stratification, diagnosis, and prediction of coronary artery disease evolution. *Annual International Conference of the IEEE Engineering in Medicine and Biology Society* 2018: 4556–4559. https://doi.org/10.1109/EMBC.2018.8513131.

**100** Sakellarios, A.I. et al. (2020). Prediction of atherosclerotic disease progression combining computational modelling with machine learning, *2020 42nd Annual International Conference of the IEEE Engineering in Medicine Biology Society (EMBC)*. In:, 2760–2763. https://doi.org/10.1109/EMBC44109.2020.9176435.

**101** Sakellarios, A., Correia, J., Kyriakidis, S. et al. (2020). A cloud-based platform for the non-invasive management of coronary artery disease. *Enterp. Inf. Syst.* 14 (8): 1102–1123. https://www.tandfonline.com/doi/abs/10.1080/17517575.2020.1746975 (accessed 29 November 2022).

**102** Kigka, V.I. et al. (2019). Site specific prediction of atherosclerotic plaque progression using computational biomechanics and machine learning. In: *2019 41st Annual International Conference of the IEEE Engineering in Medicine and Biology Society (EMBC)*, 6998–7001. https://doi.org/10.1109/EMBC.2019.8856881.

# 11

# Future Trends

## 11.1 Virtual Populations

The term "virtual population generation" relates to the development of computer techniques for generating synthetic patient data by simulating real-world distributions. Its intended use is to increase the statistical power of clinical research databases that have a major population size deficiency [1]. Currently, there are many different methodologies to generate virtual data depending mainly on the types of data to be created. In the last years, the generation of virtual populations and synthetic data has attracted the interest of many research groups. The approaches fit well with the novel concept of the *in silico* clinical trials, where some of the *in vitro* and/or *in vivo* trials are replaced by *in silico* computational modeling. The definition of the virtual population would be: "an accurate representation in terms of features and their distribution of the real population."

Mathematical techniques such as statistics and machine learning have been used to generate new virtual patients based on the real characteristics and information extracted from patients' clinical data. Hence, a part of the process of virtual patients' generation from real clinical data is related to the development of model parameters which should represent the desired patient cohort. There are two major methodological categories, the parametric and the nonparametric. The parametric methods are used mostly to resample an existing dataset and create a new one with different combinations. This approach uses distributions to extract correlations among the dataset's features. The nonparametric is a straightforward resampling method where the virtual cohort is generated by randomly selected patients from the original dataset.

*Multiscale Modelling in Biomedical Engineering*, First Edition. Dimitrios I. Fotiadis, Antonis I. Sakellarios, and Vassiliki T. Potsika.

Published 2023 by John Wiley & Sons, Inc.

### 11.1.1 Methods for Virtual Population Generation

In 2011, Marchant et al. [2] created a virtual population employing multivariate normal distributions (MVNDs) and taking into account data correlations from 2595 individuals to study the efficiency of three different models of cardiovascular risk prediction. In another study, the discrete event simulation (DES) was employed in many different datasets for the generation of a virtual population, where the realistic virtual population is considered as a function of their initial characteristics at any time $t$ over a defined time span [3]. The aim of the study was to compare the treatment effect on the virtual population with the literature-based findings. Stochastic modeling was used in [4] to generate a virtual population for metabolic simulators and glucose regulation. Their methodology included the Markov chain Monte Carlo approach to estimate model parameters.

In 2013, a study presented by Schmidt et al. [5] described the mechanistic axes population ensemble linkage (MAPEL) algorithm, to generate a cohort with statistically calibrated virtual population and develop, extend, and improve the simulation results. The MAPEL algorithm uses a generalized weight for each virtual patient to generate a new virtual population. The chosen virtual cohort was calibrated manually to achieve results which match the clinical data.

Tannenbaum et al. [6] applied a covariate distribution modeling for the creation of the "virtual" population. The covariates were continuous or categorical. This approach uses real human data to create MVNDs for each combination of covariates/parameters. If there are multiple categorical covariates (e.g. smoking status and sex), the axis of the MVND diagrams, that represents these categorical covariates, is divided into subgroups or discrete points, each of them describing unique combinations of categorical covariates (e.g. one point would be smoking status: yes, sex: woman). A log-normal multivariate distribution is employed to constrain all biological covariates to be positive. Thus, instead of covariate values, their logarithms are used to define the MVND distributions. Next, to select the "virtual" patients, the continuous values of the previous categorical covariates are mapped to discrete values. A continuous critical value function is implemented to evaluate limits of the continuous values, and therefore ranges, within which a specific categorical value is applied. Both methods were validated using simulated and real data to compare their efficiency in generating realistic "virtual" patient distributions. Teutonico et al. [7] used this method to generate realistic covariate correlations, confirming that they are comparable to a real population in a multivariate aspect.

Marchant et al. [2, 8] created a "virtual" population from a real French population of one million individuals, aged 35–64 years. They used covariates related to human cardiovascular data as input parameters to the equations that compute the Framingham and Score cardiovascular risk scores. They used the continuous method of Tannenbaum et al. [6] to create continuous covariates from the categorized ones. Using real data, "virtual" individuals from the random selection of data were created and those with impossible values beyond the range of real distributions were eliminated. Then, a discrete value was assigned to the transformed categorized covariates into continuous and completed their "virtual" population generation. Their objective was to evaluate the Framingham and Score equations for predicting cardiovascular risk.

Schmidt et al. [5] developed the MAPEL algorithm to create "virtual" populations to explore the effect of drugs. The "virtual" population is an ensemble of parameterizations. Their innovation lies in their computationally faster method of generating a realistic "virtual" population. They used statistics from real clinical trials to extract correlations of medical data. Using the detailed correlations, they developed a method to create more realistic "virtual" populations, avoiding the creation of patients with impossible combinations of medical data. Their methods enabled also the creation of specific realistic subpopulations that are used in *in silico* clinical trials, achieving a better evaluation of a drug's response to targeted patient groups. A patient is described by parameters that are medical data, called mechanical axes. The first stage of the method includes the creation of weighting equations. In the second stage, they deploy these equations along with random correlations of different parameters to create bivariate weighting equations. Each of these equations expresses the frequency of a patient relative to the others, according to a pair of parameters' values. In addition, these are developed in such a way as to form a bivariate Gaussian distribution for every combination of parameter values. In the third stage, the response to each patient's treatment is quantified and introduced to the equations achieving the expression of the response of treatment distribution to the "virtual" population. Finally, in the last stage, statistical distributions of real clinical data (NHANES database) are used to define the random probabilities that were introduced in the previous stages. The iteration of this process, using different data from real clinical trials results in the creation of different "virtual" populations or subpopulations. Through this approach, a better correlation of real patients' parameters is achieved, which is validated by the comparison of real and "virtual" clinical trial results.

Allen et al. [3] developed an algorithm that generates "virtual" patients, avoiding the creation of patients with characteristics that are not representative of realistic values. The "virtual" patients consist of a set of parameters, which account for human physiology. The algorithm uses both specific mathematical models that describe human physiology and possible ranges of human parameters. The algorithm refines the initial set of parameters by removing those that are not representative to the real ranges of human characteristics. In detail, the first step of the algorithm includes the definition of possible ranges of model parameters, which result either from real human data or from theoretical considerations. Then, an iterative process is implemented to create "virtual" patients from model simulations, using model parameters' values belonging to the defined reasonable ranges. As "virtual" patients are considered those patients whose data belong to reasonable physiological ranges, this process leads to the creation of a "virtual" population, forming a uniform distribution of patient numbers per parameter values. These "virtual" patients are then refined again in the third step, where a modified cost function is implemented, considering the remaining "virtual" population and data of real humans, resulting in a more realistic population.

Moss *et al.* [9] analyzed the dynamics of the Guyton model [10], which is a whole-body circulatory model based on human physiology. They presented the model parameters' effects on the model dynamics in the concept of cause and effect, enabling the development of a realistic "virtual" population by defining model parameters according to data from real populations. In detail, the Guyton model was initially used to explore how blood pressure and cardiac output can be controlled [11]. Moss et al. used a version of the Guyton model, which considers 219 parameters and 359 output variables and restricted their analysis to 96 parameters leading to 276 output variables. Their input to the model consisted of 32 cardiac, 21 renal, 16 autoregulation, 16 hormonal, 11 local circulation, and 4 thirst-related parameters. In their analysis, a modified Guyton formula was used to determine which parameters have a significant effect on any model output variable [12]. For this reason, Monte Carlo simulations were implemented to solve the Guyton model. The objective of each simulation was to obtain the effect of a specific parameter on a specific output variable. To achieve this, multiple simulations were performed. In each simulation, random values were assigned to the input parameters along with a small perturbation in the examined one and the Guyton model was solved. They used two input values (the random and its perturbation) of a specified parameter leading to two different solutions. That enabled the examination of the specified parameter's effect on any output variable, by evaluating the gradient of the output variable's values to the input parameter's values. The simulations that reached a steady state defined two groups of "virtual" patients since two input groups of parameters led to possible outputs, describing two human bodies. 384 000 "virtual" patients were finally created to be used in the *in silico* clinical trials. Finally, recently artificial intelligent approaches have been implemented to generate "virtual" clinical images. Generative adversarial networks (GANs) have been used for the purpose of imaging data augmentation. Kazuhiro et al. [13] employed the GANs approach to generate MRI brain images. CT images were generated using a similar approach [14] or multicontrast textured XCAT phantoms [15] were developed using a dual-discriminator conditional-GAN.

Currently, there are software commercially available to be used for virtual population generation. SimCyp® "virtual" population is used for the evaluation of drug response. Through the use of "virtual" population, the simulation of oral absorption, tissue distribution, metabolism, and excretion of drugs and drug candidates in healthy and diseased populations is enabled [16]. The inclusion of patient-specific physiology and genetics information allows the use of small molecule dynamics and biologics in SimCyp algorithms [17]. The "virtual" population is composed of medical data which are used by the simulator's underlying equations. SimCyp simulator allows the user to change the "virtual" patient parameters and creates different "virtual" populations. These parameters are: (i) demographic and genetic (age, weight, height, sex, genetics, race, and disease) and (ii) anatomical and physiological (organ size, blood flow, enzymes, transporters, plasma protein, hematocrit, transit time, and pH values) [17, 18]. Biological, physiochemical and genetic variations allowed for the creation of 19 populations of different ethnic groups, life stages (geriatric, pediatric, pregnancy, and preterm babies), and disease states (renal/hepatic impairment, obesity, inflammation, psoriasis, and cancer) [19]. Materialise® provides a specialized software tool for the evaluation of the performance of bone replacement devices on "virtual" populations. The medical image-based population constitutes geometries of human organs and tissues, along with software that modifies them, according to statistical shape modeling. Material assignment is applied to the geometries of "virtual" population and is used as input to the simulation software [20, 21]. The "virtual" population provides an understanding of the effect of the anatomical characteristics, and unlike *in vivo* clinical trials, the *in silico* clinical trials are performed faster with thousands of "virtual" patients, reducing this way the time required for the product to reach the market. A benefit of this population analysis software is the reduction of design iterations required to acquire the final shape of the device [20].

The aim of the Nova Discovery® "virtual" population is to allow the prediction of a disease's progression with or without drug administration. To achieve this, "virtual" patients and drug candidates are combined with disease models [22]. Nova Discovery's software uses statistical "virtual" populations composed of specific medical-related variables. The "virtual" population has been generated based on a representative French population of 26 million individuals (age range 35–64 years old) [23]. Among the parameters used in the "virtual" population are: the SBP, the diastolic blood pressure (DBP), the total cholesterol (TC), the HDL cholesterol (HDL-c), the glycemia (Glyc), the serum creatinine (Creat), the weight, the height, the gender, the age, and the left ventricular hypertrophy (LVH) [24].

Immunetrics focuses on the analysis of human diseases and through the development of software tools, it enables the testing of novel drugs' responses in a "virtual" population. The "virtual" population is used to evaluate the disease models (models of rheumatoid arthritis, psoriatic arthritis, ulcerative colitis, sepsis, multiorgan dysfunction, and trauma [25, 26]). The "virtual" population matches the diversity of a real-world population as it consists of patients' medical data distributed according to the real population statistics [25]. Brown et al. used Immunetrics software and evaluated the prediction of biological progress in a "virtual" population [27].

Entelos created the Physiolab Platform which is used for clinical trials in drugs. Specifically, the "virtual" populations were used for the simulation of cardiovascular diseases, metabolism, rheumatoid arthritis, hypertension, hypersensitivity, dermis, and epidermis processes [28]. Physiolab platform allows users to create a "virtual" population with medical-related data which reflect the real world's population statistics. Klinke et al. used a "virtual" population created by the Metabolism Physiolab platform to identify and evaluate biomarkers for insulin sensitivity [29].

Finally, PopGen is a simulation tool that allows the creation of a "virtual" population, which consists of realistic human anatomical and physiological data. It was developed toward providing predictions of organ volumes and blood flow of the human heart of the "virtual" individuals to be used in pharmacokinetic (PBPK) models. The "virtual" population is based on an algorithm that uses data from body weight (BW), height (H), and body mass index (BMI) from real patients. The algorithm applies perturbation on the data of the real population toward the creation of the "virtual" population. The data which are used as an input for the "virtual" population consist of data from the NHANES III study [30] and the Annals of the ICRP [31]. The NHANES III dataset includes age-, gender-, and race-dependent anthropometric data of 30 000 individuals, aged from 2 months to 90+ years. These data are considered sufficient for statistical distribution curves for BW, (H), and BMI over the whole age range of Caucasian, Afro-American, and white Mexican American populations. From the ICRP report, mean organ weights were obtained in absolute values or as a fraction of the whole BW. For the adult models, the same density was assigned to every tissue, independent of model's age since only mean values were provided by ICRP [31]. The starting point of the PopGen "virtual" population is the creation of the "mean virtual" individual. This individual is characterized by the mean body height (Hmean) and mean body mass index (BMImean) per gender. Initially, patient's height is defined to develop a "virtual" patient. The "virtual" organ weights are calculated using a scaling function, which correlates the organ weight and the height. This process is in agreement with the results of Grandmaison et al. [32], who demonstrated a better correlation between organ weights and body height compared to the BMI or BW. The scaling equation is not applied to the brain, fat, and skin tissue. It was shown that the brain weight is independent of the BW or the height for a given age and gender. Fat tissue weight is the difference between the target BW and the sum of all other organ weights [33]. The skin weight of the target individual is evaluated by another scaling function, in which skin weight is proportional to the BW. To generate a "virtual" population, the same approach is performed, with the exception that age, gender, and height are randomly created, complying with the user's selected ranges. The process is iteratively performed until the desired number of "virtual" individuals is generated. McNally et al. [21] also described briefly the methods of generating PopGen's "virtual" population. They created a "virtual" population which was initially used in the PBPK simulations and then compared with experimental data.

### 11.1.2 A Methodological Approach for a Virtual Population

The virtual population generation stage [1] licensed under CC BY-ND 4.0, which is discussed in detail later, consists of five supervised and unsupervised virtual data generation methods. 1000 virtual patients were generated. Here is a description of this strategy, along with a methodology for its implementation.

#### 11.1.2.1 Multivariate Log-Normal Distribution (log-MVND)

MVND is defined as:

$$f(\mathbf{X}) = \frac{1}{(2\pi)^{p/2}|\Sigma|^{1/2}} e^{-(\mathbf{X}-\mu)\Sigma^{-1}(\mathbf{X}-\mu)/2}, \tag{11.1}$$

where $\mathbf{X} \in R^{pxn}$, $p$ is the dimension, $\boldsymbol{\mu}$ is the mean vector of $\mathbf{X}$, $\mathbf{\Sigma}$ is the covariance matrix of $\mathbf{X}$, and $\mathbf{\Sigma}^{-1}$ is the pseudoinverse of $\mathbf{\Sigma}$. A multidimensional normal distribution is constructed from the mean vector and the covariance matrix of the input data.

#### 11.1.2.2 Supervised Tree Ensembles

A more complex approach to creating virtual populations is to train a tree ensemble on a set of training characteristics and a goal feature [34]. We generate an ensemble similar to random forests [35] during the generator's training phase, but with some additional data for the data creation phase. Based on the univariate empirical cumulative distribution function of each inner tree node, we store the generator for the splitting feature (ECDF). In each leaf node, we maintain ECDF-based generators for any variables that are not encountered on the path from the root to that leaf. To avoid the overfitting effects introduced during the training phase of the tree ensemble construction, we propose a unique technique in which one of the trees in the ensemble is randomly chosen when constructing a new instance that is passed down the tree starting at the root node. As the generator, the ensemble approximates the probability density function of regions whose properties are considered to be independent (the dependencies are likely to be resolved on the path from the root to the leaves). Due to the enormous number of different trees in the ensemble, the created samples closely approach the probability density function of the original feature space.

#### 11.1.2.3 Unsupervised Tree Ensembles

The unsupervised tree ensemble generator is identical to the supervised tree ensemble generator, with the exception that it generates a dense forest ensemble rather than a random forest ensemble [35]. Density trees are used to construct the ensemble members similarly to decision trees, but using the variance of the features as the criterion for picking the splitting feature. To avoid introducing overfitting effects during the ensemble formation process, each density tree in the ensemble is randomly picked to have the slowest convergence rate. Any details on the desired feature are redundant. The remaining components are comparable in learning and generation to those in the supervised tree ensemble.

#### 11.1.2.4 Radial Basis Function-Based Artificial Neural Networks

The radial basis function (RBF)-based artificial neural network (ANN)'s output is defined as [36]:

$$y(\mathbf{q}) = \sum_{i=1}^{N} w_i \exp\left(-\beta \left\| \boldsymbol{q} - \boldsymbol{q}_i \right\|^2\right), \tag{11.2}$$

where $y(\mathbf{q})$ is the output of the ANN, $w_i$ is the weight of the i-th neuron, $\boldsymbol{q}_i$ is the center vector of the i-th neuron, $\|\boldsymbol{q} - \boldsymbol{q}_i\|$ is the distance of each sample in $\boldsymbol{q}$ from the center vector $\boldsymbol{q}_i$ in the i-th neuron, and $\beta$ is a standard Gaussian parameter. The RBF generator is created using a standard training strategy for estimating the Gaussian parameters of the neurons. Gaussian kernels are employed by the RBF generator to account for overfitting effects and create new instances proportionate to their presence in the training set.

#### 11.1.2.5 Bayesian Networks

Bayesian networks are predicated on the assumption that during learning, relationships between variables are discovered and recorded explicitly in the form of a directed acyclic graph (DAG). The nodes indicate characteristics, whereas the edges provide a causal link between the nodes. The weights on an edge specify the likelihood of a node reaching a certain value given its predecessor. To create new instances, the structure is utilized to produce new feature values in a way compatible with the features' causal relationships. The probability distribution is uniform if the node is discrete. If the node is continuous, mean and variance are associated with each discrete parent configuration.

#### 11.1.2.6 Performance Evaluation of the Quality of the Generated Virtual Patient Data

Many approaches for generator evaluation are described [37]. The Kolmogorov-Smirnoff goodness of fit (GOF) test is used to evaluate the similarity between the real and the virtual distributions. Additional metrics is the Pearson's correlation coefficient and the Kullback–Leibler (KL) divergence.

This methodology was examined for the classification of lymphoma [1] with the results demonstrating an improved performance after the data augmentation (Figure 11.1).

### 11.1.3 A Novel Approach for a Virtual Population Combining Multiscale Modeling

Another recent technique included the creation of a virtual population of patients with obstructive (>50% stenosis) atherosclerotic disease [38]. To do this, two populations were produced first and then blended to make the final one. Virtual clinical data are created in particular utilizing an enhanced data augmentation technique that effectively replaces deficient data without impairing the distribution of the original population. From a small number of patients, fewer than 200, 10 000 virtual patients with clinical data were developed [38]. Additionally, virtual artery geometries were generated using a mechanistic plaque development model. The lumen, the outside wall, and the plaque components were all incorporated in each set of the patient's geometry. Two virtual datasets are constructed utilizing the aforementioned methodologies: (i) one with clinical data using the statistical modeling technique and (ii) one with imaging/morphological data using the plaque growth modeling methodology. The final virtual population is created by combining the two datasets and populating it with virtual patients with clinical and geometric data. We combine the two virtual datasets using the original SMARTool population of 186 patients, which include clinical and geometric data [39]. Thus, three distinct datasets were used to produce the final virtual population (the SMARTool dataset with all sorts of data, one virtual with geometrically reconstructed data, and one virtual with clinical data alone). To create a big database of a virtual population of realistic patients, the idea of probabilistic record linkage [40] will be expanded. This concept was initially designed to match two datasets. This is an expansion of the original deterministic matching notion, which is based on the precise matching of specified fields. Previously established algorithms were mostly used for clinical or scientific reasons, where match precision is critical. On the other hand, with virtual populations, a certain amount of "error" or "variance" is necessary to produce a high number of examples while maintaining the biological validity of all created cases. In our scenario, we have three datasets: A, B, and C; dataset C (186 patients from the SMARTool

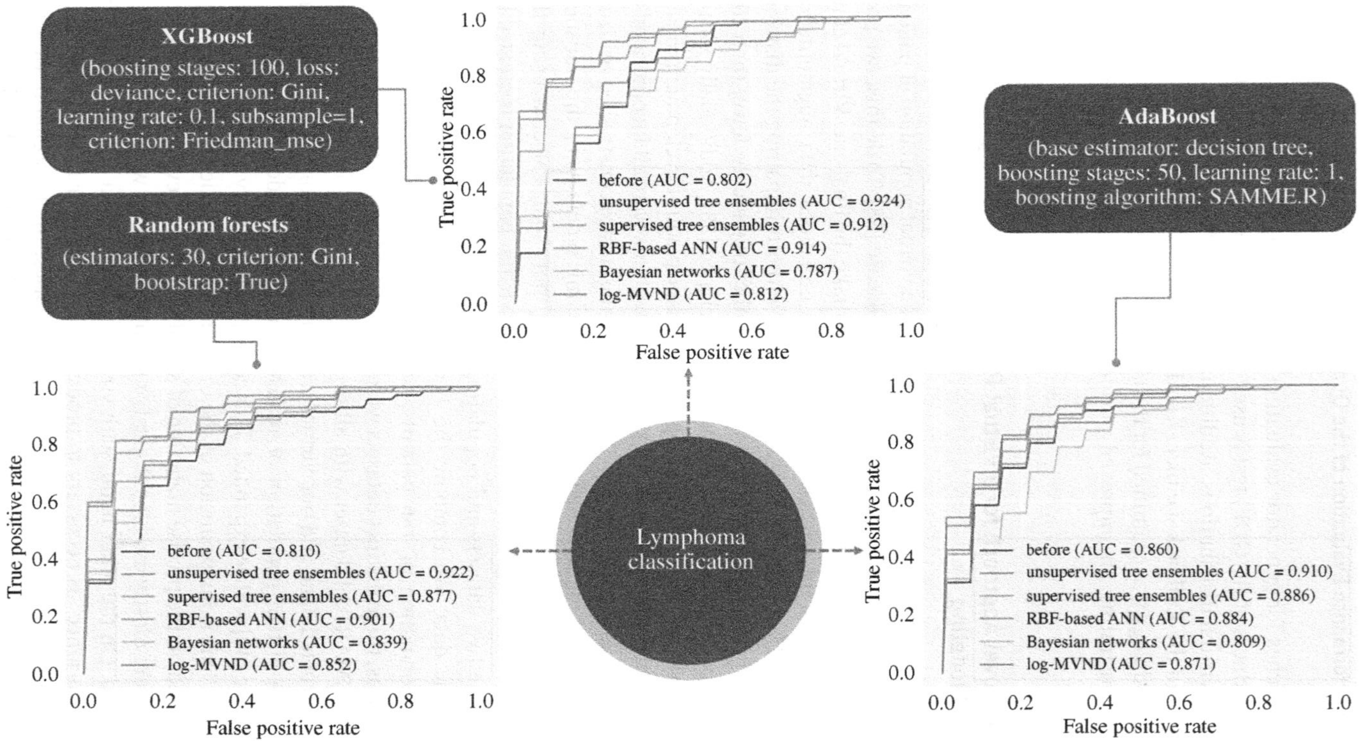


**Figure 11.1** ROC curves depicting the classification performance of the XGBoost, the AdaBoost, and the Random Forests for lymphoma classification with and without data augmentation. *Source:* Pezoulas et al. [1]/Elsevier/CC BY.

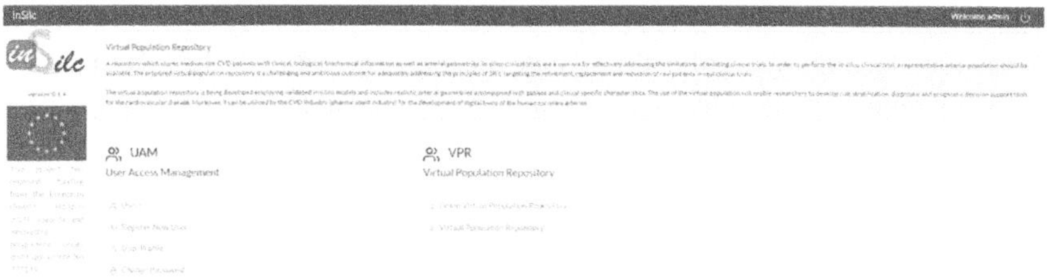


**Figure 11.2** Demo page of the cardiovascular virtual population.

population) has all sorts of attributes from both A (virtual dataset with clinical information) and B (virtual dataset without clinical information) (virtual dataset with geometrical information). Dataset C is included in the virtual population as a whole (after deidentification). To include the other two datasets (A and B), they must be merged with C or used in conjunction with C as a "linkage" table. The results were integrated into an online database which can be used for research purposes upon request (Figure 11.2).

## 11.2 Digital Twins

The digital twin idea has been applied to healthcare in recent years. A digital twin is a fictitious depiction of a real thing or system that exists throughout its existence. Digital twins in healthcare provide personalized therapy by integrating an individual's genetic, biochemical, physiological, and behavioral characteristics. The digital twin may be thought of as a revolutionary approach to incorporating multiscale modeling into healthcare and using it for diagnostic, prognostic, or preventative reasons while also including an interpretable workflow.

A benefit of digital twins in healthcare is that they provide a more accurate representation of any deviations from the norm. Thus, every human at any level (body, organ, tissue, etc.) may be represented *in silico* with a normal phenotype as well as a variety of diseased phenotypes, allowing for comparison and testing of any research ideas. This "normal" or healthy condition may be described at a high resolution and in numerous data dimensions during a person's lifetime using genetic, phenotypic, and behavioral data [41]. Natural diversity between people, which makes defining what is really normal challenging, may be represented in this high-dimensional space. Heterogeneity in data collecting is replaced by continuous monitoring of metrics during the course of one's life. This technique enables a far more precise statistical description of the "normal" or healthy condition, as well as illness states or susceptibilities to disease [41]. Additionally, the patient is characterized at a high resolution utilizing a huge quantity of data, which makes digital twins a method of customized therapy by allowing for the transition of the health condition to the illness state for the same patient. The fact that digital twins represent an individual's condition and enable transparent comparison of these people results in a conceptual shift in how health and sickness are defined. The transparency of what is "normal" in its variability begs the issue of whether natural levels are optimum and amenable to engineering [42]. When compared to other members of the population, the healthy state may suddenly be seen as unsatisfactory; a situation that demands correction. Additionally, the health condition may morph into a state of "symptomless sickness" as a result of the data's ability to infer disease probabilities.

### 11.2.1 Ecosystem of the Digital Twin for Health

This framework encapsulates the process of collecting data from personal health devices, analyzing it, and providing feedback to the user in a closed-loop cycle. By integrating noncompliant health devices with an X73 wrapper module, the framework presents a way to include not only X73 compliant devices but also noncompliant health devices. The needs may be summarized in four points using this framework: (i) data gathering through sensory devices, (ii) data standardization, (iii) data analytics to find trends, and (iv) sending feedback to actual twins and caretakers via hard and soft actuation.

On the basis of the ecosystem shown in Figure 11.3 and the X73 standards [44], an ISO/IEEE 11073 standardized digital twin framework architecture for health and well-being was suggested. Figure 11.4 illustrates how this sort of digital twin architecture may be simply implemented on cloud platforms for healthcare. This design incorporates a layer for data collection through sensors or wearable devices, which have become more popular in recent years. Additionally, a communication level is specified in terms of the X73 standards' applicability or nonapplicability. The next level concerns data analysis. This level often comprises, in addition to data preprocessing, the development of tailored risk algorithms. This level requires the use of mechanistic multiscale modeling techniques to supplement data-driven models. The last level is concerned with the result post-processing in the form of reports and visualization capabilities. Particularly for multiscale modeling in which tissue or organs are replicated, visualization enables greater comprehension and explanation of the results. On the other hand, data-driven models, such as those based on machine learning, should be presented as explainable AI, as detailed in Chapter 10.

### 11.2.2 An Example Workflow of a Digital Twin

A seminal publication published in the European Heart Journal [45] posits the digital twin as a method of reaching precision cardiology, proposing two distinct

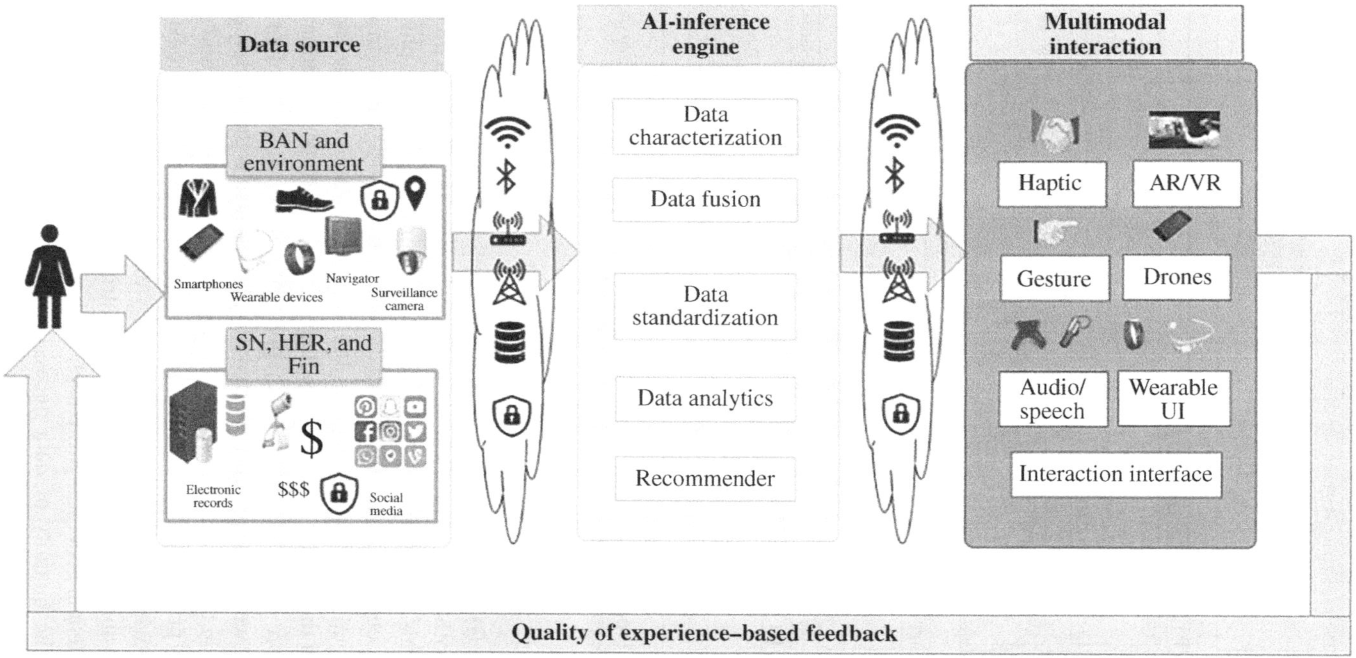


**Figure 11.3** Ecosystem of the Digital Twin for Health and Well-being. *Source:* Adapted from El Saddik et al. [43]/IEEE COMMUNICATIONS SOCIETY/CC BY-ND 4.0.

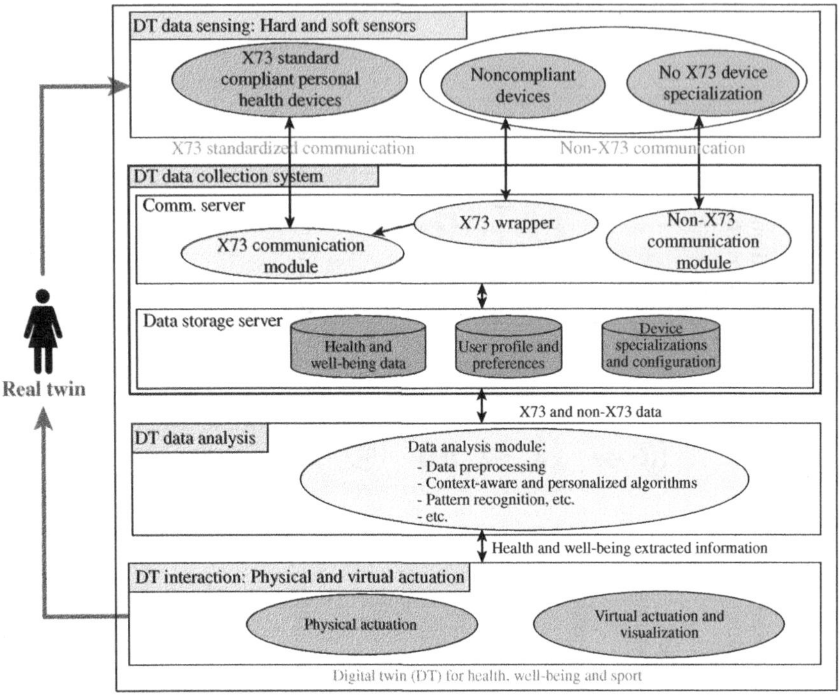


**Figure 11.4** Proposed X73-based Digital Twin (DT) System Architecture. *Source:* Adapted from El Saddik et al. [43] licensed under CC BY-ND 4.0.

approaches: (i) a mechanistic computational approach and (ii) a data-driven approach. Both arms are predicated on the study of patient data and their interplay. To do this, a multidisciplinary strategy including cardiologists, physicians, radiologists, engineers, and bioinformaticians is necessary. The accuracy of a possible digital twin is dependent on the amount of data available for model building and the quality of analysis applied to mechanistic computational models. The more data that is gathered, the more customized the digital twin becomes. However, the primary shortcoming of this idea is that it excludes preclinical and *in vitro* data. The existing digital twins in the area of cardiovascular illness focus primarily on the heart's physiology, as shown by the construction of a digital twin of human electrophysiology that matches clinical findings as nearly as feasible [46]. Another research constructed a digital twin of the cardiovascular system to generate synthetic photoplethysmogram data with the goal of increasing the quantity of data necessary to train the established machine learning models [47]. Another recently described digital twin offers noninvasive diagnosis of carotid stenosis by combining computational biomechanics with video-captured head motion [48]. The findings of this research are encouraging in terms of enabling the widespread deployment of the digital twin in practice. Petrini and coworkers used the notion of the digital twin to construct a digital twin of a coronary artery stent in order to determine the stent's material qualities when these are unavailable [49].

This work combined experimental data with computational findings in order to increase the digital twin's accuracy. To summarize, tremendous effort is being made to construct digital twins for cardiovascular disease. Additionally, it is self-evident that a multidisciplinary approach is required to create an appropriate digital twin. In particular, machine learning and artificial intelligence technologies are required for the processing of various forms of data, while mechanistic computational models will accurately recreate the physiology. Simultaneously, not only clinical data but also *in vitro* and preclinical investigations will provide the required parameters for the models. An active form of the digital twin may be constructed under this idea, allowing for continual updating and calibration with fresh data, allowing for progressive illness treatment and monitoring. Figure 11.5 depicts a typical and comprehensive process. The initial phase in the process, as described there, is data gathering. To construct a highly accurate digital twin, several sorts of data from clinical research and, in certain situations, *in vitro* tests are necessary. Medical imaging is a distinct category of obtained data since it not only permits precise illness evaluation but also the 3D depiction of the organ or tissue under examination in recent years. The next level of the process is composed of two arms: (i) the first is a mechanistic modeling approach. Typically, this section involves modeling the physiologic system on several scales and levels. These models facilitate clinical interpretation and decision-making. A multiscale model of atherosclerotic plaque development was described in Chapter 7 as an example of this approach. This model combines differential equations such as Navier–Stokes and convection–diffusion equations, as well as structural dynamics. Additionally, it makes use of patient data and parameters derived from experimental results. (ii) the second limb is comprised of data-driven models. Historically, this section relied heavily on statistical modeling, particularly binary logistic modeling when used for predictive purposes.

Technical, ethical, and economical restrictions hinder the collection of data necessary to aid therapeutic decision-making. Synergy between mechanistic and statistical models has been demonstrated to be beneficial for diagnosing, treating, and evaluating prognosis. A fully developed digital twin will blend population and individual representations to provide the most accurate information possible for clinical decision-making (Figure 11.6).

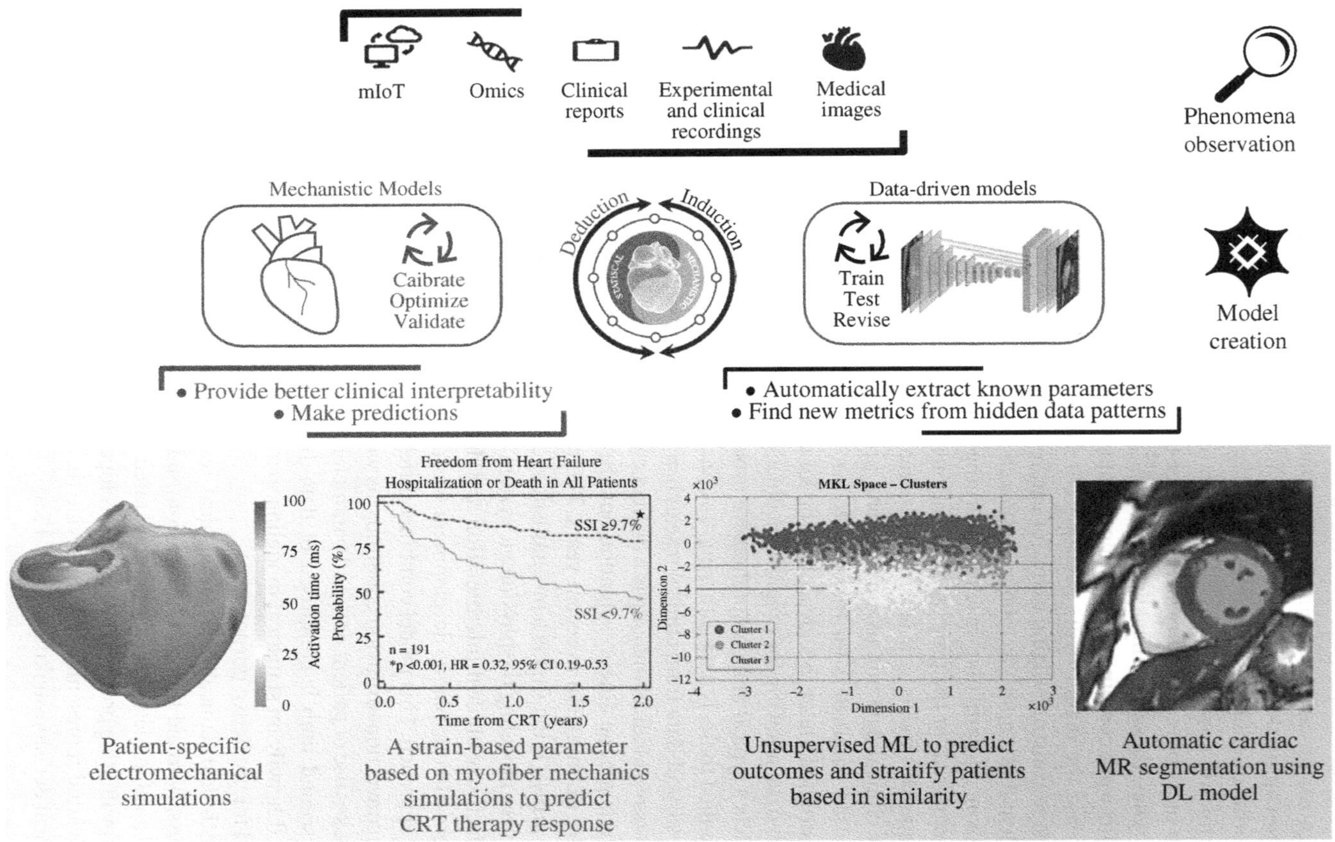


**Figure 11.5** The two pillars of the digital twin, mechanistic and statistical models, illustrating its construction and four examples of use (bottom layer). *Source:* Corral-Acero et al. [45]/Oxford University Press/CC BY-ND 4.0.

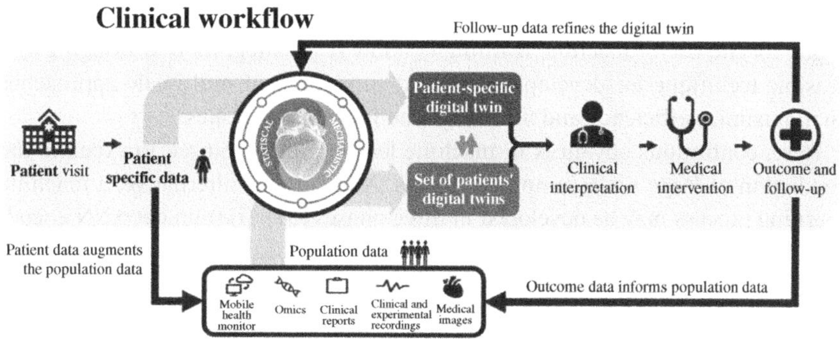


**Figure 11.6** Clinical workflow using the fully developed digital twin concept. *Source:* Corral-Acero et al. [45] licensed under CC BY-ND 4.0.

## 11.3 Integrating Multiscale Modeling and Machine Learning

Parallel developments are occurring in the disciplines of multiscale modeling, data science, and machine learning. Multiscale modeling enables comprehension of the simulated systems, regardless of whether the modeling is conducted at the micro- or macroscale. However, the integrity of such models is always contingent on the parameters set and used. These parameters are derived from the literature, experimental data, or clinical data, and are often calibrated and adjusted throughout the postprocessing stage. On the other hand, although solely data-driven models of molecular and cellular systems are generated by data science techniques [50], particularly machine learning methods, including deep learning methods [51], are simple to train and deploy, the underlying model manifests as a black box. While the overall approach followed by the machine learning community is well suited for classification, learning, and regression issues, it suffers from interpretability and explainability restrictions, despite recent efforts to build more explainable approaches. Thus, there is an area in which the two disciplines may complement one another.

Multiscale analysis has developed as a potential approach for developing *in silico* prediction models during the last two decades by systematically combining information at the tissue, cellular, and molecular levels. Machine learning-based approaches may be used to accelerate, improve, and autotune a number of current solutions for multiphysics simulations [52].

As previously stated, one of the primary goals of multiscale modeling is to connect the physics at multiple sizes using bridge algorithms that transmit information between them. However, implementing this process is often hard and time-consuming. To overcome this substantial implementation constraint, we argue for the use of machine learning to allow scale integration or bridging as a feasible technique for developing the next generation of multiscale approaches that maximize efficiency and flexibility when combining scales.

Here, continuous advances in machine learning may be used to expedite the prediction of large-scale computer models. As a suitable alternative, a machine learning process may be developed in three phases [53]: (i) train deep NN encoders that relate attributes of the multiscale model at Scale 1 (e.g. MD or MC) to those at Scale 2 (e.g. continuum model), utilizing explicit multiscale calculations to cover all conceivable situations; (ii) replace the coupling methods with encoders to connect Scales 1 and 2; (iii) verify that the property profiles acquired from the two scales match by designing a cost function that constrains the training of NNs in both scales (i). Scale coupling using NNs should be stable and computationally efficient for multiscale algorithms.

One problem is to develop interpretable models from heterogeneous data with varying degrees of accuracy and to direct the prudent collection of fresh knowledge toward revealing how biological systems acquire function. This difficulty may be overcome by exposing the whole multiscale model to modern data science and statistical approaches, including sensitivity, evolvability, robustness analysis, uncertainty quantification, multifidelity modeling, pattern identification, and model reduction.

### 11.3.1 Physics-Informed NN (PINN)

Can we use prior physics-derived knowledge to avoid overfitting or creating non-physical predictions? From a conceptual approach, is it possible to enhance machine learning with a set of well-known physics-based equations, as is done in engineering fields using MSM models? While data-driven methodologies may provide answers that are not constrained by past assumptions or models, their predictions must follow fundamental physical laws. There are well-known instances of deep learning NNs that seem to be extremely accurate yet generate highly erroneous predictions when exposed to data outside their training regime, and others that provide highly wrong predictions when exposed to apparently modest changes to the target data [54]. To circumvent this prevalent limitation of machine learning-only methodologies, several solutions exist for integrating machine learning with MSM in order to a priori adhere to fundamental physical laws while avoiding data overfitting.

One possibility is to combine deterministic and stochastic models. Combining the deterministic governing equations of multiscale models, e.g. the Navier–Stokes equations in a fluid dynamics problem with the stochastic equations of systems biology such as chemokinetics may aid in the development of computational models for otherwise ill-posed problems. A possible method that leverages deep NNs' well-known ability to approximate universal functions is the physics-informed NNs [55]. In this way, nonlinear difficulties can be handled without making important assumptions, which will affect significantly the results. These NNs must obey any symmetry, invariance, or conservation requirements derived from the physical laws that control the observed data, as defined by generic time-dependent and nonlinear partial differential equations. Using such approaches and strategies, it is possible to solve efficiently complex and difficult problems or develop novel machine learning algorithms or solvers of differential equations based on machine learning.

### 11.3.2 Deep NN Algorithms Inspired by Statistical Physics and Information Theory

Massive volumes of data, low-cost computing, and computational methods are propelling robust deep learning systems' outstanding performance and acceptance. However, developing, maintaining, and growing these systems remains an art that involves much trial and error. Learning and inference techniques have historically been motivated by and drawn from numerical methods [56]. NN algorithms can be created from a synthesis of statistical approaches and information theory methodologies and feed them back as input into multiscale models through the prescription of new computational strategies for deep NNs. Generalization in deep NNs is using algebraic topology to describe the space of attainable functions through stochastic dynamics on data in order to develop computationally efficient designs and training algorithms [57]. Recent work has proposed to use representation learning principles to formulate variational approaches for evaluating the performance of deep learning algorithms that provide guarantees about the final model's performance.

## 11.4 Conclusion and Future Trends

Multiscale modeling needs increased computational requirements. For example, for each dimension of the musculoskeletal system, separate and specialized computational techniques are utilized. Novel approaches for parallel processing using high-performance computing will improve the effectiveness of multiscale modeling in healthcare. In particular, real-time predictive modeling or diagnosis of a disease can be achieved when modeling is performed using cloud services. Another important asset of multiscale modeling is the need for accurate boundary conditions. For this reason, the computational field has to interact closely with basic science in a way that both disciplines will benefit from the findings. A novel way to do this nowadays is using 3D printing and bioplotting approaches. For example, in the field of cardiovascular disease, bioplotting of vascular scaffolds and their use in fluidic devices will provide evidence about the reality and the physiology of cells' interaction under flow conditions, which are then simulated with accuracy. Going a step further, using such experimental settings, complex biological mechanisms such as atherosclerotic plaque growth and inflammation in vasculitis or cancer can be used to provide input to computational multiscale modeling. Finally, it is a necessary need that the clinicians and physicians should interact with the engineers and data scientists. All can together build an ecosystem which will be updated regularly with new input. In this way, the accuracy of the developed models will increase, while the use of explainable ways of artificial intelligence will provide evidence to the physicians to utilize the models in clinical practice.

## References


1 Pezoulas, V.C., Grigoriadis, G.I., Gkois, G. et al. (2021). A computational pipeline for data augmentation towards the improvement of disease classification and risk stratification models: a case study in two clinical domains. *Comput. Biol. Med.* 134: 104520. https://doi.org/10.1016/j.compbiomed.2021.104520.

2 Marchant, I., Nony, P., Cucherat, M. et al. (2011). The global risk approach should be better applied in french hypertensive patients: a comparison between simulation and observation studies. *PLoS One* 6 (3): e17508. https://doi.org/10.1371/journal.pone.0017508.

3 Allen, R., Rieger, T. and Musante, C. (2016), Efficient Generation and Selection of Virtual Populations in Quantitative Systems Pharmacology Models. CPT Pharmacometrics Syst. Pharmacol., 5: 140–146. https://doi.org/10.1002/psp4.12063

4 Haidar, A., Wilinska, M.E., Graveston, J.A., and Hovorka, R. (2013). Stochastic virtual population of subjects with type 1 diabetes for the assessment of closed-loop glucose controllers. *IEEE Trans. Biomed. Eng.* 60 (12): 3524–3533. https://doi.org/10.1109/TBME.2013.2272736.

5 Schmidt, B.J., Casey, F.P., Paterson, T., and Chan, J.R. (2013). Alternate virtual populations elucidate the type I interferon signature predictive of the response to rituximab in rheumatoid arthritis. *BMC Bioinf.* 14 (1): 221. https://doi.org/10.1186/1471-2105-14-221.

6 Tannenbaum, S.J., Holford, N.H.G., Lee, H. et al. (2006). Simulation of correlated continuous and categorical variables using a single multivariate distribution. *J. Pharmacokinet. Pharmacodyn.* 33 (6): 773–794. https://doi.org/10.1007/s10928-006-9033-1.

7 Teutonico, D., Musuamba, F., Maas, H.J. et al. (2015). Generating virtual patients by multivariate and discrete re-sampling techniques. *Pharm. Res.* 32 (10): 3228–3237. https://doi.org/10.1007/s11095-015-1699-x.

8 Marchant, I., Boissel, J.P., Kassaï, B. et al. (2009). SCORE should be preferred to Framingham to predict cardiovascular death in French population. *Eur. J. Cardiovasc. Prev. Rehabil.* 16 (5): 609–615. https://doi.org/10.1097/HJR.0b013e32832da006.

9 Moss, R., Grosse, T., Marchant, I. et al. (2012). Virtual patients and sensitivity analysis of the guyton model of blood pressure regulation: towards individualized models of whole-body physiology. *PLoS Comput. Biol.* 8 (6): e1002571. https://doi.org/10.1371/journal.pcbi.1002571.

10 Guyton, A.C., Coleman, T.G., and Granger, H.J. (1972). Circulation: overall regulation. *Annu. Rev. Physiol.* 34: 13–46. https://doi.org/10.1146/annurev.ph.34.030172.000305.

11 Montani, J.-P. and Van Vliet, B.N. (2009). Understanding the contribution of Guyton's large circulatory model to long-term control of arterial pressure: Contribution of Guyton's model. *Exp. Physiol.* 94 (4): 382–388. https://doi.org/10.1113/expphysiol.2008.043299.

12 Morris, M.D. (1991). Factorial sampling plans for preliminary computational experiments. *Technometrics* 33 (2): 161. https://doi.org/10.2307/1269043.

13 Kazuhiro, K., Werner, R.A., Toriumi, F. et al. (2018). Generative adversarial networks for the creation of realistic artificial brain magnetic resonance images. *Tomography* 4 (4): 159–163. https://doi.org/10.18383/j.tom.2018.00042.

14 Russ, T., Goerttler, S., Schnurr, A.K. et al. (2019). Synthesis of CT images from digital body phantoms using CycleGAN. *Int. J. Comput. Assist. Radiol. Surg.* 14 (10): 1741–1750. https://doi.org/10.1007/s11548-019-02042-9.

15 Chang, Y., Lafata, K., Segars, W.P. et al. (2020). Development of realistic multi-contrast textured XCAT (MT-XCAT) phantoms using a dual-discriminator conditional-generative adversarial network (D-CGAN). *Phys. Med. Biol.* 65 (6): 065009. https://doi.org/10.1088/1361-6560/ab7309.

16 Younis, I.R., Robert Powell, J., Rostami-Hodjegan, A. et al. (2017). Utility of model-based approaches for informing dosing recommendations in specific populations: report from the Public AAPS Workshop: model-based approaches to dosing recommendations. *J. Clin. Pharmacol.* 57 (1): 105–109. https://doi.org/10.1002/jcph.787.

17 Virtual Populations. Certara. https://www.certara.com/software/ (accessed March 26 2018).

18 Jamei, M., Marciniak, S., Feng, K. et al. (2009). The Simcyp® population-based ADME simulator. *Expert Opin. Drug Metab. Toxicol.* 5 (2): 211–223. https://doi.org/10.1517/17425250802691074.

19 Using Virtual Cancer Patients to Probe the Mechanisms of Oncology Drug Disposition. Certara. https://www.certara.com/2018/01/30/using-virtual-cancer-patients-to-probe-the-mechanisms-of-oncology-drug-disposition/?ap%5B0%5D=PBPK (accessed 29 November 2022).

**20** 3D Population Driven Design. Materialise Medical 3D Printing. http://www.materialise.com/en/medical/3d-population-driven-design (accessed 29 November 2022).

**21** McNally, K., Cotton, R., Hogg, A., and Loizou, G.D. (2014). PopGen: A virtual human population generator. *Toxicology* 315: 70–85.

**22** The Effect Model. Novadiscovery. http://www.novadiscovery.com/how-we-do-it/the-effect-model/ (accessed 29 November 2022).

**23** Boissel, J.-P., Auffray, C., Noble, D. et al. (2015). Bridging systems medicine and patient needs: systems medicine and patient needs. *CPT Pharmacomet. Syst. Pharmacol.* 4 (3): 135–145. https://doi.org/10.1002/psp4.26.

**24** Kahoul, R., Gueyffier, F., Amsallem, E. et al. (2014). Comparison of an effect-model-law-based method versus traditional clinical practice guidelines for optimal treatment decision-making: application to statin treatment in the French population. *J. R. Soc. Interface* 11 (100): 20140867. https://doi.org/10.1098/rsif.2014.0867.

**25** About Us. Immunetrics. https://www.immunetrics.com/about-us/ (accessed 29 November 2022).

**26** Disease Specific Models. Immunetrics. https://www.immunetrics.com/products/disease-specific-models/ (accessed 29 November 2022).

**27** Brown, D., Namas, R.A., Almahmoud, K. et al. (2015). Trauma in silico: Individual-specific mathematical models and virtual clinical populations. *Sci. Transl. Med.* 7 (285): 285ra61. https://doi.org/10.1126/scitranslmed.aaa3636.

**28** FDA Receives License to Entelos PhysioLab Modeler® Software. FierceBiotech. https://www.fiercebiotech.com/it/fda-receives-license-to-entelos-physiolab-modeler%C2%AE-software (accessed 29 November 2022).

**29** Klinke, D.J. (2008). Integrating epidemiological data into a mechanistic model of type 2 diabetes: validating the prevalence of virtual patients. *Ann. Biomed. Eng.* 36 (2): 321–334. https://doi.org/10.1007/s10439-007-9410-y.

**30** NHANES III (1988–1994) – Data Files. https://wwwn.cdc.gov/nchs/nhanes/nhanes3/DataFiles.aspx (accessed 29 November 2022).

**31** Valentin, J. (2002). Basic anatomical and physiological data for use in radiological protection: reference values: ICRP Publication 89: approved by the Commission in September 2001. *Ann. ICRP* 32 (3, 4): 1–277. https://doi.org/10.1016/S0146-6453(03)00002-2.

**32** de la Grandmaison, G.L., Clairand, I., and Durigon, M. (2001). Organ weight in 684 adult autopsies: new tables for a Caucasoid population. *Forensic Sci. Int.* 119 (2): 149–154. https://doi.org/10.1016/S0379-0738(00)00401-1.

**33** Willmann, S., Höhn, K., Edginton, A. et al. (2007). Development of a physiology-based whole-body population model for assessing the influence of individual variability on the pharmacokinetics of drugs. *J. Pharmacokinet. Pharmacodyn.* 34 (3): 401–431. https://doi.org/10.1007/s10928-007-9053-5.

**34** Ram, P. and Gray, A.G. (May 2011). Density estimation trees. In: *Proceedings of the 17th ACM SIGKDD International Conference on Knowledge Discovery and Data Mining*, New York, NY, USA, 627–635. https://doi.org/10.1145/2020408.2020507.

**35** Breiman, L. (2001). Random forests. *Mach. Learn.* 45 (1): 5–32. https://doi.org/10.1023/A:1010933404324.

**36** Robnik-Šikonja, M. (2016). Data generators for learning systems based on RBF networks. *IEEE Trans. Neural Netw. Learn. Syst.* 27 (5): 926–938. https://doi.org/10.1109/TNNLS.2015.2429711.

**37** Robnik-Sikonja, M. (2018). Dataset comparison workflows. *Int. J. Data Sci.* 3: 126. https://doi.org/10.1504/IJDS.2018.10013385.

**38** Pleouras, D.S., Sakellarios, A., Rigas, G. et al. (2021). A novel approach to generate a virtual population of human coronary arteries for in silico clinical trials of stent design. *IEEE Open J. Eng. Med. Biol.* 2: 201–209. https://doi.org/10.1109/OJEMB.2021.3082328.

**39** Sakellarios, A.I., Rigas, G., Kigka, V. et al. (2017). SMARTool: a tool for clinical decision support for the management of patients with coronary artery disease based on modeling of atherosclerotic plaque process. In: *Annual International Conference of the IEEE Engineering in Medicine and Biology Society. IEEE Engineering in Medicine and Biology Society. Annual International Conference*, Jeju, Korea, vol. 2017, 96–99.

**40** Dusetzina, S.B., Tyree, S., Meyer, A.-M. et al. (2014). *Linking Data for Health Services Research: A Framework and Instructional Guide*. Rockville (MD): Agency for Healthcare Research and Quality (US) http://www.ncbi.nlm.nih.gov/books/NBK253313/ (accessed 29 November 2022).

**41** Bruynseels, K., Santoni de Sio, F., and van den Hoven, J. (2018). Digital twins in health care: ethical implications of an emerging engineering paradigm. *Front. Genet.* 9: https://www.frontiersin.org/article/10.3389/fgene.2018.00031 (accessed 29 November 2022).

**42** Kahane, G. and Savulescu, J. (2015). Normal human variation: refocussing the enhancement debate. *Bioethics* 29 (2): 133–143. https://doi.org/10.1111/bioe.12045.

**43** El Saddik, A., Badawi, H.F., Martinez, R. et al. (2019). Dtwins: a digital twins ecosystem for health and well-being. *IEEE COMSOC MMTC Commun. – Front.* 14: 39–46.

**44** Laamarti, F., Badawi, H.F., Ding, Y. et al. (2020). An ISO/IEEE 11073 standardized digital twin framework for health and well-being in smart cities. *IEEE Access* 8: 105950–105961. https://doi.org/10.1109/ACCESS.2020.2999871.

**45** Corral-Acero, J., Margara, F., Marciniak, M. et al. (2020). The 'Digital Twin' to enable the vision of precision cardiology. *Eur. Heart J.* 41 (48): 4556–4564. https://doi.org/10.1093/eurheartj/ehaa159.

**46** Gillette, K., Gsell, M.A.F., Prassl, A.J. et al. (2021). A Framework for the generation of digital twins of cardiac electrophysiology from clinical 12-leads ECGs. *Med. Image Anal.* 71: 102080. https://doi.org/10.1016/j.media.2021.102080.

**47** Mazumder, O., Roy, D., Bhattacharya, S. et al. (2019). Synthetic PPG generation from haemodynamic model with baroreflex autoregulation: a Digital twin of cardiovascular system. In: *Annual International Conference of the IEEE Engineering in Medicine and Biology Society. IEEE Engineering in Medicine and*

*Biology Society*, vol. 2019, 5024–5029. https://doi.org/10.1109/EMBC.2019.8856691.

**48** Chakshu, N.K., Carson, J., Sazonov, I., and Nithiarasu, P. (2019). A semi-active human digital twin model for detecting severity of carotid stenoses from head vibration – a coupled computational mechanics and computer vision method. *Int. J. Numer. Methods Biomed. Eng.* 35 (5): https://doi.org/10.1002/cnm.3180.

**49** Antonini, L., Berti, F., Isella, B. et al. (2021). From the real device to the digital twin: a coupled experimental-numerical strategy to investigate a novel bioresorbable vascular scaffold. *PLoS One* 16 (6): e0252788. https://doi.org/10.1371/journal.pone.0252788.

**50** Lee, M.J., Ye, A.S., Gardino, A.K. et al. (2012). Sequential application of anticancer drugs enhances cell death by rewiring apoptotic signaling networks. *Cell* 149 (4): 780–794. https://doi.org/10.1016/j.cell.2012.03.031.

**51** Sachs, K., Perez, O., Pe'er, D. et al. (2005). Causal protein-signaling networks derived from multiparameter single-cell data. *Science* 308 (5721): 523–529. https://doi.org/10.1126/science.1105809.

**52** Balaprakash, P., Dongarra, J., Gamblin, T. et al. (2018). Autotuning in high-performance computing applications. *Proc. IEEE* 106 (11): 2068–2083. https://doi.org/10.1109/JPROC.2018.2841200.

**53** Alber, M., Buganza Tepole, A., Cannon, W.R. et al. (2019). Integrating machine learning and multiscale modeling-perspectives, challenges, and opportunities in the biological, biomedical, and behavioral sciences. *NPJ Digit. Med.* 2: 115. https://doi.org/10.1038/s41746-019-0193-y.

**54** Lillicrap, T. P. and Kording, K. P. (2019). What does it mean to understand a neural network? https://arxiv.org/abs/1907.06374v1 (accessed 29 November 2022).

**55** Raissi, M., Perdikaris, P., and Karniadakis, G.E. (2019). Physics-informed neural networks: a deep learning framework for solving forward and inverse problems involving nonlinear partial differential equations. *J. Comput. Phys.* 378: 686–707. https://doi.org/10.1016/j.jcp.2018.10.045.

**56** Lachmann, A., Giorgi, F.M., Lopez, G., and Califano, A. (2016). ARACNe-AP: gene network reverse engineering through adaptive partitioning inference of mutual information. *Bioinforma. Oxf. Engl.* 32 (14): 2233–2235. https://doi.org/10.1093/bioinformatics/btw216.

**57** Chaudhari, P., Oberman, A., Osher, S. et al. (2018). Deep relaxation: partial differential equations for optimizing deep neural networks. *Res. Math. Sci.* 5 (3): 30. https://doi.org/10.1007/s40687-018-0148-y.

# Index



*Multiscale Modelling in Biomedical Engineering*, First Edition. Dimitrios I. Fotiadis, Antonis I. Sakellarios, and Vassiliki T. Potsika.

Published 2023 by John Wiley & Sons, Inc.

## d

## IEEE Press Series in Biomedical Engineering

The focus of our series is to introduce current and emerging technologies to biomedical and electrical engineering practitioners, researchers, and students. This series seeks to foster interdisciplinary biomedical engineering education to satisfy the needs of the industrial and academic areas. This requires an innovative approach that overcomes the difficulties associated with traditional textbooks and edited collections.

**Series Editor, Metin Akay,** University of Houston, TX

1. *Time Frequency and Wavelets in Biomedical Signal Processing*
   Metin Akay
2. *Neural Networks and Artificial Intelligence for Biomedical Engineering*
   Donna L. Hudson and Maurice E. Cohen
3. *Physiological Control Systems: Analysis, Simulation, and Estimation*
   Michael C.K. Khoo
4. *Principles of Magnetic Resonance Imaging: A Signal Processing Perspective*
   Zhi-Pei Liang and Paul C. Lauterbur
5. *Nonlinear Biomedical Signal Processing, Volume 1, Fuzzy Logic, Neural Networks, and New Algorithms*
   Metin Akay
6. *Fuzzy Control and Modeling: Analytical Foundations and Applications*
   Hao Ying
7. *Nonlinear Biomedical Signal Processing, Volume 2, Dynamic Analysis and Modeling*
   Metin Akay
8. *Biomedical Signal Analysis: A Case-Study Approach*
   Rangaraj M. Rangayyan
9. *System Theory and Practical Applications of Biomedical Signals*
   Gail D. Baura
10. *Introduction to Biomedical Imaging*
    Andrew Webb
11. *Medical Image Analysis*
    Atam P. Dhawan

12. *Identification of Nonlinear Physiological Systems*
David T. Westwick and Robert E. Kearney

13. *Electromyography: Physiology, Engineering, and Non-Invasive Applications*
Roberto Merletti and Philip A. Parker

14. *Nonlinear Dynamic Modeling of Physiological Systems*
Vasilis Z. Marmarelis

15. *Genomics and Proteomics Engineering in Medicine and Biology*
Metin Akay

16. *Handbook of Neural Engineering*
Metin Akay

17. *Medical Image Analysis, Second Edition*
Atam P. Dhawan

18. *Advanced Methods of Biomedical Signal Processing*
Sergio Cerutti and Carlo Marchesi

19. *Epistemology of the Cell: A Systems Perspective on Biological Knowledge*
Edward R. Dougherty and Michael L. Bittner

20. *Micro and Nanotechnologies in Engineering Stem Cells and Tissues*
Murugan Ramalingam, Esmaiel Jabbari, Seeram Ramakrishna, and Ali Khademhosseini

21. *Introduction to Neural Engineering for Motor Rehabilitation*
Dario Farina, Winnie Jensen, and Metin Akay

22. *Introduction to Tissue Engineering: Applications and Challenges*
Ravi Birla

23. *Handbook of Biomedical Telemetry*
Konstantina S. Nikita

24. *Biomedical Signal Analysis, Second Edition*
Rangaraj M. Rangayyan

25. *Models and Algorithms for Biomolecules and Molecular Networks*
Bhaskar Das Gupta and Jie Liang

26. *Surface Electromyography: Physiology, Engineering and Applications*
Roberto Merletti and Dario Farina

27. *m-Health: Fundamentals and Applications*
Robert Istepanian and Bryan Woodward

28. *Physiological Control Systems: Analysis, Simulation, and Estimation Second Edition*
    Michael C.K. Khoo

29. *Computational Modeling and Simulation Examples in Bioengineering*
    Nenad D. Filipovic

30. *Introduction to Biomedical Imaging, Second Edition*
    Andrew Webb

31. *Multiscale Modelling in Biomedical Engineering*
    Dimitrios I. Fotiadis, Antonis I. Sakellarios, and Vassiliki T. Potsika

Printed and bound by CPI Group (UK) Ltd, Croydon, CR0 4YY

17/09/2026

14964317-0006